Boundary Value Problems for Elliptic Equations and Systems

Pitman Monographs and
Surveys in Pure and Applied Mathematics 46

Boundary Value Problems for Elliptic Equations and Systems

Guo Chun Wen

Peking University and Yantai University

Heinrich G W Begehr

Free University of Berlin

Longman Scientific & Technical

Copublished in the United States with
John Wiley & Sons, Inc., New York

Longman Scientific & Technical
Longman Group UK Limited
Longman House, Burnt Mill, Harlow
Essex CM20 2JE, England
and Associated Companies throughout the world.

Copublished in the United States with
John Wiley & Sons Inc., 605 Third Avenue, New York, NY 10158

First published in 1990

AMS Subject Classification:
(Main) 35J60, 35J55, 30C60
(Subsidiary) 30E25, 30G20, 35J25
ISSN 0269-3666

British Library Cataloguing in Publication Data

Wen, Guo Chun
Boundary value problems for elliptic equations
and systems.
1. Elliptic partial differential equations.
Boundary value problems
I. Title II. Begehr, Heinrich G.W.
515.3′53

ISBN 0-582-03159-1

Libary of Congress Cataloguing-in-Publication Data

Wen, Guo Chun.
Boundary value problems for elliptic equations
and systems
(Pitman monographs and surveys in pure and applied mathematics ; 46)
Bibliography: p.
Includes index.
1. Boundary value problems. 2. Differential
equations, Elliptic. I. Begehr, Heinrich G.W.
II. Title. III. Series.
QA379.W46 1989
515.3′53 88-31511
ISBN 0-470-21304-3

Set in 10/12 Times New Roman
Printed and Bound in Great Britain
at The Bath Press, Avon

Contents

Preface

This book deals mainly with various boundary value problems for linear and nonlinear elliptic equations and systems of first and second order. The canonical forms of linear uniformly elliptic systems of two first order equations were considered in detail by L. Bers and I.N. Vekua. They established systematic mathematical theories related to one another and called the theory of "pseudoanalytic functions" and "generalized analytic functions", respectively. In this book these theories are extended to general nonlinear uniformly elliptic systems. Simultaneously, more complete results for linear uniformly elliptic systems are given. Hence, the content of this book gives a continuation and development of Bers' and Vekua's theory.

It differs from the theory of pseudoanalytic and generalized analytic functions in the considered systems as well as in the methods used. Because we mainly discuss nonlinear elliptic equations and systems with weak conditions, they cannot be reduced to the above mentioned canonical form. We will give some representation theorems for solutions to general nonlinear elliptic equations and systems, establish the relations between their solutions and analytic and harmonic functions, prove some properties of the solutions, and then consider the solvability of corresponding boundary value problems. In order to prove the existence of solutions to the boundary value problems, the following methods are used.

1. Transformation of the general real elliptic equation and system with some conditions into complex forms.
2. Establishing integral representations of solutions to various boundary value problems and their properties.

3. Deducing a priori estimates for solutions to the boundary value problems from the representation theorems for solutions.
4. Using the Fredholm theory suitably, the method of continuity, the Schauder fixed-point theorem and the Leray-Schauder theorem.

There are four characteristics of this book: the nonlinear elliptic equations and systems are rather general, the considered regions are almost always multiply connected domains, the boundary conditions are quite general, and various methods are used. Moreover, as applications of our methods, we obtain some existence theorems for nonlinear quasiconformal mappings from arbitrary multiply connected domains onto canonical domains.

Most of the content originates in investigations of the first named author, mainly published in Chinese; moreover, some results are published here for the first time. It is clear that the boundary value problems investigated in this book have many applications in hydromechanics, elasticity, and many other fields.

We are planning to continue this presentation of results and methods on elliptic equations and systems in a second volume. This first part leans heavily on the classical work of L. Bers, R.P. Gilbert, I.N. Vekua, W.L. Wendland, and others. However, most of the topics discussed are distinct from those considered in the above mentioned books.

For technical reasons this manuscript was typed twice, both times by Mrs Barbara Wengel, secretary in the I. Math. Institute of the FU Berlin. While her colleagues switched to wordprocessors she had to stay with her typewriter until the manuscript was retyped. We thus have to thank her not only for her good work but also for her patience. We are grateful to Gabriele Ledworuski and Ji Shu-hua for preparing the index of this volume.
The preparation of this book was supported by the Deutsche Forschungsgemeinschaft and the Educational State Commission of P.R. of China on the basis of their scientific exchange program as well as through the Freie Universität Berlin and the Peking University.

Beijing and Berlin, June 1989 G.C. Wen, H. Begehr

I Boundary value problems for simple complex equations

In this chapter, we shall discuss some boundary value problems for simple complex equations which include the Cauchy-Riemann system of equations and the Poisson equation as special cases. We mainly consider the complex equations $w_{\bar{z}} = f(z)$ and $u_{z\bar{z}} = f(z)$, where

$$(\ )_{\bar{z}} = \frac{1}{2}[(\ )_x + i(\ )_y] \ , \ (\ )_z = \frac{1}{2}[(\ )_x - i(\ )_y] \ .$$

In particular, $w_{\bar{z}} = 0$ is the complex form of the Cauchy-Riemann system. In addition

$$u_{z\bar{z}} = \frac{1}{4}\Delta u \ , \ u_{z\bar{z}} = 0$$

is the complex form of the Poisson equation. In this book, we use the same notation and definitions as in [10]1), [97]2). We give the proper proposition for the Riemann-Hilbert boundary value problem of $w_{\bar{z}} = f(z)$ in a multiply connected domain, and prove the existence and uniqueness of the solution for the problem by a new method. Moreover, we consider the regular and irregular oblique derivative boundary value problems for $u_{z\bar{z}} = f(z)$. Besides, we obtain integral representations for solutions of the boundary value problems and prove some properties of the integral operators involved.

§ 1. Some boundary value problems for analytic functions

In this section, we mainly consider the Riemann-Hilbert boundary value problem for analytic functions in a multiply connected domain. In the books written by I.N. Vekua and F.D. Gakhov (cf. [97]2), [33]), applying the method of integral equations, the solvability of the above boundary value problem is discussed. We shall use a different method to handle the boundary value problem, and give the solvability conditions and the total number of solutions for the problem.

In addition, we also introduce the compound boundary value problem for analytic functions.

1. Formulation of the Riemann-Hilbert boundary value problem

Let D be an $(N+1)$-connected domain in the z-plane $\mathbb{C}$ with boundary $\Gamma \in C_\mu (0 < \mu \le 1)$. Without loss of generality, we may assume that D is a circular domain in the unit disc $|z| < 1$, which is bounded by $N + 1$ circles $\Gamma_j = \{|z-z_j| = \gamma_j\}$, $j = 1,\ldots,N$ and $\Gamma_{N+1} = \Gamma_0 = \{|z| = 1\}$, $z = 0 \in D$. Otherwise, through a conformal mapping of z, we can map the general domain onto a circular domain.

Problem A. The Riemann-Hilbert boundary value problem for analytic functions may be formulated as follows: Find an analytic function $\Phi(z)$ in D, continuous on $\overline{D}$ and satisfying the boundary condition

$$Re[\overline{\lambda(t)}\Phi(t)] = r(t)\ , \quad t \in \Gamma\ , \tag{1.1}$$

where $|\lambda(t)| = 1, C_\alpha[\lambda,\Gamma] = C[\lambda,\Gamma] + H_\alpha[\lambda,\Gamma] \le \ell, C_\alpha[r,\Gamma] \le \ell$, and $\alpha(0 < \alpha < 1)$, $\ell(0 \le \ell < \infty)$ are constants. The above boundary value problem will be denoted Problem A.

Problem A with $r(t) = 0$ is called Problem A_0, and $K = 1/(2\pi)\Delta_\Gamma \arg \lambda(t)$ is called the index of Problem A and Problem A_0. When the index $K < N$, the problem A may not be solvable and when $K \ge N$, the solutions of Problem A are not unique. Hence, we consider the modified boundary value problem:

Problem B. Find an analytic function $\Phi(z)$ in D, continuous on $\overline{D}$ and satisfying the modified boundary condition

$$Re[\overline{\lambda(t)}\Phi(t)] = r(t) + h(t)\ , \quad t \in \Gamma\ . \tag{1.2}$$

(1) If $K \ge N$, we take $h(t) = 0$, and suppose that $\Phi(z)$ satisfies $2K - N + 1$ point conditions:

$$Im[\overline{\lambda(a_j)}\Phi(a_j)] = b_j \ , \ j = 1,\ldots,2K-N+1 \ , \tag{1.3}$$

where a_j is a point on Γ_j , $j = 1,\ldots,N$, and $a_j (j=N+1,\ldots,2K-N+1)$ are distinct points on Γ_0 , we may put $a_{N+1} = a_0 = 1$, and $b_j (j=1,\ldots,2K-N+1 \ , b_0=b_{N+1})$ are all real constants with the conditions $|b_j| \leq \ell$, $j=1,\ldots,2K-N+1$.

(2) If $0 \leq K < N$, we take

$$h(t) = \begin{cases} h_j \ , \ t \in \Gamma_j \ , \ j = 1,\ldots N-K \ , \\ 0 \ \ , \ t \in \Gamma_j \ , \ j = N-K+1,\ldots,N+1 \ , \end{cases} \tag{1.4}$$

where $h_j (j=1,\ldots,N-K)$ are all unknown real constants to be determined appropriately, and suppose that $\Phi(z)$ satisfies $K + 1$ point conditions:

$$Im[\overline{\lambda(a_j)}\Phi(a_j)] = b_j \ , \ j = N-K+1,\ldots,N+1 \ , \tag{1.5}$$

in which a_j , b_j $(j=N-K+1,\ldots,N+1)$ are as stated in connection with (1.3).

(3) If $K < 0$, we take

$$h(t) = h_j - Re[\overline{\lambda(t)} \sum_{m=1}^{|K|-1} (H_m^+ + iH_m^-)t^{-m}] \ , \ t \in \Gamma_j \ , \ j = 0,1,\ldots,N \ , \tag{1.6}$$

where $h_j (j = 0,1,\ldots,N)$, $H_m^{\pm} (m=1,\ldots,|K|-1)$ are appropriate real constants.

In particular, if $\lambda(t) = 1$, $t \in \Gamma$, then the problem B is the modified Dirichlet problem, which is called Problem D. Moreover, Problem B with $r(t) = 0$ is called Problem B_0.

Later, we shall need the canonical form of boundary condition (1.2). For this purpose, we find an analytic function $S(z)$, which is a solution for Problem D with the boundary condition

$$Re\, S(t) = S_1(t) - \theta(t) \ , \ t \in \Gamma \ , \ Im\, S(1) = 0 \ , \tag{1.7}$$

where $S_1(t) = \begin{cases} arg\, \lambda(t) - K\, arg\, t + arg\, \Pi(t) \ , \ t \in \Gamma_0 \ , \\ arg\, \lambda(t) + arg\, \Pi(t) \ , \ t \in \Gamma_j \ , \ j = 1,\dots,N \ , \end{cases}$

$$\theta(t) = \begin{cases} 0 \ , \ t \in \Gamma_0 \ , \\ \theta_j \ , \ t \in \Gamma_j \ , \ j = 1,\dots,N \ , \end{cases}$$

$$\Pi(t) = \prod_{j=1}^{N} (t-z_j)^{K_j} \ , \ K_j = \frac{1}{2\pi} \Delta_{\Gamma_j} arg\, \lambda(t) \ , \ j = 0,\dots,N,$$

$\theta_j (j=1,\dots,N)$ are all appropriate constants (cf. Theorem 1.5). Let

$$\Psi(z) = e^{-iS(z)}\, \Pi(z)\, \Phi(z) \ , \ z \in D \ . \tag{1.8}$$

Then the boundary condition (1.2) and the point conditions (1.3), (1.5) can be transformed into the canonical form

$$Re[\overline{\Lambda(t)}\Psi(t)] = R(t) + H(t) \ , \ t \in \Gamma \ , \tag{1.9}$$

$$Im[\overline{\Lambda(a_j)}\Psi(a_j)] = B_j \ , \ j \in \{j\} = \begin{cases} 1,\dots,2K-N+1 \ , \text{ if } \ K \ge N \ , \\ N-K+1,\dots,N+1 \ , \text{ if } \ 0 \le K < N \ , \end{cases} \tag{1.10}$$

where

$$\Lambda(t) = \begin{cases} t^K, t \in \Gamma_0 \\ e^{i\theta_j}, t \in \Gamma_j \ , \ j=1,\dots,N \ , \end{cases}$$

$$R(t) = r(t)e^{S_2(t)} |\Pi(t)| \ , \ H(t) = h(t)e^{S_2(t)} |\Pi(t)| \ ,$$

$$B_j = b_j e^{S_2(a_j)} |\Pi(a_j)| \ , \ S_2(t) = Im\, S(t) \ .$$

The boundary value problem with conditions (1.9) and (1.10) is called Problem C.

2. Uniqueness of the solution for Problem B

First of all, let us recall a lemma, which has been proved in § 4 of Chapter 4 in [97]2).

Lemma 1.1 Let $\Phi(z)$ be a nontrivial solution of Problem B_0 $(K \geq -1)$ on $\overline{D}$ and denote by N_D and N_Γ the total number of zero-points on D and Γ, respectively. Then the following equality holds:

$$2N_D + N_\Gamma = 2K . \tag{1.11}$$

Furthermore, the total number of zeros on $\Gamma_j (j=0,1,\ldots,n)$ are even numbers, and when $K = -1$, Problem B_0 only has the trivial solution. If $\Phi(z)$ has isolated poles on D and Γ, then

$$2N_D + N_\Gamma - 2P_D - P_\Gamma = 2K , \tag{1.12}$$

where P_D and P_Γ denote the total number of poles on D and Γ, respectively.

Theorem 1.2 The solution of Problem B is unique.

Proof. Let $\Phi_1(z)$ and $\Phi_2(z)$ be two solutions of Problem B. It is easy to see that $\Phi(z) = \Phi_1(z) - \Phi_2(z)$ is an analytic function, and satisfies the boundary condition

$$Re[\overline{\lambda(t)}\Phi(t)] = h(t) , \quad t \in \Gamma , \tag{1.13}$$

$$Im[\overline{\lambda(a_j)}\Phi(a_j)] = 0 , \quad j \in \{j\} . \tag{1.14}$$

For the case of $0 \leq K < N$, noting $\Phi(a_j) = 0$, $j = N-K+1,\ldots,N+1$ and using Lemma 1.1, if $\Phi(z) \not\equiv 0$ in D, then $\Phi(z)$ on $\Gamma \setminus \Gamma_0$ has at least $2(K+1)$ zeros. From (1.11) the absurd inequality follows

$$2K + 2 \leq 2N_D + N_\Gamma = 2K .$$

Therefore, $\Phi(z) \equiv 0$, $z \in D$, i.e. $\Phi_1(z) \equiv \Phi_2(z)$.

Similarly, we can prove $\Phi_1(z) \equiv \Phi_2(z)$, when $K \geq N$.

As for the case of $K < 0$, due to

$$h(t) = h_j - Re[\overline{\lambda(t)} \sum_{m=1}^{|K|-1} (H_m^+ + iH_m^-)t^{-m}] ,$$

we know that

$$\Psi_*(z) = z^{|K|-1}[\Phi(z) + \sum_{m=1}^{|K|-1} (H_m^+ + iH_m^-)z^{-m}]$$

is analytic in D and satisfies the boundary condition

$$Re[\overline{\lambda_1(t)}\Psi_*(t)] = h(t) = h_j , \; t \in \Gamma_j , \; j = 0,\dots,N , \tag{1.15}$$

where the index of $\lambda_1(t) = \lambda(t)\overline{t}^{1-|K|}$ equals -1 . By Lemma 1.1, it follows that $\Psi_*(z) \equiv 0$, $z \in D$, thus $\Phi(z) \equiv 0$, i.e. $\Phi_1(z) \equiv \Phi_2(z)$. This completes the proof. □

3. A priori estimates of solutions for Problem B

If the domain D is the unit disc $|z| < 1$ and $\lambda(t) = 1$, $t \in \Gamma = \{|t|=1\}$, then Problem B is the Dirichlet problem D on $\overline{\mathbb{D}} = \{|z|\le 1\}$. In this case, the Privalov theorem gives an a priori estimate for solutions to the Dirichlet problem D (cf. [10]1)).

Lemma 1.3 Suppose that $\Phi(z) = u(z) + iv(z)$ is analytic in $\mathbb{D} = \{|z|<1\}$, $|v(0)| \le M_1 < \infty$, and $u(z)$ is continuous on $\overline{\mathbb{D}}$ satisfying $C_\alpha[u,\Gamma] \le M_2 < \infty$, where $\alpha(0 < \alpha < 1)$, M_1 and M_2 are all constants. Then $\Phi(z)$ is continuous on $\overline{\mathbb{D}}$ and satisfies the estimate

$$C_\alpha[\Phi,\overline{\mathbb{D}}] \le M_3 , \tag{1.16}$$

where $M_3 = M_3(\alpha,M_1,M_2)$, i.e. M_3 is a constant depending only on α , M_1 and M_2 .

Applying Theorem 1.2 and Lemma 1.3, we can derive the estimate for solutions of Problem B in the (N+1)-connected circular domain.

Theorem 1.4 Suppose that $\Phi(z)$ is a solution of Problem B. Then $\Phi(z)$ satisfies the following estimate

$$C_\alpha[\Phi,\overline{D}] = C[\Phi,\overline{D}] + H_\alpha[\Phi,\overline{D}] \le M_4 , \tag{1.17}$$

where $M_4 = M_4(\alpha,\ell,K,D)$.

Proof. We discuss the three cases for the index, $K \ge N$, $-1 \le K < N$ and $K < -1$, successively.

(1) $K \ge N$. We first prove the boundedness of the solution:

$$C[\Phi,\overline{D}] = \max_{z\in\overline{D}} |\Phi(z)| \le M_5 = M_5(\alpha,\ell,K,D) . \tag{1.18}$$

(For the sake of convenience, sometimes we shall denote the constant $M(\alpha,\ell,K,D)$ by $M(\alpha,\ell,D)$.)

Otherwise, there exist two sequences of functions $\{\lambda_n(t)\}$, $\{r_n(t)\}$ and a sequence of numbers $\{b_j^{(n)}\}$ $(j=1,\ldots,2K-N+1)$, $|\lambda_n(t)| = 1$, $C_\alpha[\lambda_n,\Gamma] \le \ell$, $C_\alpha[r_n,\Gamma] \le \ell$, $|b_j^{(n)}| \le \ell$ $(j=1,\ldots,2K-N+1)$ so that when $n \to \infty$, $\{\lambda_n(t)\}$, $\{r_n(t)\}$ converge uniformly to $\lambda_0(t)$, $r_0(t)$ on Γ , respectively and $\{b_j^{(n)}\}$ converges to $b_j^{(n)}$ $(j=1,\ldots,2K-N+1)$, and there exists a sequence of solutions $\{\Phi_n(z)\}$ for Problem B which satisfy the boundary conditions

$$Re[\overline{\lambda_n(t)}\Phi_n(t)] = r_n(t) , \quad t \in \Gamma , \tag{1.19}$$

$$Im[\overline{\lambda_n(a_j)}\Phi_n(a_j)] = b_j^{(n)} , \quad j = 1,\ldots 2K-N+1 , \tag{1.20}$$

and when $n \to \infty$,

$$C[\Phi_n,\overline{D}] = H_n \to \infty . \tag{1.21}$$

We may assume $H_n \ge 1$ $(n=1,2,\ldots)$, it then is obvious that $\psi_n(z) = \Phi_n(z)/H_n$ satisfies the boundary conditions

$$Re[\overline{\lambda_n(t)}\psi_n(t)] = r_n(t)/H_n , \quad t \in \Gamma , \tag{1.22}$$

$$Im[\overline{\lambda_n(a_j)}\Psi_n(a_j)] = b_j^{(n)}/H_n \ , \ j = 1,\ldots,2K-N+1 \ , \tag{1.23}$$

and $|\Psi_n(z)| \le 1$, $z \in D$. On the basis of the Cauchy formula for analytic functions, the estimate of $\Psi(z)$ on any closed set D_* in D is derived easily

$$C^1[\Psi_n,D_*] = C[\Psi_n,D_*] + C[\Psi_n',D_*] \le M_6 = M_6(D,D_*) \ . \tag{1.24}$$

Let $2d$ denote the least distance between $\Gamma_j(j=0,1,\ldots,N)$ and $z = 0$, and $K_0 = 1/(2\pi)\Delta_{\Gamma_0} arg\,\lambda_n(t)$. First we see that the analytic function $\Psi_n^*(z) = z^{-K_0}\Psi_n(z)$ in $D_0 = \{1-d<|z|<1\}$ satisfies

$$Re[\overline{\lambda_n^*(t)}\Psi_n^*(t)] = r_n(t)/H_n \ , \ \lambda_n^*(t) = \lambda_n(t)t^{-K_0} \ , \ t \in \Gamma_0 \tag{1.25}$$

where $1/(2\pi)\Delta_\Gamma arg\,\lambda_n^*(t) = 0$. By Lemma 1.3, the analytic function in $\mathbb{D} = \{|z|<1\}$

$$S_n(z) = \frac{1}{2\pi i}\int_{\Gamma_0}\frac{(t+z)arg\lambda_n^*(t)}{(t-z)t}\,dt \tag{1.26}$$

satisfies the estimate

$$C_\alpha[S_n,\overline{\mathbb{D}}] \le M_7 = M_7(\alpha,\ell) \ . \tag{1.27}$$

Hence the function $e^{-iS_n(z)}\Psi_n^*(z)$ is analytic in D_0 , and satisfies the condition on Γ_0:

$$Re[e^{-iS_n(t)}\Psi_n^*(t)] = r_n(t)e^{ImS_n(t)}/H_n = r_n^*(t) \ , \ t \in \Gamma_0 \ . \tag{1.28}$$

Similarly to (1.26) and (1.27), the analytic function in $\mathbb{D}$

$$T_n(z) = \frac{1}{2\pi i}\int_{\Gamma_0}\frac{(t+z)r_n^*(t)}{(t-z)t}\,dt \tag{1.29}$$

satisfies the estimate

$$C_\alpha[T_n, \mathbb{D}] \le M_8 = M_8(\alpha,\ell) . \qquad (1.30)$$

Because the analytic function $T_n^*(z) = e^{-iS_n(z)} \Psi_n^*(z) - T_n(z)$ satisfies a homogeneous boundary condition on Γ_0,

$$Re[T_n^*(t)] = 0 , \ t \in \Gamma_0 , \qquad (1.31)$$

we can define an analytic function on $\tilde{D} = D_0 \cup D_0^*$, $D_0^* = \{z: 1 \le |z| < \frac{1}{1-d}\}$

$$\Phi_n^*(z) = \begin{cases} T_n^*(z) , & z \in D_0 , \\ -\overline{T_n^*(\frac{1}{\bar{z}})} , & z \in D_0^* . \end{cases} \qquad (1.32)$$

Taking account of the boundedness of $\Phi_n^*(z)$, i.e.

$$C[\Phi_n^*, \tilde{D}] \le M_9 = M_9(\alpha,\ell,D) , \qquad (1.33)$$

and using the Cauchy formula, we can obtain the estimate

$$C^1[\Phi_n^*, \tilde{D}_0] \le M_{10} = M_{10}(\alpha,\ell,D) , \qquad (1.34)$$

where $\tilde{D}_0 = \{1-\frac{d}{2} \le |z| \le \frac{2}{2-d}\}$. From (1.27), (1.30), and (1.34), it follows that

$$C_\alpha[\Psi_n, \hat{D}_0] \le M_{11} = M_{11}(\alpha,\ell,D) , \qquad (1.35)$$

where $\hat{D}_0 = \{1-\frac{d}{2} \le |z| \le 1\}$. As for the estimates of $\Psi_n(z)$ on $\hat{D}_j = \{\gamma_j \le |z-z_j| \le \gamma_j + \frac{d}{2}\}$, we may use the conformal mapping $\zeta = \frac{\gamma_j}{z-z_j}$, which maps $D_j = \{\gamma_j < |z-z_j| < \gamma_j + d\}$ onto $G_j = \{\frac{\gamma_j}{\gamma_j+d} < |\zeta| < 1\}$ $(j=1,\dots,N)$. Then using the same argument as in the proof of (1.35), we can derive that $\Psi_n(z)$ satisfies the estimate

$$C_\alpha[\Psi_n, \hat{D}_j] \le M_{12} = M_{12}(\alpha,\ell,D) , \qquad (1.36)$$

where $\hat{D}_j = \{\gamma_j \le |z-z_j| \le \gamma_j + \frac{d}{2}\}$, $j = 1,\dots,N$. Thus combining

(1.24), (1.35), and (1.36), we obtain

$$C_\alpha[\Psi_n,\overline{D}] \le M_{13} = M_{13}(\alpha,\ell,D) . \qquad (1.37)$$

Therefore, we can choose a subsequence of $\{\Psi_n(z)\}$, which converges uniformly to an analytic function $\Psi_0(z)$ in D, satisfying the estimate (1.37) and the boundary conditions

$$Re[\overline{\lambda_0(t)}\Psi_0(t)] = 0 , \quad t \in \Gamma , \qquad (1.38)$$

$$Im[\overline{\lambda_0(a_j)}\Psi_0(a_j)] = 0 , \quad j = 1,\ldots,2K-N+1 . \qquad (1.39)$$

On the basis of Theorem 1.2, it is easily shown that $\Psi_0(z) = 0 , z \in D$. However on the other side, from

$$\max_{z\in\overline{D}} |\Phi_n(z)|/H_n = 1 ,$$

we have

$\max_{z\in\overline{D}} |\Psi_0(z)| = 1$. This contradiction proves the estimate (1.18). Using the same method as in the proof of (1.37), we obtain also the estimate (1.17).

(2) $-1 \le K < N$. Substitute the solution $\Phi(z)$ for Problem B $(-1 \le K < N)$ into the boundary condition (1.2). Then $h(t)$ in (1.2) is determined. Using the reductio ad absurdum and Theorem 1.2, we can prove

$$|h(t)| \le M_{14} = M_{14}(\alpha,\ell,K,D) . \qquad (1.40)$$

After that, similarly to (1), the estimate (1.17) can be derived.

(3) $K < -1$. Let $\Phi(z)$ be the solution for Problem B $(K < -1)$ and denote

$$\Psi_*(z) = z^{|K|-1}\,[\Phi(z) + \sum_{m=1}^{|K|-1} H_m z^{-m}]\ . \qquad (1.41)$$

Then $\Psi_*(z)$ is analytic in D and satisfies the boundary condition

$$Re[\overline{\lambda_1(t)}\Psi_*(t)] = r(t) + h(t)\ ,\ t \in \Gamma\ , \qquad (1.42)$$

where $h(t) = h_j$, $t \in \Gamma_j$, $h_j (j=0,1,\ldots,N)$ are all constants and the index of $\lambda_1(t) = \lambda(t)\overline{t}^{\,1-|K|}$ is equal to -1 . According to (2), we conclude the estimate (1.17). □

4. Solvability of Problem B for analytic functions

Applying the results of conformal mappings of multiply connected domains, or the theory of subharmonic functions, we can prove the existence of solutions to the Dirichlet problem for harmonic functions in the multiply connected domain D (cf. [103]24)). Now using the above result, we shall derive the following theorem for the modified Dirichlet problem D for analytic functions in D .

Theorem 1.5 The Problem D for analytic functions has a unique solution.

Proof. Let $\Phi_1(z)$ and $\Phi_2(z)$ be two solutions of Problem D and denote $\Phi(z) = \Phi_1(z) - \Phi_2(z) = u(z) + iv(z)$. By the Green formula, we have

$$\iint_D [(u_x)^2 + (u_y)^2]dxdy = \int_\Gamma [uu_x dy - uu_y dx] =$$

$$= \int_\Gamma u\frac{\partial u}{\partial n}\, ds = \int_\Gamma u\frac{\partial v}{\partial s}\, ds = \sum_{j=1}^{N} h_j \int_{\Gamma_j} dv = 0\ . \qquad (1.43)$$

Consequently, $u_x = u_y = 0$, $z \in D$, and $u(z) = v(z) = 0$, i.e. $\Phi_1(z) = \Phi_2(z)$, $z \in D$.

To prove the existence of the solution of Problem D, we first seek a harmonic function $u(z)$ with the boundary condition

$$u(t) = \Phi(t) \ , \ t \in \Gamma \ , \tag{1.44}$$

and the harmonic measure $\omega_j(z)$ in D with the boundary condition

$$\omega_j(t) = \delta_{jk} = \begin{cases} 1, t \in \Gamma_j \ , k=j \\ 0, t \in \Gamma_k \ , k \neq j \ , \end{cases} \quad j = 0,1,\ldots,N \ . \tag{1.45}$$

By using reductio ad absurdum and the uniqueness of the solution of the Dirichlet problem for harmonic functions in D , we can obtain

$$J = \begin{vmatrix} \gamma_{11} & \cdots & \gamma_{1N} \\ \vdots & & \vdots \\ \gamma_{N1} & \cdots & \gamma_{NN} \end{vmatrix} \neq 0, \gamma_{jk} = \frac{1}{2\pi} \int_{\Gamma_j} \frac{\partial \omega_k}{\partial n} ds \ , \ j,k=1,\ldots,N \ , \tag{1.46}$$

where $\vec{n}$ is the outward normal vector on Γ . Hence the system of linear algebraic equations

$$\sum_{k=1}^{N} h_k \gamma_{jk} = \beta_j = -\frac{1}{2\pi} \int_{\Gamma_j} \frac{\partial u}{\partial n} ds \ , \ j = 1,\ldots,N \tag{1.47}$$

has a solution $h_k (k=1,\ldots,N)$. Furthermore the conjugate harmonic function $V(z)$ of the harmonic function

$$U(z) = u(z) + \sum_{k=1}^{N} h_k \omega_k(z) \tag{1.48}$$

satisfies the equalities

$$\frac{1}{2\pi} \int_{\Gamma_j} dV = \frac{1}{2\pi} \int_{\Gamma_j} \frac{\partial U}{\partial n} ds = -\beta_j + \sum_{k=1}^{N} h_k \gamma_{jk} = 0 \ , \ j=1,\ldots,N \ . \tag{1.49}$$

This shows that $\Phi(z) = U(z) + iV(z)$ is a single-valued in D , and $\Phi(z)$ is a solution of Problem D.

<u>Lemma 1.6</u> Let ℓ_1 be a given positive constant. If $C_\alpha[r,\Gamma] \leq \ell$, $|b_j| \leq \ell (j \in \{j\})$ in (1.2), (1.3) and (1.5) are replaced by

$$C_\alpha[r,\Gamma] \le \ell_1 \ , \ |b_j| \le \ell_1 \ , \ j \in \{j\} \ , \tag{1.50}$$

then the solution $\Phi(z)$ of Problem B satisfies the estimate

$$C_\alpha[\Phi,\overline{D}] \le \ell_1 M_{15}, \ M_{15} = M_{15}(\alpha,\ell,K,D) \ . \tag{1.51}$$

Proof. Put $\Psi(z) = \Phi(z)/\ell_1$. The analytic function $\Psi(z)$ in D satisfies the boundary condition

$$Re[\overline{\lambda(t)}\Psi(t)] = r(t)/\ell_1 + h(t) \ , \ t \in \Gamma \ , \tag{1.52}$$

$$Im[\overline{\lambda(a_j)}\Psi(a_j)] = b_j/\ell_1 \ , \ j \in \{j\} \ . \tag{1.53}$$

On account of $C_\alpha[r/\ell_1,\Gamma] \le 1$, $|b_j/\ell_1| \le 1$, $j \in \{j\}$, it follows from Theorem 1.4 that $\Psi(z)$ satisfies the estimate

$$C_\alpha[\Psi,\overline{D}] \le M_{15} \ . \tag{1.54}$$

Hence the estimate (1.51) is true. □

Theorem 1.7 If the index $K = 0$, then Problem B for analytic functions has a solution.

Proof. Due to the relation between Problem B and Problem C, we shall discuss only Problem C. We can suppose that $Im\ \Psi(1) = 0$. Setting $\theta(t) = \frac{1}{i}\ ln\ \Lambda(t)$, i.e. $\Lambda(t) = e^{i\theta(t)}$, we have $C_\alpha[\theta,\Gamma] \le \ell_2 = \ell_2(\alpha,\ell,D)$. Now using the method of parameter extension, we introduce the boundary value problem D* with the boundary condition

$$Re[e^{ik\theta(t)}\Psi(t)] = s(t) + H(t) \ , \ t \in \Gamma \ , \ Im\ \Psi(1) = 0 \ , \tag{1.55}$$

where k is a parameter $(0 \le k \le 1)$ and $s \in C_\alpha(\Gamma)$. It is evident that when $k = 0$, Problem D^* is the modified Problem D and so there exists a solution. Suppose that when $k = k_0 (0 \le k_0 < 1)$, Problem D* is solvable. Then we can find a neighbourhood

$K_\delta = \{|k-k_0| \le \delta, 0 \le k \le 1\}$ so that for every $k \in K_\delta$, Problem D* is solvable. In fact, the boundary condition (1.55) can be written in the form

$$Re[e^{ik_0\theta(t)}\Psi(t)]$$
$$= Re[(e^{ik_0\theta(t)}-e^{ik\theta(t)})\Psi(t)] + s(t) + H(t), \quad Im\,\Psi(1) = 0 . \tag{1.56}$$

Replacing $\Psi(z)$ on the right-hand side of (1.56) by a function $\Psi_0(z) \in C_\alpha(\overline{D})$, especially by $\Psi_0(z) = 0$, obviously, the boundary value problem (1.56) has a solution $\Psi_1(z)$. By using successive iteration, we obtain a sequence of solutions $\Psi_n(z)$, $n = 1,2,\ldots$, which satisfy for $t \in \Gamma$

$$\left.\begin{aligned} &Re[e^{ik_0\theta(t)}\Psi_{n+1}(t)] = Re[(e^{ik_0\theta(t)}-e^{ik\theta(t)})\Psi_n(t)] + s(t) + H(t), \\ &Im\,\Psi_{n+1}(1) = 0, \quad n = 1,2,\ldots . \end{aligned}\right\} \tag{1.57}$$

It follows from the above formulae that

$$\left.\begin{aligned} &Re[e^{ik_0\theta(t)}(\Psi_{n+1}(t)-\Psi_n(t))] = g_n(t) + H(t), \quad Im[\Psi_{n+1}(1)-\Psi_n(1)] = 0 \\ &g_n(t) = Re[(e^{ik_0\theta(t)}-e^{ik\theta(t)})(\Psi_n(t)-\Psi_{n-1}(t))], \quad t \in \Gamma . \end{aligned}\right\} \tag{1.58}$$

Taking

$$e^{ik\theta(t)}-e^{ik_0\theta(t)} = \int_{k_0}^{k} i\theta(t)e^{is\theta(t)}ds$$

into account, it is easy to see that

$$C_\alpha[e^{ik\theta(t)}-e^{ik_0\theta(t)},\Gamma] \le |k-k_0|M_{16}, \quad M_{16} = M_{16}(\alpha,\ell,D) .$$

We choose $|k-k_0| < \delta = 1/(2M_{15}M_{16}+1)$, so that $|k-k_0|M_{15}M_{16} < \frac{1}{2}$. From Lemma 1.6, we have

$$C_\alpha[\Psi_{n+1}-\Psi_n,\overline{D}] \le |k-k_0|M_{15}\,M_{16}\,C_\alpha[\Psi_n-\Psi_{n-1},\overline{D}] < \frac{1}{2}\,C_\alpha[\Psi_n-\Psi_{n-1},\overline{D}] \tag{1.59}$$

and when $n,m > N_0 + 1$ (N_0 is a positive integer),

$$C_\alpha[\Psi_n-\Psi_m,\overline{D}] \le \frac{1}{2^{N_0}} \sum_{j=1}^{\infty} \frac{1}{2^j} C_\alpha[\Psi_2-\Psi_1,\overline{D}] \le \frac{1}{2^{N_0}} C_\alpha[\Psi_2-\Psi_1,\overline{D}] . \tag{1.60}$$

Hence $\{\Psi_n(z)\}$ is a Cauchy sequence. Due to the completeness of the Banach space $C_\alpha(\overline{D})$, there exists a function $\Psi_0(z) \in C_\alpha(\overline{D})$, so that $C_\alpha[\Psi_n-\Psi_0,\overline{D}] \to 0$, for $n \to \infty$. From (1.57), we can obtain that $\Psi_0(z)$ satisfies the condition (1.55). It is shown that $\Psi_0(z)$ is a solution of Problem D* for every $k \in K_\delta$. Because the constant δ is independent of $k_0 (0 \le k_0 < 1)$, therefore from the solvability of Problem D* when $k_0 = 0$, we can derive the solvability of Problem D* when $t = \delta, 2\delta, \ldots, [\frac{1}{\delta}]\delta, 1$. In particular, when $t = 1$ and $s(t) = R(t)$, Problem D* is Problem C with the index $K = 0$. This completes the proof. □

Theorem 1.8 When the index $K \neq 0$, Problem B for analytic functions is solvable.

Proof. (1) $0 < K \le N$. We first discuss the case $K = 1$. Using Theorem 1.7 solutions $\Psi_0(z)$, $\Psi_1(z)$, $\Psi_2(z)$ of Problem B can be found, which satisfy the boundary conditions

$$\left.\begin{aligned} &Re[\lambda_2(t)\Psi_0(t)] = r(t) + h(t), \overline{\lambda_2(t)} = \overline{\lambda(t)}t, t \in \Gamma, \Delta_\Gamma arg\, \lambda_2(t) = 0, \\ &Im[\overline{\lambda_2(a_0)}\Psi_0(a_0)] = b_0 = b_{N+1}, a_0 = a_{N+1} \in \Gamma_0 = \Gamma_{N+1}, \end{aligned}\right\} \tag{1.61}$$

$$\left.\begin{aligned} &Re[\overline{\lambda_2(t)}\Psi_1(t)] = Re[\overline{\lambda_2(t)}\frac{1}{t}] + h(t), t \in \Gamma, \\ &Im[\overline{\lambda_2(a_0)}\Psi_1(a_0)] = Im[\overline{\lambda_2(a_0)}\frac{1}{a_0}], \end{aligned}\right\} \tag{1.62}$$

$$\left.\begin{aligned} &Re[\overline{\lambda_2(t)}\Psi_2(t)] = Re[\overline{\lambda_2(t)}\frac{i}{t}] + h(t), t \in \Gamma, \\ &Im[\overline{\lambda_2(a_0)}\Psi_2(a_0)] = Im[\overline{\lambda_2(a_0)}\frac{i}{a_0}], \end{aligned}\right\} \tag{1.63}$$

respectively. Putting $\Phi_1(z) = \Psi_1(z) - 1/z$, $\Phi_2(z) = \Psi_2(z) - i/z$, it is evident that $\Phi_1(z)$ and $\Phi_2(z)$ are linearly independent over $\mathbb{R}$, so that

$$J = \begin{vmatrix} Re\ \Phi_1(a_N) & Im\ \Phi_1(a_N) \\ Re\ \Phi_2(a_N) & Im\ \Phi_2(a_N) \end{vmatrix} \neq 0 . \tag{1.64}$$

Hence the following algebraic system has a solution C_1 , C_2 :

$$\left.\begin{aligned} C_1 Re\ \Phi_1(a_N) + C_2 Re\ \Phi_2(a_N) &= Re\frac{\lambda_2(a_N)}{|a_N|^2}(h_N+ib_N^*-ib_N) \\ C_1 Im\ \Phi_1(a_N) + C_2 Im\ \Phi_2(a_N) &= Im\frac{\lambda_2(a_N)}{|a_N|^2}(h_N+ib_N^*-ib_N) \end{aligned}\right\} \tag{1.65}$$

where $h_N + ib_N^* = \overline{\lambda_2(a_n)}\ \Psi_0(a_N) - r(a_N)$. Thus the function $\Phi(z) = z\ \Psi(z) = z[\Psi_0(z)-C_1\Phi_1(z)-C_2\Phi_2(z)]$ is a solution of Problem B with $K = 1$. Similarly, we can derive the solvability of Problem B with the index $K(1 < K \leq N)$.

(2) $K > N$. In order to prove the solvability of Problem B with $K = N+1$, we find solutions $\Psi_0(z)$, $\Psi_1(z)$, $\Psi_2(z)$ of the boundary value problems with the boundary conditions:

$$\left.\begin{aligned} &Re[\overline{\lambda_3(t)}\Psi_0(t)] = r(t), \overline{\lambda_3(t)} = \overline{\lambda(t)}t, t \in \Gamma, \frac{1}{2\pi}\Delta_\Gamma\, arg\, \lambda_3(t) = N, \\ &Im[\overline{\lambda_3(a_j)}\Psi_0(a_j)] = b_j\ , \ a_j \in \Gamma_j\ , \ j = 1,\ldots,N+1\ , \end{aligned}\right\} \tag{1.66}$$

$$\left.\begin{aligned} &Re[\overline{\lambda_3(t)}\Psi_1(t)] = Re[\overline{\lambda_3(t)}\frac{1}{t}]\ , \ t \in \Gamma\ , \\ &Im[\overline{\lambda_3(a_j)}\Psi_1(a_j)] = Im[\overline{\lambda_3(a_j)}\frac{1}{a_j}]\ , \ j = 1,\ldots,N+1, \end{aligned}\right\} \tag{1.67}$$

$$\left.\begin{aligned} &Re[\overline{\lambda_3(t)}\Psi_2(t)] = Re[\overline{\lambda_3(t)}\frac{i}{t}]\ , \ t \in \Gamma\ , \\ &Im[\overline{\lambda_3(a_j)}\Psi_2(a_j)] = Im[\overline{\lambda_3(a_j)}\frac{i}{a_j}]\ , \ j = 1,\ldots,N+1, \end{aligned}\right\} \tag{1.68}$$

corresponding to (1.61), (1.62) and (1.63), respectively and proceed as in (1). Similarly, we can prove the solvability of Problem B with the index $K(N < K < \infty)$.

(3) If $K = -1$, we consider the boundary value problems with the boundary conditions

$$\left.\begin{aligned} &Re[\overline{\lambda_4(t)}\Psi_0(t)] = r(t) + h(t) \ , \ \overline{\lambda_4(t)} = \overline{\lambda(t)}/t \ , \ t \in \Gamma \ , \\ &\frac{1}{2\pi}\Delta_\Gamma \, arg \ \lambda_4(t) = 0 \ , \end{aligned}\right\} \tag{1.69}$$

$$\left.\begin{aligned} &Re[\overline{\lambda_4(t)}\Psi_1(t)] = 1 + h(t) \ , \ t \in \Gamma \ , \\ &Im[\overline{\lambda_4(a_0)}\Psi_1(a_0)] = 0 \ , \end{aligned}\right\} \tag{1.70}$$

$$\left.\begin{aligned} &Re[\overline{\lambda_4(t)}\Psi_2(t)] = h(t) \ , \ t \in \Gamma \ , \\ &Im[\overline{\lambda_4(a_0)}\Psi_2(a_0)] = 1 \ , \end{aligned}\right\} \tag{1.71}$$

respectively and can find the solution C_1 , C_2 of the algebraic system

$$\left.\begin{aligned} &C_1 Re \ \Psi_1(0) + C_2 Re \ \Psi_2(0) = Re \ \Psi_0(0) \\ &C_1 Im \ \Psi_1(0) + C_2 Im \ \Psi_2(0) = Im \ \Psi_0(0) \end{aligned}\right\}. \tag{1.72}$$

Thus, the function $\Phi(z) = [\Psi_0(z) - C_1\Psi_1(z) - C_2\Psi_2(z)]/z$ is a solution of Problem B with $K = -1$.

In the case $K < -1$, similarly to (1.15), the function

$$\Psi(z) = z^{|K|-1}[\Phi(z) + \sum_{m=1}^{|K|-1} (H_m^+ + iH_m^-)z^{-m}] \ , \ z \in D \ ,$$

is an analytic function satisfying the boundary condition (1.15), and then the function

$$\Phi(z) = \Psi(z)z^{1-|K|} - \sum_{m=1}^{|K|-1} (H_m^+ + iH_m^-)z^{-m}$$

is a solution of Problem B with the index $K < -1$. □

Theorem 1.9 (1) If the index $K \geq N$, Problem A is solvable and the general solution $\Phi(z)$ admits the representation

$$\Phi(z) = \Phi_0(z) + \sum_{k=1}^{J} c_k \Phi_k(z) \ , \tag{1.73}$$

where $\Phi_0(z)$ is a special solution to Problem A, $\Phi_k(z)(k=1,\ldots,J\ ,\ J = 2K-N+1)$ are over $\mathbb{R}$ linearly independent solutions of the corresponding homogeneous boundary value problem.

(2) If $0 \le K < N$, the general solution $\Phi(z)$ of Problem A has to satisfy $N - K$ solvability conditions. If the solvability conditions are satisfied, then its general solutions $\Phi(z)$ can be expressed by (1.73), where $J = K+1$.

(3) If $K < 0$, Problem A has $N-2K-1$ solvability conditions.

Proof. (1) In this case, Problem A coincides with Problem B. Let us denote by $\Phi_k(z)(k=1,\ldots,2K-N+1)$ the linearly independent solutions of Problem B_0 defined by the point conditions

$$\overline{\lambda(a_j)}\Phi(a_j) = \begin{cases} i\ , & j = k\ , \\ 0\ , & j \neq k\ , \end{cases} \quad k=1,\ldots,2K-N+1\ . \tag{1.74}$$

Then every solution $\Phi(z)$ of Problem A can be expressed in the form (1.73), where the constants $c_k(k=1,\ldots 2K-N+1)$ are determined by the following formulae

$$ic_k = \overline{\lambda(a_k)}[\Phi(a_k)-\Phi_0(a_k)]\ ,\ k=1,\ldots,2K-N+1\ . \tag{1.75}$$

From Theorem 1.2, it is easy to see that

$$\Phi(z) - \Phi_0(z) = \sum_{k=1}^{2K-N+1} c_k\Phi_k(z)\ , \tag{1.76}$$

consequently, we have the expression (1.73).

(2) By using a method similar to that of the case (1), we can obtain the expression (1.73) with $J = K+1$.

(3) Substituting the solution $\Phi(z)$ of Problem B into (1.2) where $\Phi(z)$ in the case $h(t) = 0$ is a solution of Problem A, too it

follows that Problem A has $N-2K-1$ solvability conditions.

5. The compound boundary value problem for analytic functions

Let D be an $(N+1)$-connected domain as stated above. In the interior of D, there are n mutually disjoint contours $L_j (j=1,\ldots,m)$. Set

$$L = \bigcup_{j=1}^{m} L_j \in C_\mu (0 < \mu < 1)$$

and denote

$$D^+ = D \setminus D^- \ , \ D^- = \bigcup_{j=1}^{m} D_j^- \ , \ D_j^- = D \cap d_j \ ,$$

where d_j is the bounded domain surrounded by L_j. Without loss of generality, we assume $0 \in D$.

Problem F. The compound boundary value problem for analytic functions is to find a sectionally analytic function

$$\phi(z) = \begin{cases} \phi^+(z) \ , \ z \in D^+ \ , \\ \phi^-(z) \ , \ z \in D^- \ , \end{cases}$$

satisfying the boundary conditions

$$\phi^+(t) = G(t)\phi^-(t) + g(t) \ , \ t \in L \tag{1.77}$$

$$Re[\overline{\lambda(t)}\phi(t)] = r(t) \ , \ t \in \Gamma \ , \tag{1.78}$$

where $\lambda(t)$, $r(t)$ are stated in (1.1), and $G(t)$, $g(t)$ satisfy

$$C_\nu[G,L] \le d < \infty \ , \ G(t) \neq 0 \ , \ t \in L \ , \ C_\nu[g,L] \le d \ , \ 0 < \nu < 1 \ . \tag{1.79}$$

The above boundary value problem is called Problem F, and Problem F with $g(t) = 0$, $r(t) = 0$ is called Problem F_0. The index of Problem F and Problem F_0 is defined as

$$K = \frac{1}{2\pi} [\Delta_\Gamma \ arg \ \lambda(t) + \Delta_L \ arg \ G(t)] \ . \tag{1.80}$$

By using the method of elimination by $X\Phi = \phi - \Phi_0$ (cf.[71]2)), the boundary conditions (1.77), (1.78) can be transformed into the following Riemann-Hilbert boundary condition

$$Re[\overline{\Lambda(t)}\Phi(t)] = R(t) \ , \ t \in \Gamma \tag{1.81}$$

where

$$\left.\begin{aligned} &\Lambda(t) = \lambda(t)\overline{X(t)} \ , \ R(t) = r(t) - Re[\overline{\lambda(t)}\Phi_0(t)] \ , \ t \in \Gamma \ , \\ &\Phi_0(z) = X(z)\Psi(z) \ , \ \Psi(z) = \frac{1}{2\pi i}\int_L \frac{g(t)}{X^+(t)}\frac{dt}{t-z} \ , \\ &X(z) = \begin{cases} X^+(z) = e^{\Gamma^+(z)/\Pi(z)} , z \in D^+ , \\ X^-(z) = e^{\Gamma^-(z)} , \ z \in D^- , \end{cases} \quad \Gamma(z) = \frac{1}{2\pi i}\int_L \frac{\ln[G(t)\Pi(t)]}{t-z}\,dt \ , \end{aligned}\right\} \tag{1.82}$$

in which $\Pi(z) = \prod_{j=1}^{m} (z-s_j)^{K_j}$, $s_j \in d_j$, and

$K_j = \frac{1}{2\pi}\Delta_{L_j} \arg G(t)$, $j = 1,\ldots,m$.

Since Problem F with $K < N$ may not be solvable, we discuss the modified boundary value problem, where the boundary condition (1.78) is replaced by

$$Re[\overline{\lambda(t)}\phi(t)] = r(t) + h(t) \ , \ t \in \Gamma \ , \tag{1.83}$$

$$h(t) = \begin{cases} 0 \ , \ K \ge N \ , \\ \left.\begin{aligned} &h_j \ , \ t \in \Gamma_j \ , \ j = 1,\ldots,N-K \ , \\ &0 \ , \ t \in \Gamma_j \ , \ j = N-K+1,\ldots,N+1 , \end{aligned}\right\} 0 \le K < N \ , \\ h_j - Re[\overline{\lambda(t)}X(t) \displaystyle\sum_{m=1}^{|K|-1} (H_m^+ + iH_m^-)t^{-m}] , t \in \Gamma_j , j=0,\ldots,N, K < 0 \ , \end{cases} \tag{1.84}$$

where $h_j\,(j=0,1,\ldots,N)$ and $H_m^{\pm}\,(m=1,\ldots,|K|-1)$ are appropriate real constants. Besides we suppose that the solution satisfies the point conditions

$$Im[\overline{\lambda(a_j)}\phi(a_j)] = b_j \ , j \in \{j\} = \begin{cases} 1,\ldots,2K-N+1 \ , \ K \geq N \ , \\ N-K+1,\ldots,N+1 \ , \ 0 \leq K < N \ , \end{cases} \tag{1.85}$$

where a_j , b_j are as stated in (1.3) and(1.5). The above modified boundary value problem is called Problem G, and Problem G with $g(t) = 0$, $r(t) = 0$ is called Problem G_0 .

Theorem 1.10 Let $\phi(t)$ be a solution of Problem G for analytic functions. Then the function

$$\Phi(z) = [\phi(z)-\Phi_0(z)]/X(z) \tag{1.86}$$

is a solution of Problem B for analytic functions, which satisfies the boundary condition

$$Re[\overline{\Lambda(t)}\Phi(t)] = R(t) + h(t) \ , \ t \in \Gamma \ . \tag{1.87}$$

Conversely, let $\Phi(z)$ be a solution of Problem B. Then

$$\phi(z) = \Phi_0(z) + X(z)\Phi(z) \tag{1.88}$$

is a solution of Problem G.

Proof. Let $\phi(z)$ be a solution of Problem G. The functions $\Phi_0(z)$, $X(z)$ are given by (1.82), and $X(z)$ satisfies the boundary condition

$$X^+(z) = G(t)X^-(t) \ , \ t \in L \ . \tag{1.89}$$

Therefore $\phi(z)$ satisfies the following boundary condition

$$\begin{aligned} \phi^+(t) &= \Phi_0^+(t) + X^+(t)\Phi^+(t) = G(t)[\Phi_0^-(t) + X^-(t)\Phi^+(t)] + g(t) \\ &= G(t)\phi^-(t) + g(t) = G(t)[\Phi_0^-(t) + X^-(t)\Phi^-(t)] + g(t) \ , \ t \in L \ . \end{aligned} \tag{1.90}$$

From $X^-(t) \neq 0$, $t \in L$, it follows $\Phi^+(t) = \Phi^-(t)$, $t \in L$. Hence

$$\Phi(z) = \begin{cases} \Phi^+(z), z\in D^+ \\ \Phi^-(z), z\in D^- \end{cases}$$

is an analytic function in D and is a solution of Problem B.

Conversely, it is clear that the second part of the result is true, too. □

On the basis of Theorem 1.2, Theorem 1.9 and the above theorem, we can obtain the following theorem.

Theorem 1.11 Problem G for analytic functions has a unique solution.

(1) When $K \geq N$, the general solution of Problem F contains $2K-N+1$ arbitrary real constants.

(2) When $0 \leq K < N$, Problem F has $N-K$ solvability conditions. If the solvability conditions are satisfied, then its general solution includes $K+1$ arbitrary real constants.

(3) When $K < 0$, Problem F has $N-2K-1$ solvability conditions.

§ 2. Some boundary value problems for harmonic functions

Now, we shall discuss some basic boundary value problems for harmonic functions, i.e. the Dirichlet problem, the Neumann problem, the regular oblique derivative problem and the irregular oblique derivative problem in a multiply connected domain. We shall establish relations between the above problems for harmonic functions and the Riemann-Hilbert boundary value problem for analytic functions and use some theorems from § 1 to prove the solvability of the above problems for harmonic functions, which include some new results.

1. Formulation and uniqueness of the regular oblique derivative problem

As stated in § 1, there is no harm in assuming that the multiply connected domain D is a circular domain.

Problem IV. The so-called regular oblique derivative problem for harmonic functions is formulated as follows: Find a harmonic function $u(z)$ in D, continuously differentiable on $\overline{D}$ and satisfying the boundary condition

$$\frac{\partial u}{\partial \nu} + 2\sigma(t)u(t) = 2\tau(t) + h(t)\ ,\ t \in \Gamma\ ,\ u(a_0) = b_0\ ,\ a_0 = 1 \in \Gamma_0\ , \qquad (2.1)$$

where $\vec{\nu}$ can be arbitrarily prescribed provided that $\cos(\nu,n) \geq 0$ on Γ, $\vec{n}$ being the outward normal vector. We note that

$$\begin{aligned}\frac{\partial u}{\partial \nu} &= \frac{\partial u}{\partial x}\cos(\nu,x) + \frac{\partial u}{\partial y}\sin(\nu,x) = (u_z+u_{\bar{z}})\cos(\nu,x) \\ &\quad + i(u_z-u_{\bar{z}})\sin(\nu,x) = 2Re[\overline{\lambda(z)}u_z]\ ,\ z \in \Gamma\ ,\end{aligned} \qquad (2.2)$$

where $\lambda(z) = \cos(\nu,x) + i\sin(\nu,x) = e^{i(\nu,x)}$, (ν,x) is the angle between $\vec{\nu}$ and the x-axis. The boundary condition (2.1) can be written in the following form

$$Re[\overline{\lambda(t)}u_t] + \sigma(t)u(t) = \tau(t) + h(t)\ ,\ t \in \Gamma\ ,\ u(a_0) = b_0\ . \qquad (2.3)$$

Suppose $\lambda(t)$, $\sigma(t)$, $\tau(t)$ and b_0 satisfy

$$C_\alpha[\lambda,\Gamma] \leq \ell < \infty\ ,\ \sigma(t) \geq 0\ ,\ C_\alpha[\sigma,\Gamma] \leq \ell\ ,\ C_\alpha[\tau,\Gamma] \leq \ell\ ,\ |b_0| \leq \ell\ , \qquad (2.4)$$

where $\alpha(0 < \alpha < 1)$ is a constant. If $\cos(\nu,n) = 0$, and $\sigma(t) = 0$ for every point t on Γ_j, we assume

$$\int_{\Gamma_j} \tau(t)ds = 0\ ,\ u(a_j) = b_j\ ,\ 1 \leq j \leq N\ , \qquad (2.5)$$

where $a_j \in \Gamma_j$, b_j is a real constant, $|b_j| \leq \ell$, $1 \leq j \leq N$. Without loss of generality, we assume that $\cos(\nu,n) = 0$, $\sigma(t) = 0$, on $\Gamma_* = \Gamma_1 \cup \ldots \cup \Gamma_{N_0}$ $(N_0 \leq N)$ and that conditions do not hold on $\Gamma_j (N_0 < j \leq N, \Gamma_{**} = \Gamma \setminus (\Gamma_* \cup \Gamma_0)$. Besides,

$$h(t) = \begin{cases} h_0 , & t \in \Gamma_0 , \\ 0 , & t \in \Gamma \setminus \Gamma_0 , \end{cases} \tag{2.6}$$

where h_0 is an appropriate real constant.

If $\cos(\nu,n) > 0$ on Γ, Problem IV is the third boundary value Problem III. If $\cos(\nu,n) = 1$, $\sigma(t) = 0$ on Γ, Problem IV is the second boundary value Problem II (Neumann problem). If $\cos(\nu,n) = 0$, $\sigma(t) = 0$ on Γ, Problem IV is equivalent to the first boundary value Problem I (Dirichlet problem). In fact, the first boundary value problem possesses the boundary condition

$$u(t) = r(t) , \quad t \in \Gamma . \tag{2.7}$$

If $r \in C^1_\alpha(\Gamma)$, then $\tau(t) = \frac{dr(t)}{ds}$ satisfies (2.5), where $b_j = r(a_j)$, $j = 1,\dots,N$. Moreover, it is clear that

$$\int_{\Gamma_0} \tau(t)ds = 0 ,$$

and we set $b_0 = r(a_0)$ and $h_0 = 0$. The reverse is also true, if we provide that

$$r(t) = \int_{a_j}^{t} \tau(t)ds + r(a_j) , \quad r(a_j) = b_j , \quad j = 1,\dots,N , \tag{2.8}$$

and select h_0 appropriately, so that

$$\int_{\Gamma_0} [\tau(t)+h_0]ds = 0 , \quad r(t) = \int_{a_0}^{t} \tau(t)ds + r(a_0) , \quad r(a_0) = b_0 . \tag{2.9}$$

Lemma 2.1 Let G be the disc $\{|z-z_0| < R \ (<\infty)\}$, and $u(z)$ be continuously differentiable on $\overline{G}$ and not a constant. If the function $u(z)$ is a subharmonic function, and attains its maximum at z_1, $|z_1-z_0| = R$, then

$$\frac{\partial}{\partial \nu} u(z)\Big|_{z=z_1} > 0 ,$$

where $\vec{\nu}$ is a vector in z_1, so that $cos(\nu,n) > 0$.

This is the Hopf lemma, cf. § 1, Chapter 6 in [103]24).

Theorem 2.2 The solution of Problem IV is unique.

Proof. Let $u_1(z)$ and $u_2(z)$ be two solutions of Problem IV. Then $u(z) = u_1(z) - u_2(z)$ is a harmonic function satisfying the boundary condition

$$\frac{\partial u}{\partial \nu} + 2\sigma(t)u(t) = h(t) , \ t \in \Gamma , \ u(a_j) = 0 , \ j = 0,1,\ldots,N_0 . \qquad (2.10)$$

Suppose that $u(z) \not\equiv 0$ on $\bar{D}$. From the maximum and minimum principle we have

$$M = u(t_*) = \max_{z \in \bar{D}} u(z) \geq 0 , \quad m = u(t_{**}) = \min_{z \in \bar{D}} u(z) \leq 0 ,$$

where t_* and $t_{**} \in \Gamma$. If $M = u(t_*) > 0$, $t_* \in \Gamma_*$, due to $cos(\nu,n) = 0$, $\sigma(t) = 0$, $t \in \Gamma_*$, then $\partial u/\partial \nu = 0$, $u(t) = 0$, $t \in \Gamma_*$. This is impossible. If $t_* \in \Gamma_{**}$, it is clear that

$$\frac{\partial u}{\partial \nu}\Big|_{t=t_*} \geq 0$$

and $\sigma(t_*) = 0$, and so $cos(\nu,n) = 0$, for $t = t_*$. Let $\tilde{\Gamma}$ be the longest circular arc of Γ including the point t_* so that $cos(\nu,n) = 0$ and $\sigma(t) = 0$ on $\tilde{\Gamma}$. Thereby, $\partial u/\partial \nu = 0$, $u(t) = M$ for $t \in \tilde{\Gamma}$. However, there exists a point $t_*' \in \Gamma \setminus \tilde{\Gamma}$, such that $cos(\nu,n) > 0$, $cos(\nu,s) > 0$, $\partial u/\partial n > 0$, $\frac{\partial u}{\partial s} \geq 0$ at $t = t_*'$, and

$$\frac{\partial u}{\partial \nu}\Big|_{t=t_*'} = \left[\frac{\partial u}{\partial n} cos(\nu,n) + \frac{\partial u}{\partial s} cos(\nu,s)\right]\Big|_{t=t_*'} > 0 ;$$

this is a contradiction to (2.10). Or there exists $t_*' \in \Gamma \setminus \tilde{\Gamma}$, so that $cos(\nu,n) = 0$, $\sigma(t) > 0$ at $t = t_*'$. This is a contradiction to (2.10), too. This shows that $u(z)$ does not attain its maximum on $\Gamma \setminus \Gamma_0$. By a similar method, we can prove that $u(z)$ does not attain a negative minimum on $\Gamma \setminus \Gamma_0$. Hence t_* and $t_{**} \in \Gamma_0$. From (2.10), we derive

$$[\frac{\partial u}{\partial \nu} + 2\sigma(t)u(t)]_{t=t_*} = h_0 \geq 0 \, , \, [\frac{\partial u}{\partial \nu} + 2\sigma(t)u(t)]_{t=t_{**}} = h_0 \leq 0 \quad . \quad (2.11)$$

Thus $h_0 = 0$, and according to the same reason as above, we find that $M = m = 0$, i.e. $u(z) = 0$, $u_1(z) \equiv u_2(z)$, $z \in \bar{D}$.

Finally, we mention that if $\sigma(t) \not\equiv 0$ on Γ , we may let $h(t) \equiv 0$ instead of $u(a_0) = b_0$, and then the proposition is also true. □

2. A priori estimates of solutions for the regular oblique derivative problem

Making use of the estimate of solutions to the Riemann-Hilbert boundary value problem for analytic functions from § 1, we shall give an estimate for solutions of the regular oblique derivative problem for harmonic funcitons.

Theorem 2.3 Any solution $u(z)$ of Problem IV satisfies the estimate

$$\begin{aligned} C^1_\alpha[u,\bar{D}] &= C[u,\bar{D}] + H_\alpha[u,\bar{D}] + \\ &+ C[u_z,\bar{D}] + H_\alpha[u_z,\bar{D}] \leq M_1 = M_1(\alpha,\ell,D) \; . \end{aligned} \quad (2.12)$$

Proof. We first prove that

$$C^1[u,\bar{D}] = C[u,\bar{D}] + C[u_z,\bar{D}] \leq M_2 = M_2(\alpha,\ell,\bar{D}) \; . \quad (2.13)$$

Otherwise, there exist three sequences of functions $\{\lambda_n(t)\}$, $\{\sigma_n(t)\}$, $\{\tau_n(t)\}$ *) and a sequence of numbers $\{b^n_j\}$ which

---

*) If $\lambda(t)$, $\sigma(t)$ satisfy the conditions $-Re[\frac{\gamma_j\overline{\lambda(t)}}{t-z_j}] = cos(\nu,n) \geq \eta > 0$ or $\sigma(t) \geq \eta > 0$ on a circular arc $\Gamma^*_j \subset \Gamma_j (N_0 < j \leq N)$, then we require that $-Re[\frac{\gamma_j\overline{\lambda_n(t)}}{t-z_j}] \geq \eta > 0$ or $\sigma_n(t) \geq \eta > 0$ on Γ^*_j , where η is a constant (cf.[103]24)).

satisfy the same conditions as $\lambda(t)$, $\sigma(t)$, $\tau(t)$, and b_j in (2.4), so that when $n \to \infty$, $\{\lambda_n(t)\}$, $\{\sigma_n(t)\}$, $\{\tau_n(t)\}$ converge uniformly to $\lambda_0(t)$, $\sigma_0(t)$, $\tau_0(t)$ on Γ , respectively, and $\{b_j^n\}$ converges to b_j^0 , $j = 0,1,\ldots,N_0$, and there exists a sequence of solutions $\{u_n(z)\}$ of Problem IV for harmonic functions which satisfy the boundary conditions

$$\left.\begin{array}{l} Re[\overline{\lambda_n(t)u_{nt}}] + \sigma_n(t)u_n(t) = \tau_n(t) + h_n(t) \ , \ t \in \Gamma \ , \\ \int_{\Gamma_j} \tau_n(t)ds = 0 \ , \ j=1,\ldots,N_0 \ , \ u_n(a_j) = b_j^n \ , \ j=0,1,\ldots,N_0 \ , \end{array}\right\} \tag{2.14}$$

so that when $n \to \infty$,

$$C^1[u_n,\bar{D}] = Q_n \to \infty \ , \ Q_n \geq 1 \ . \tag{2.15}$$

Obviously, $U_n(z) = u_n(z)/Q_n$ is a harmonic function in D satisfying the boundary condition

$$\left.\begin{array}{l} Re[\overline{\lambda_n(t)U_{nt}}] + \sigma_n(t)U_n(t) = [\tau_n(t)+h_n(t)]/Q_n \ , \ t \in \Gamma \ , \\ U_n(a_j) = b_j^n/Q_n \ , \ j=0,1,\ldots,N_0 \ . \end{array}\right\} \tag{2.16}$$

Noting $(U_{nz})_{\bar{z}} = 0$ in D , we see that $W_n(z) = U_{nz}$ is an analytic function in D , satisfying the boundary condition

$$\left.\begin{array}{l} Re[\overline{\lambda_n(t)}W_n(t)] = r_n(t) \ , \ t \in \Gamma \ , \\ r_n(t) = -\sigma_n(t)U_n(t) + [\tau_n(t)+h_n(t)]/Q_n \ . \end{array}\right\} \tag{2.17}$$

Moreover, from (2.15) it is not difficult to see that

$$C[W_n,\bar{D}] \leq 1 \ , \ C_\alpha[r_n,\Gamma] \leq M_3 = M_3(\alpha,\ell,D) \ . \tag{2.18}$$

According to Theorem 1.4 in § 1, we obtain

$$C_\alpha[W_n,\bar{D}] \leq M_4 \ , \ C_\alpha^1[U_n,\bar{D}] \leq M_5 \tag{2.19}$$

where $M_j = M_j(\alpha,\ell,D)$, $j = 4,5$. Therefore, we may choose a subsequence of $\{U_n(z)\}$, which converges uniformly to $U_0(z)$ on $\overline{D}$ and $U_0(z)$ satisfies the boundary condition

$$\left.\begin{aligned} &Re[\overline{\lambda_0(t)U_{0t}}] + \sigma_0(t)U_0(t) = h_0(t) \ , \ t \in \Gamma \ , \\ &\int_{\Gamma_j} \tau_0(t)ds = 0 \ , \ j=1,\ldots,N_0 \ , \ U_0(a_j) = 0 \ , \ j=0,1,\ldots,N_0 \ . \end{aligned}\right\} \tag{2.20}$$

By Theorem 2.2, we have $U_0(z) \equiv 0$, $z \in D$. However, from (2.15) we obtain $C^1[U_0,\overline{D}] = 1$. This contradiction proves the estimate (2.13). Using the same method as in the proof of (2.19), we conclude also the estimate (2.12).

3. Solvability of the regular oblique derivative problem

In order to prove the existence of a solution of Problem IV, we introduce some of its special cases. If the vector $\vec{\nu}$ in the boundary condition (2.1) satisfies

$$cos(\nu,n) = -Re[\frac{\gamma_j\overline{\lambda(t)}}{t-z_j}] \geq \eta > 0 \quad \text{on} \quad \Gamma_j \subset \Gamma_{**} \ ,$$

where η is a constant, then Problem IV with this condition is called Problem L. Further, Problem L with $\sigma(t) \equiv 0$ on $\Gamma \setminus \Gamma_*$ is called Problem M. Besides, the conditions $u(a_j) = b_j$, $j = 1,\ldots,N_0$ are replaced by

$$Im[\overline{\lambda(t)}u_t]_{t=a_j} = c_j \ , \ j = 1,\ldots,N_0 \ , \tag{2.21}$$

where $c_j(j=1,\ldots,N_0)$ are all constants; the modified Problem M is called Problem N. In the following we shall prove the solvability of Problem N, Problem M, Problem L, and Problem IV, successively.

Theorem 2.4 Problem N for harmonic functions is solvable.

Proof. On the basis of Theorem 1.8, we first find three solutions $\Phi_0(z)$, $\Phi_1(z)$, and $\Phi_2(z)$ of Problem B for analytic functions with boundary conditions

$$\left.\begin{array}{l} Re[\overline{\lambda_0(t)}\Phi_0(t)] = \tau(t) \ , \ \overline{\lambda_0(t)} = \overline{\lambda(t)}/t \ , \ t \in \Gamma \ , \\ Im[\overline{\lambda_0(a_j)}\Phi_0(a_j)] = c_j \ , \ j = 0,\ldots,N \ , \end{array}\right\} \tag{2.22}$$

$$\left.\begin{array}{l} Re[\overline{\lambda_0(t)}\Phi_1(t)] = 0 \ , \ t \in \Gamma \ , \\ Im[\overline{\lambda_0(a_j)}\Phi_1(a_j)] = \begin{cases} 1, j=0 \ , \\ 0, j=1,\ldots,N \ , \end{cases} \end{array}\right\} \tag{2.23}$$

$$\left.\begin{array}{l} Re[\overline{\lambda_0(t)}\Phi_2(t)] = \begin{cases} 1, t \in \Gamma_0 \ , \\ 0, t \in \Gamma \setminus \Gamma_0 \ , \end{cases} \\ Im[\overline{\lambda_0(a_j)}\Phi_2(a_j)] = 0 \ , \ j=0,1,\ldots,N \ , \end{array}\right\} \tag{2.24}$$

respectively, where $a_j \in \Gamma_j$, $j = 0,\ldots,N$, $c_j (j=0,\ldots,N)$ are all real constants, and the index of $\lambda_0(t)$ is N . From Lemma 1.1, we can derive

$$\begin{vmatrix} Re\ \Phi_1(0) & Im\ \Phi_1(0) \\ Re\ \Phi_2(0) & Im\ \Phi_2(0) \end{vmatrix} \neq 0 \ ,$$

and so there exist two real constants C_1' , C_2' , which satisfy $C_1'\ \Phi_1(0) + C_2'\ \Phi_2(0) = -\Phi_0(0)$. Hence the function

$$\Psi_0(z) = [\Phi_0(z) + C_1'\ \Phi_1(z) + C_2'\ \Phi_2(z)]/z$$

is a solution of Problem B with the boundary condition

$$\left.\begin{array}{l} Re[\overline{\lambda(t)}\Psi_0(t)] = \tau(t) + h(t) \ , \ t \in \Gamma \ , \\ Im[\overline{\lambda(a_j)}\Psi_0(a_j)] = c_j \ , \ j = 1,\ldots,N \ . \end{array}\right\} \tag{2.25}$$

Due to (2.5), it follows that

$$Re\int_{\Gamma_j} \Psi_0(z)dz = -\int_0^{2\pi} Re[i(z-z_j)\Psi_0(z)]d\theta = \int_0^{2\pi} \tau(t)\gamma_j d\theta = 0, \ j=1,\ldots,N_0 \ . \tag{2.26}$$

If

$$Re \int_{\Gamma_j} \Psi_0(z)dz = 0 \ , \ j = N_0+1,\dots,N \ , \tag{2.27}$$

then the function

$$u(z) = 2Re \int_1^z \Psi_0(\zeta)d\zeta + b_0 \tag{2.28}$$

is a single valued solution of Problem N for harmonic functions. Otherwise, we seek $N - N_0$ solutions $\Psi_{N_0+1},\dots,\Psi_N$ of Problem B with boundary conditions

$$\left.\begin{aligned} &Re[\overline{\lambda(t)}\Psi_k(t)] = 0 \ , \ t \in \Gamma \ , \\ &Im[\overline{\lambda(a_j)}\Psi_k(a_j)] = \delta_{jk} = \begin{cases} 1 \ , \ j=k \ , \ k=N_0+1,\dots,N \ , \\ 0 \ , \ j\neq k \ , \ j=1,\dots,N \ . \end{cases} \end{aligned}\right\} \tag{2.29}$$

It is evident that $\Psi_{N_0+1}(z),\dots,\Psi_N(z)$ are linearly independent. We shall prove

$$J = \begin{vmatrix} I_{N_0+1,N_0+1} & \dots & I_{N_0+1,N} \\ \vdots & & \vdots \\ I_{N,N_0+1} & \dots & I_{N,N} \end{vmatrix} \neq 0 \ , \ I_{jk} = Re \int_{\Gamma_j} \Psi_k(z)dz, j,k=N_0+1,\dots,N \ . \tag{2.30}$$

Suppose $J = 0$. Then there exist constants

$$c_{N_0+1},\dots,c_N \ (|c_{N_0+1}|+\dots+|c_N|\neq 0)$$

so that

$$\Psi^*(z) = \sum_{k=N_0+1}^{N} c_k \Psi_k(z)$$

satisfies

$$Re \int_{\Gamma_j} \Psi^*(z)dz = Re \sum_{k=N_0+1}^{N} c_k \int_{\Gamma_j} \Psi_k(z)dz = 0 \ , \ j=N_0+1,\ldots,N \ . \qquad (2.31)$$

Hence, there are points $a_j^* \in \Gamma_j$ satisfying

$$Re[i(z-z_j)\Psi^*(z)]\Big|_{z=a_j^*} = -Im[(a_j^*-z_j)\Psi^*(a_j^*)] = 0 \ , \ j=N_0+1,\ldots,N \ . \quad (2.32)$$

Noting

$$Re[\frac{\overline{\lambda(t)}}{t-z_j}(t-z_j)\Psi^*(t)]\Big|_{t=a_j^*} = Re[\overline{\lambda(a_j^*)}\Psi^*(a_j^*)] = 0 \ , \ j=N_0+1,\ldots,N \ ,$$

and $Re[\frac{\overline{\lambda(t)}}{t-z_j}] \neq 0$, $j = N_0+1,\ldots,N$, it is easy to see that

$$Re[(a_j^*-z_j)\Psi^*(a_j^*)] = 0 \ , \ j = N_0+1,\ldots,N \ .$$

Hence $(a_j^*-z_j)\Psi^*(a_j^*) = 0$, i.e. $\Psi^*(a_j^*) = 0$, $j = N_0+1,\ldots,N$. Similarly, from (2.29), we can derive $\Psi^*(a_j) = 0$, $a_j \in \Gamma_j$, $j = 1,\ldots,N_0$. This shows that $\Psi^*(z)$ has at least $2N$ zeros on Γ . By Lemma 1.1, we derive the absurd inequality

$$2N \le 2N_D + N_\Gamma = 2(N-1) \ . \qquad (2.33)$$

Thus $J \neq 0$. Afterwards, we seek the solution $C_1,\ldots,C_n$ of the linear system of algebraic equations

$$\begin{aligned} C_{N_0+1}I_{N_0+1,N_0+1}+\ldots+C_N I_{N_0+1,N} &= -Re \int_{\Gamma_{N_0+1}} \Psi_0(z)dz \ , \\ \ldots & \\ C_{N_0+1}I_{N,N_0+1}+\ldots+C_N I_{N,N} &= -Re \int_{\Gamma_N} \Psi_0(z)dz \ , \end{aligned} \qquad (2.34)$$

and substitute

$$\Psi(z) = \Psi_0(z) + \sum_{k=N_0+1}^{N} C_k \Psi_k(z) \tag{2.35}$$

into the position of $\Psi_0(z)$ in (2.28). Then the function $u(z)$ determined by integration in (2.28) is a solution of Problem N for harmonic functions. □

<u>Theorem 2.5</u> Problem M for harmonic functions is solvable.

<u>Proof.</u> Let $u_0(z)$ be a solution of Problem N and denote $u_0(a_j) = b_j^*$, $j = 1,\dots,N_0$. If $b_j^* = b_j$, $j = 1,\dots,N_0$, then $u_0(z)$ is also a solution of Problem M. Otherwise, we find N_0 solutions $u_1(z),\dots,u_{N_0}(z)$ of Problem N with the boundary conditions

$$\left.\begin{aligned} &Re[\overline{\lambda(t)}u_{kt}] = h(t) \ , \ t \in \Gamma \ , \ u_k(a_0) = 0 \ , \\ &Im[\overline{\lambda(t)}u_{kt}]_{t=a_j} = \delta_{jk} = \begin{cases} 1, j=k \ , \\ 0, j \neq k \ , \end{cases} j,k=1,\dots,N_0 \ , \end{aligned}\right\} \tag{2.36}$$

which are linearly independent. On account of Theorem 2.2, we can verify

$$J = \begin{vmatrix} u_1(a_1) & \dots & u_{N_0}(a_1) \\ \vdots & & \vdots \\ u_1(a_{N_0}) & \dots & u_{N_0}(a_{N_0}) \end{vmatrix} \neq 0 \ ,$$

and then seek the solution $c_1,\dots,c_{N_0}$ of the algebraic system

$$u_1(a_j)c_j + \dots + u_{N_0}(a_j)c_{N_0} = b_j^* - b_j \ , \ j=1,\dots,N_0 \ . \tag{2.37}$$

Thus the function

$$u(z) = u_0(z) - \sum_{k=1}^{N_0} c_k u_k(z) \tag{2.38}$$

is a solution of Problem M . □

Theorem 2.6 Problem L for harmonic functions has a solution.

Proof. We use the imbedding and discuss the boundary value problem L_* with the boundary condition

$$\left.\begin{aligned} &\frac{\partial u}{\partial \nu} + 2k\,\sigma(t)u(t) = s(t) + h(t)\ ,\ t \in \Gamma\ , \\ &\int_{\Gamma_j} s(t)ds = 0\ ,\ 1 \le j \le N_0\ ,\ u(a_j) = b_j\ ,\ j=0,1,\ldots,N_0\ , \end{aligned}\right\} \tag{2.39}$$

where $s(t) \in C_\alpha(\Gamma)$, k is the parameter $(0 \le k \le 1)$. When $k = 0$, Problem L_* is Problem M, and so it is solvable. According to the method in the proof of Theorem 1.7, we may prove that if Problem L_* with some parameter $k_0 (0 \le k_0 \le 1)$ has a solution, then there exists a positive constant δ which is independent of k_0 , so that Problem L_* is solvable for any parameter k satisfying $|k-k_0| \le \delta$ and $0 \le k \le 1$. Hence when $k = \delta$, $2\delta,\ldots,[\frac{1}{\delta}]\delta, 1$, Problem L_* is sovable. Especially, when $k = 1$ and $s(t) = 2\tau(t)$, Problem L_*, i.e. Problem L has a solution. □

Theorem 2.7 Problem IV for harmonic functions is solvable.

Proof. We can choose ν_m corresponding to $\lambda_m(t)$, so that

$$cos(\nu_m, n) \ge \frac{1}{m} > 0\ ,\ C_\alpha[\lambda_m, \Gamma] \le \ell$$

and $\lambda_m(t)$ converges uniformly to $\lambda(t)$ on Γ . From Theorem 2.6, Problem L with the boundary condition

$$\left.\begin{aligned} &\frac{\partial u}{\partial \nu_m} + 2\sigma(t)u(t) = 2\tau(t) + h(t)\ ,\ t \in \Gamma\ , \\ &\int_{\Gamma_j} \tau(t)ds = 0\ ,\ j=1,\ldots,N_0\ ,\ u(a_j) = b_j\ ,\ j=0,1,\ldots,N_0\ , \end{aligned}\right\} \tag{2.40}$$

has a solution $u_m(z)$. For all positive integers, the corresponding solutions $u_m(z)$ $(m=1,2,\ldots)$ satisfy the estimate (2.12). Therefore, we may select a subsequence of $\{u_m(z)\}$, so that $\{u_m(z)\}$, $\{u_{mz}\}$ converge uniformly to $u_0(z)$, u_{0z} on $\overline{D}$, respectively. The harmonic function $u_0(z)$ satisfies (2.1)-(2.6), so $u_0(z)$ is a solution of Problem IV. □

4. The irregular oblique derivative problem for harmonic functions

Problem P The so-called irregular oblique derivative problem for harmonic functions is to find a harmonic function $u(z)$ in D, continuously differentiable on $\overline{D}$ and satisfying the boundary condition

$$\frac{\partial u}{\partial \mu} + 2\sigma(t)u(t) = 2\tau(t), \quad t \in \Gamma, \quad u(a_0) = b_0 \tag{2.41}$$

where $\vec{\mu}$ is a given vector at the point $t \in \Gamma$, $a_0 = 1 \in \Gamma_0$, b_0 is an arbitrary real constant. Noting

$$\frac{\partial u}{\partial \mu} = \frac{\partial u}{\partial x} \cos(\mu,x) + \frac{\partial u}{\partial y} \cos(\mu,y) =$$

$$= 2Re\,[\overline{\lambda(t)}u_t], \quad \overline{\lambda(t)} = \cos(\mu,x) + i\sin(\mu,x),$$

(2.41) can be written in the complex form

$$Re[\overline{\lambda(t)}u_t] + \sigma(t)u(t) = \tau(t), \quad t \in \Gamma, \quad u(a_0) = b_0. \tag{2.42}$$

We assume that $\lambda(t)$, $\sigma(t)$, $\tau(t)$ satisfy

$$\left.\begin{aligned} &C_\alpha[\lambda,\Gamma] \le \ell < \infty, \quad C_\alpha[\sigma,\Gamma] \le \ell, \\ &C_\alpha[\tau,\Gamma] \le \ell, \end{aligned}\right\} \tag{2.43}$$

where $\alpha (0 < \alpha < 1)$, $\ell\ (0 < \ell < \infty)$ are positive constants.

This boundary value problem is also called Poincaré problem.

If $\sigma(t) = 0$, $t \in \Gamma$, the solvability of Problem P is not difficult to derive from Theorem 1.9. In fact, putting $\Phi(z) = u_z$, the boundary condition (2.42) can be transformed into the boundary condition of the Riemann-Hilbert problem A:

$$Re[\overline{\lambda(t)}\Phi(t)] = \tau(t) \ , \ t \in \Gamma \ . \qquad (2.44)$$

If we find a solution $\Phi(z)$ of this Problem A and the function $u(z)$ determined by the integral

$$u(z) = 2Re\int_1^z \Phi(z)dz + b_0 \qquad (2.45)$$

is single-valued, then $u(z)$ is a solution of Problem P. However, if D is a multiply connected domain, $u(z)$ must not be single-valued. It is clear that

$$u(z) = 2Re\int_1^z [\Phi(z) + \sum_{j=1}^{N} \frac{id_j}{z-z_j}]dz + b_0 \ , \ z \in D \ , \qquad (2.46)$$

with properly chosen real constants $d_j (j=1,\ldots,N)$ is a single-valued function in D. Hence, if the solution $\Phi(z)$ of Problem A satisfies the following N conditions

$$\int_{\Gamma_j} \Phi(z)dz = 0 \ , \ j=1,\ldots,N \ , \qquad (2.47)$$

then we may take $d_j = 0 \ (j=1,\ldots,N)$ in (2.46). This shows that under the N conditions (2.47) the solution of Problem A gives the solution of Problem P for harmonic functions, too. In addition, by Theorem 1.9 we have

Theorem 2.8 Suppose $\sigma(t) = 0$ in (2.41) or (2.42). Then Problem P possesses the following results on solvability.

(1) If the index $K \geq N$, Problem P has N solvability conditions, its general solution depends on $2K-N+2$ arbitrary real constants.

(2) If $0 \leq K < N$, Problem P has $2N-K$ solvability conditions, its general solution depends on $K+2$ arbitrary real constants.

(3) If $K < 0$, Problem P has $2N-2K-1$ solvability conditions, its general solution depends on one arbitrary real constant.

In Chapter 3, we shall consider the solvability of the general Problem P.

Problem P^*. This is a special Problem P with the proper modified boundary condition

$$\left.\begin{aligned} &\frac{\partial u}{\partial \mu} + 2\,[\sigma(t)+h(t)]u(t) = \tau(t)\ ,\ t \in \Gamma\ , \\ &\text{i.e. } Re[\overline{\lambda(t)}u_t] + [\sigma(t)+h(t)]u(t) = \tau(t)\ ,\ t \in \Gamma\ , \end{aligned}\right\} \tag{2.48}$$

where $\vec{\mu}$, $\sigma(t)$ and $\tau(t)$ are as stated in (2.41) - (2.43) satisfying the condition

$$cos(\mu,n)\ \sigma(t) \geq 0\ ,\ t \in \Gamma\ . \tag{2.49}$$

Now we explain in detail the above conditions. We denote by Γ^+ and Γ^- those circular arcs on Γ on which $cos(\mu,n) \geq 0$ and $cos(\mu,n) \leq 0$, respectively (there is at least a point on each of those arcs such that $cos(\mu,n) \neq 0$), so that $\Gamma^+ \cup \Gamma^- = \Gamma$, $\Gamma^+ \cap \Gamma^- = \emptyset$, and $\overline{\Gamma^+} \cap \overline{\Gamma^-}$ contains a finite number of points: $a_j (j=1,\ldots,m)$ and $a'_j (j=1,\ldots,m')$, such that every circular arc includes its initial point and does not include the terminal point, and $a_j \in \Gamma^+$, $a'_j \in \Gamma^-$, when the direction of $\vec{\mu}$ at a_j , a'_j is equal to the direction of Γ , and $a_j \in \Gamma^-$, $a'_j \in \Gamma^+$, when the direction of $\vec{\mu}$ at a_j , a'_j is opposite to the direction of Γ , and $cos(\mu,n)$ changes sign once on the two circular arcs with the end point a_j or a'_j . Moreover, when $m > 0$, we take $h(t) = 0$, and when $m = 0$,

$$h(t) = \begin{cases} h_0, & t \in \Gamma_0\ , \\ 0\ , & t \in \Gamma \setminus \Gamma_0\ , \end{cases}$$

h_0 is an appropriate real constant. In addition, we assume that

$$\sigma(t) \geq 0\ ,\ \text{when } t \in \Gamma^+\ ;\ \sigma(t) \leq 0\ ,\ \text{when } t \in \Gamma^-\ . \tag{2.50}$$

If $cos(\mu,n) = 0$, $\sigma(t) = 0$, $t \in \Gamma_j$, $1 \leq j \leq n$, then we assume

$$\int_{\Gamma_j} \tau(t)ds = 0 \;,\; u(a_j^*) = b_j^* \;,\; a_j^*(\neq a_j') \in \Gamma \;,\; |b_j^*| \le \ell \tag{2.51}$$

and

$$u(a_j) = b_j \;,\; j=1,\ldots,J, J = \begin{cases} 1 \;, \text{ for } m = 0 \;, \\ m \;, \text{ for } m \ge 1 \;, \end{cases} |b_j| \le \ell \;. \tag{2.52}$$

The index K of Problem P_* is given by

$$K = \frac{1}{2\pi} \Delta_\Gamma \, arg \, \lambda(t) = N-1 + \frac{m-m'}{2} \;. \tag{2.53}$$

In § 4, Chapter 6 of [103],24), the following result is proved.

<u>Theorem 2.9</u> Under m' conditions, Problem P_* possesses a continuously differentiable solution $u(z)$ on $\overline{D}$.

<u>§ 3. Some boundary value problems for simple first order complex equations</u>

In this section, we shall discuss the simple complex equation

$$w_{\overline{z}} = \omega(z) \tag{3.1}$$

where $\omega(z) \in L_p(\overline{D})$, $2 < p < \infty$, i.e.

$$L_p[\omega,\overline{D}] := [\iint_D |\omega(z)|^p dxdy]^{1/p} < \infty$$

and shall give some proper modified propositions for the Riemann-Hilbert problem for (3.1). Besides, we shall give an integral representation for solutions of the above modified problems, and prove some properties of the integral operators.

<u>1. Some proper modified propositions of the Riemann-Hilbert problem for (3.1)</u>

In § 1, we have given a proper proposition of the Riemann-Hilbert problem for analytic functions. Now we shall give two other properly modified propositions for the Riemann-Hilbert problem for (3.1).

Problem B^*. One of the well posed Riemann-Hilbert problems for (3.1) is to find a solution $w(z)$ of (3.1) in the (N+1)-connected circular domain D, continuous on $\overline{D}$ and satisfying the modified boundary conditions

$$\left.\begin{aligned} &Re[\overline{\lambda(t)}w(t)] = r(t) + h(t) \ , \ t \in \Gamma \ , \\ &h(t) = \begin{cases} 0 \ , \ t \in \Gamma \ , \ \text{if} \quad K \geq N \ , \\ \left.\begin{array}{l} h_j \ , \ t \in \Gamma_j \ , \ j=1,\ldots,N-K \ , \\ 0 \ , \ t \in \Gamma_j \ , \ j=N-K+1,\ldots,N+1 \ , \end{array}\right\} \text{if} \quad 0 \leq K < N \ , \\ \left.\begin{array}{l} h_j \ , \ t \in \Gamma_j \ , \ j=1,\ldots,N \ , \\ h_0 + \sum_{m=1}^{|K|-1} Re[(H_m^+ + iH_m^-)t^m] \ , \ t \in \Gamma_0 \end{array}\right\} \text{if} \quad K < 0 \ , \end{cases} \end{aligned}\right\} \tag{3.2}$$

$$Im[\overline{\lambda(a_j)}w(a_j)] = b_j \ , \ j \in \{j\} = \begin{cases} 1,\ldots,2K-N+1, \ \text{if} \ K \geq N \ , \\ N-K+1,\ldots,N+1, \ \text{if} \ 0 \leq K < N \ , \end{cases} \tag{3.3}$$

where $h_j (j=1,\ldots,N+1)$, $H_m^{\pm}(m=1,\ldots,|K|-1)$ are all unknown real constants to be determined appropiately, and $a_j \in \Gamma_j$, $j=1,\ldots,N$, $a_j \in \Gamma_0 (j=N+1,\ldots,2K-N+1)$ are distinct points, $b_j (j=1,\ldots,2K-N+1)$ are all real constants, where $\lambda(t)$, $r(t)$, $b_j (j \in \{j\})$ satisfy the conditions stated in (1.1) - (1.5). The difference between Problem B and Problem B^* besides (3.1) and (3.3) is only in the condition (3.2) if $K < 0$.

Through the transformation of the function $w(z)$ into

$$W(z) = e^{-iS(z)} \Pi(z)w(z) = w(z)/\Delta(z) \ , \ z \in D \ , \tag{3.4}$$

where $S(z)$, $\Pi(z)$ are as stated in (1.7), the complex equation (3.1) can be reduced to the following equation

$$W_{\bar{z}} = \Omega(z) \ , \ \Omega(z) = e^{-iS(z)} \Pi(z) \ \omega(z) = \omega(z)/\Delta(z) \ , \tag{3.5}$$

and the boundary conditions (3.2) and (3.3) likewise to the boundary conditions

$$Re[\overline{\Lambda(t)}W(t)] = R(t) + H(t)\ , \quad t \in \Gamma\ , \tag{3.6}$$

$$Im[\overline{\Lambda(a_j)}W(a_j)] = B_j\ , \ j \in \{j\} = \begin{cases} 1,\dots,2K-N+1\ , & \text{if } K \geq N \\ N-K+1,\dots,N+1\ , & \text{if } 0 \leq K < N\ , \end{cases} \tag{3.7}$$

where $\Lambda(t)$, $R(t)$, $H(t)$, and B_j are as stated in (1.9) and (1.10), in which $h(t)$ is as stated in (3.2). The boundary value problem with boundary conditions (3.6) and (3.7) for (3.5) is called <u>Problem C^*</u>, and Problem C^* with $R(t) = 0$ is called <u>Problem C_0^*</u>.

<u>Problem C^{**}.</u> Another properly modified Riemann-Hilbert problem for (3.5) in D possesses the boundary condition (3.6), but the point conditions (3.7) are replaced by the integral conditions

$$\left.\begin{array}{l} L_j(W) = Im\displaystyle\int_{\Gamma_j} \overline{\Lambda(t)}W(t)d\theta = B_j\ , \ j = \begin{cases} N-K+1,\dots,N+1\ , & \text{if } 0 \leq K < N\ , \\ 1,\dots,N+1\ , & \text{if } K \geq N\ , \end{cases} \\ L_j(W) = \begin{cases} Re\displaystyle\int_{\Gamma_0} t^{j-N-1}\overline{\Lambda(t)}W(t)d\theta = B_j\ , \\ Im\displaystyle\int_{\Gamma_0} t^{j-K-1}\overline{\Lambda(t)}W(t)d\theta = B_j\ , \end{cases} \quad j = \begin{cases} N+2,\dots,K+1\ , \\ K+2,\dots,2K-N+1\ , \end{cases} \quad \text{if } K \geq N \end{array}\right\} \tag{3.8}$$

where $B_m (m=1,\dots,2K-N+1)$ are all real constants.

Problem C^{**} with $R(t) = 0$ is called <u>Problem C_0^{**}</u>.

Essentially, Problem C^* is equivalent to Problem C^{**}. If Problem C^* for (3.5) is solvable, and $W_0(z)$ is a solution to this problem, we denote

$$L_jW_0 = B_j^*\ , \ j \in \{j\} = \begin{cases} N-K+1,\dots,N+1\ , & \text{if } 0 \leq K < N\ , \\ 1,\dots,2K-N+1\ , & \text{if } K \geq N\ . \end{cases} \tag{3.9}$$

If $B_j^* = B_j$, $j \in \{j\}$, then $W_0(z)$ is also a solution of Problem C^{**}. Otherwise, we can find the solutions $\Psi_1(z),\ldots,\Psi_{2K-N+1}(z)$ $(K \geq N)$ of Problem C_0^* with

$$Im[\overline{\Lambda(a_j)}\Psi_k(a_j)] = \delta_{jk} = \begin{cases} 1 , & j=k , \\ 0 , & j \neq k , \end{cases} \quad j,k=1,\ldots,2K-N+1$$

for analytic functions. Using the same method as in the proof of Theorem 3.1, we can prove

$$J = \begin{vmatrix} I_{1,1} & \cdots & I_{1,2K-N+1} \\ \vdots & & \vdots \\ I_{2K-N+1,1} & \cdots & I_{2K-N+1,2K-N+1} \end{vmatrix} \neq 0 , \; I_{j,k} = L_j\Psi_k , \; j,k=1,\ldots,2K-N+1 ,$$

so that we can seek the solution $c_1,\ldots,c_{2K-N+1}$ to the algebraic system

$$c_1I_{j,1} + c_2I_{j,2} + \ldots + c_{2N-K+1}I_{j,2K-N+1} = B_j - B_j^* , \; j=1,\ldots,2K-N+1 ,$$

and

$$W(z) = W_0(z) + \sum_{k=1}^{2K-N+1} c_k\Psi_k(z) \tag{3.10}$$

is just a solution of Problem C^{**} for (3.5). Similarly, we may discuss the case when $0 \leq K < N$.

Conversely, it is not difficult to derive the solution to Problem C^* from that to Problem C^{**} for (3.5).

2. Properties of solutions to Problem B* for analytic functions

We first prove the existence and uniqueness of the solution to Problem B^*, and then give an integral representation for the solution to Problem B^* for analytic functions.

Theorem 3.1 Problem B^* for analytic functions has a unique solution.

Proof. If the index $K \geq 0$, the corresponding results were given in Theorem 1.2, Theorem 1.7, and Theorem 1.8. Now we discuss the case where $K < 0$. To prove the existence of the solution of Problem B^* for $K < 0$, we need only to find the appropriate real constants $h_j\,(j=0,1,\ldots,N)$ and $H_m^{\pm}(m=1,\ldots,|K|-1)$ so that the following solvability conditions of Problem B^* are satisfied

$$\int_\Gamma \lambda(t)\Psi_n(t)[r(t)+h(t)]t'(s)ds = 0 \;,\; n=1,\ldots,N-2K-1 \;, \tag{3.11}$$

where $\Psi(z)$ $(n=1,\ldots,N-2K-1)$ is a complete system of linearly independent solutions of the conjugate homogeneous Problem B_0' with the boundary condition

$$Re[\lambda(t)\Psi_n(t)t'(s)] = 0 \;,\; t \in \Gamma \;,\; n=1,\ldots,N-2K-1 \tag{3.12}$$

(cf. [97]2)). From (3.11) and (3.12), it follows that

$$\begin{aligned}&\sum_{j=0}^{N} h_j \int_{\Gamma_j} Im[\lambda(t)\Psi_n(t)t'(s)]ds \\ &+ \sum_{m=1}^{|K|-1} \int_{\Gamma_0} Re[(H_m^+ + iH_m^-)t^m]\; Im[\lambda(t)\Psi_n(t)t'(s)]ds \\ &= -\int_\Gamma \lambda(t)\Psi_n(t)t'(s)r(t)ds \;,\; n=1,\ldots,N-2K-1 \;.\end{aligned} \tag{3.13}$$

We shall prove that the coefficients determinant of the system of algebraic equations (3.13) is not equal to zero, i.e.

$$J = \begin{vmatrix} a_{1,1} & \cdots & a_{1,N-2K-1} \\ \vdots & & \vdots \\ a_{N-2K-1,1} & \cdots & a_{N-2K-1,N-2K-1} \end{vmatrix} \neq 0 , \tag{3.14}$$

where

$$a_{n,j} = \begin{cases} \frac{1}{i} \int_{\Gamma_j} \lambda(t)\Psi_n(t)dt , & j=1,\ldots,N+1 , \\ \frac{1}{i} \int_{\Gamma_0} \lambda(t)\Psi_n(t)\cos(j-N-1)sdt , & j=N+2,\ldots,N-K , \\ i \int_{\Gamma_0} \lambda(t)\Psi_n(t)\sin(j-N+K)sdt , & j=N-K+1,\ldots,N-2K-1 , \\ & n=1,\ldots,N-2K-1 . \end{cases} \tag{3.15}$$

Suppose that $J = 0$, we can seek real constants $c_1,\ldots,c_{N-2K-1}$, which are not all equal to zero, so that

$$\left. \begin{aligned} & \int_{\Gamma_j} Im[\lambda(t)\Psi(t)t'(s)]ds = 0 , \; j=0,1,\ldots,N , \\ & \int_{\Gamma_0} Im[\lambda(t)\Psi(t)t'(s)]\{{\cos ms \atop \sin ms}\}ds = 0 , \; m=1,\ldots,|K|-1 , \end{aligned} \right\} \tag{3.16}$$

where

$$\Psi(z) = \sum_{n=1}^{N-2K-1} c_n\Psi_n(z)$$

is a solution of the conjugate homogeneous Problem B_0' and $\Psi(z) \not\equiv 0$ on $\overline{D}$. By the first formula of (3.16), there are points $a_j^* \in \Gamma_j$ satisfying $Im[\lambda(t)\Psi(t)t'(s)]|_{t=a_j^*} = 0$, $j=1,\ldots,N$, and so

$$\lambda(t)\Psi(t)t'(s)|_{t=a_j^*} = 0 , \text{ i.e. } \Psi(a_j^*) = 0 , \; j=1,\ldots,N . \tag{3.17}$$

In addition, let $U(z)$ be a harmonic function in $\mathbb{D} = \{|z|<1\}$, which satisfies the boundary condition $U(t) = Im[\lambda(t)\Psi(t)t'(s)]$ for $|t| = 1$. From (3.16) we obtain

$$U(z) = \frac{1}{2\pi}\int_0^{2\pi} U(t)\, Re\,\frac{t+z}{t-z}\,d\theta = Re\,\frac{z^{|K|}}{\pi}\int_0^{2\pi}\frac{U(t)d\theta}{t^{|K|-1}(t-z)},\ z \in \mathbb{D}\ . \qquad (3.18)$$

It is not difficult to see that if $U(z) \not\equiv 0$ in $\mathbb{D}$, there exist at least $2|K|$ level curves $U(z) = 0$ passing through $z = 0$ and so $U(z)$ possesses at least $2|K|$ zeros $a^*_{N+1},\ldots,a^*_{N+2|K|} \in \Gamma_0$. Hence

$$\lambda(t)\Psi(t)t'(s)\Big|_{t=a^*_j} = 0\ ,\ \text{i.e.}\ \Psi(a^*_j) = 0\ ,\ j=N+1,\ldots,N-2K\ . \qquad (3.19)$$

From (3.17) and (3.19), we know that $\Psi(z)$ possesses at least 2N-2K zeros on Γ. Using Lemma 1.1 we get the absurd inequality

$$2N-2K \leq 2N_D + N_\Gamma = 2N-2K-2\ , \qquad (3.20)$$

where $N-K-1$ is the index of Problem B_0'. This contradiction proves that (3.14) is true. Hence, we can find a solution $h_j\,(j=0,1\ldots,N)$, $H^{\pm}_m(m=1,\ldots,|K|-1)$ to (3.13). This shows that Problem B^* is solvable.

Next, we prove the uniqueness of the solution to Problem B^*. Let $\Phi_1(z)$ and $\Phi_2(z)$ be two solutions of Problem B^* with $K < 0$. Then $\Phi(z) = \Phi_1(z) - \Phi_2(z)$ satisfies the boundary condition

$$Re[\overline{\lambda(t)}\Phi(t)] = h(t)\ ,\ t \in \Gamma\ . \qquad (3.21)$$

As in (3.11), it is clear that $h(t)$ satisfies the conditions

$$\int_\Gamma \lambda(t)\Psi(t)h(t)t'(s)ds = 0\ ,\ n=1,\ldots,N-2K-1\ , \qquad (3.22)$$

where $\Psi_n(t)\ (n=1,\ldots,N-2K-1)$ are stated in connection with (3.11). Because of (3.14), we see

$$h_j = 0(j=0,1,\ldots,N) \; , \; H_m^{\pm} = 0 \; (m=1,\ldots,|K|-1) \; . \tag{3.23}$$

It follows from Theorem 1.2 that $\Phi(z)=0$, i.e. $\Phi_1(z)=\Phi_2(z)$, $z\in D$. □

Theorem 3.2 Let $\Phi(z)$ be a solution of Problem B^* for analytic functions. Then $\Phi(z)$ satisfies the estimate

$$C_\alpha[\Phi,\bar{D}] \le M_1 \; , \tag{3.24}$$

where $\alpha(0<\alpha<1)$, $M_1 = M_1(\alpha,\ell,K,D)$ are constants.

Proof. We discuss only the case $K<0$. According to the method in the proof of Theorem 1.4 and using the uniqueness result in Theorem 3.1, we can prove the estimate (3.24). □

Theorem 3.3 Under the condition in Theorem 3.2, the solution $\Phi(z)$ can be expressed as

$$\Phi(z) = \frac{1}{2\pi}\int_\Gamma T(z,t)r(t)d\theta + \Phi_0(z) \tag{3.25}$$

where $T(z,t) = P(z,t)e^{iS(z)+\mathit{Im}S(t)}|\Pi(t)|/\Pi(z)$ is the Schwarz kernel, $S(z)$, $\Pi(z)$ are as stated in (1.7), and $P(z,t)$ is

$$\left.\begin{aligned} &P(z,t) = \sum_{j=0}^{N} P_j(z,t) + P_*(z,t) \; , \; P_j(z,t) = \begin{cases} \dfrac{e^{i\theta_j}(t+z-2z_j)}{t-z} \; , & t\in\Gamma_j \; , \\ 0, t\in\Gamma\setminus\Gamma_j, j=1,\ldots,N \; , & \end{cases} \\ &P_0(z,t) = \begin{cases} z^K(t+z)/(t-z) \; , \; t\in\Gamma_0 \; , \; K\ge 0 \; , \\ 2/[t^{|K|-1}(t-z)] \; , \; t\in\Gamma_0 \; , \; K<0 \; , \\ 0 \; , \; t\in\Gamma\setminus\Gamma_0 \; , \end{cases} \end{aligned}\right\} \tag{3.26}$$

in which $P_*(z,t)$ is an analytic function in $z\in D$ satisfying the boundary condition

$$Re[\overline{\Lambda(z)}P_*(z,t)] = -Re[\overline{\Lambda(z)}Q(z,t)] + h(z,t) , \tag{3.27}$$

$$Q(z,t) = \sum_{\substack{m=0 \\ m\neq j}}^{N} P_m(z,t) , \quad z \in \Gamma_j , \quad 0 \le j \le N ,$$

$$Im[\overline{\Lambda(a_j)}P_*(a_j,t)] = -Im[\overline{\Lambda(a_j)}Q(a_j,t)] , \; a_j \in \Gamma_j , \; j \in \{j\} , \; t \in \Gamma , \tag{3.28}$$

and $\Phi_0(z)$ is an analytic function satisfying the corresponding homogeneous boundary condition. (Here we replace $h(t)$ in (3.2) by $h(t)e^{-ImS(t)}/|\Pi(t)|$. The revised Problem is still called Problem B^*.)

Proof. We first give the expression for the solution to Problem C_0^* for analytic functions. Due to the estimate

$$C_\alpha[\overline{\Lambda}Q(z,t),\Gamma] \le M_2 = M_2(\alpha,\ell,D) \tag{3.29}$$

we can prove the existence and uniqueness of the solution $P_*(z,t)$ and that $P_*(z,t)$ satisfies the estimate

$$C_\alpha[P_*(z,t),\overline{D}] \le M_3 = M_3(\alpha,\ell,D) \tag{3.30}$$

Hence for every $t_0 \in \Gamma$, $P_*(z,t)$ and $P'_{*z}(z,t)$ converge uniformly to $P_*(z,t_0)$ and $P'_{*z}(z,t_0)$ on $\overline{D}$, respectively, when $t \to t_0$. This shows that $P_*(z,t)$ and $P'_{*z}(z,t)$ are continuous in $t \in \Gamma$. Thus, we obtain the expression for the solution $\Psi(z)$ to Problem C^* for analytic functions

$$\Psi(z) = \frac{1}{2\pi}\int_\Gamma P(z,t)R(t)d\theta + \Psi_0(z) , \; R(t) = r(t)e^{ImS(t)}|\Pi(t)| , \tag{3.31}$$

where $\Psi_o(z)$ is an analytic function satisfying the homogeneous boundary condition

$$Re[\overline{\Lambda(t)}\Psi_0(t)] = h(t) , \quad t \in \Gamma . \tag{3.32}$$

From (3.4) and (3.31), the integral expression (3.25) is derived. □

3. The homogeneous Problem C_0^* for simple complex equations

In the following, we first give the integral expression for solutions to Problem C_0^* for the first order simple complex equation (3.5) and then discuss properties of solutions of Problem C_0^* above.

Theorem 3.4 The solution $W(z)$ of Problem C_0^* with the boundary condition

$$Re[\overline{\Lambda(t)}W(t)] = h(t) \ , \ t \in \Gamma \tag{3.33}$$

for (3.5) possesses the following integral representation

$$W(z) = \tilde{T}\Omega = -\frac{1}{\pi}\iint_D [G_1(z,\zeta)Re\Omega(\zeta) + iG_2(z,\zeta)Im\Omega(\zeta)]d\sigma_\zeta \tag{3.34}$$

$$= T\Omega + \sum_{k=0}^{N} T_k\Omega + T_*\Omega \ ,$$

where $\Omega(z) \in L_p(\overline{D})$, $p > 2$,

$$\left.\begin{aligned} &G_1(z,\zeta) = \frac{1}{\zeta-z} + \sum_{k=0}^{N} g_k(z,\zeta) \ , \ G_2(z,\zeta) = \frac{1}{\zeta-z} - \sum_{k=0}^{N} g_k(z,\zeta) \ , \\ &g_0(z,\zeta) = \begin{cases} z^{2K+1}/(1-\overline{\zeta}z) \ , \ K \geq 0 \\ \overline{\zeta}^{-2K-1}/(1-\overline{\zeta}z) \ , \ K < 0 \ , \end{cases} \quad g_k(z,\zeta) = \frac{e^{2i\theta_k}(z-z_k)}{\gamma_k^2-(\overline{\zeta}-\overline{z_k})(z-z_k)} \ , \ k > 0 \end{aligned}\right\} \tag{3.35}$$

$$\left.\begin{aligned} &T\Omega = -\frac{1}{\pi}\iint_D \frac{\Omega(\zeta)}{\zeta-z}d\sigma_\zeta \ , \ T_0\Omega = -\frac{1}{\pi}\iint_D g_0(z,\zeta)\overline{\Omega(\zeta)}d\sigma_\zeta \ , \\ &T_k\Omega = -\frac{1}{\pi}\iint_D g_k(z,\zeta)\overline{\Omega(\zeta)}d\sigma_\zeta \ , \ k=1,\dots,N \ , \end{aligned}\right\} \tag{3.36}$$

and

$$T_*\Omega = -\frac{1}{\pi}\iint_D g_*(z,\zeta)\overline{\Omega(\zeta)}d\sigma_\zeta \ ,$$

$g_*(z,\zeta)$ is an analytic function, which satisfies the boundary condition

$$\left.\begin{array}{l} Re[\overline{\Lambda(z)}g_*(z,\zeta)] = -Re[\overline{\Lambda(z)}g(z,\zeta)] + h(z) \ , \\ \qquad g(z,\zeta) = \sum\limits_{\substack{m=0 \\ m\neq k}}^{N} g_m(z,\zeta) \ , \ z \in \Gamma_k, \ k=0,1,\ldots,N \ , \\ \qquad Im[\overline{\Lambda(a_j)}g_*(a_j,\zeta)] = -Im[\overline{\Lambda(a_j)}g(a_j,\zeta)] \ , \ j \in \{j\} \ . \end{array}\right\} \tag{3.37}$$

Proof. On the basis of the properties of the integrals $T\Omega$, $T_k\Omega$ and $T_*\Omega$ we obtain

$$(T\Omega)_{\bar{z}} = \Omega(z) \ , \ (T_k\Omega)_{\bar{z}} = 0 \ , \ (T_*\Omega)_{\bar{z}} = 0 \ , \ z \in D \ . \tag{3.38}$$

Hence, the function $W(z) = \tilde{T}\Omega$ is a solution to the equation (3.5). Noticing

$$\left.\begin{array}{l} Re\left[\dfrac{e^{-i\theta_k}\Omega(\zeta)}{\zeta-z} + \dfrac{e^{i\theta_k}(z-z_k)\overline{\Omega(\zeta)}}{\gamma_k^2-(\bar{\zeta}-\bar{z}_k)(z-z_k)}\right] = \\ = Re\left[\dfrac{e^{-i\theta_k}\Omega(\zeta)}{\zeta-z_k-(z-z_k)} + \dfrac{e^{i\theta_k}\overline{\Omega(\zeta)}}{\bar{z}-\bar{z}_k-(\bar{\zeta}-\bar{z}_k)}\right] = 0 \ , \text{ for } z\in\Gamma_k \ , \ k=1,\ldots,N \ , \end{array}\right\} \tag{3.39}$$

$$Re[\frac{\bar{z}^{K}\Omega(z)}{\zeta-z} + \frac{z^{K+1}\overline{\Omega(z)}}{1-\bar{\zeta}z}] = Re[\frac{\bar{z}^{K}\Omega(z)}{\zeta-z} + \frac{z^{K}\overline{\Omega(z)}}{\bar{z}-\bar{\zeta}}] = 0 \ , \text{ for } z\in\Gamma_0 \ , \ K\geq 0 \ , \tag{3.40}$$

and if $K < 0$,

$$\left.\begin{array}{l} Re\Big\{-\dfrac{1}{\pi}\displaystyle\iint_D [\dfrac{z^{|K|}\Omega(\zeta)}{\zeta-z} + \dfrac{z^{|K|}\bar{\zeta}^{2|K|-1}\overline{\Omega(\zeta)}}{1-\bar{\zeta}z} \\ \qquad + (\zeta^{|K|-1} + z\zeta^{|K|-2}+\ldots+z^{|K|-1})\Omega(\zeta) \\ \qquad + (z\bar{\zeta}^{|K|} + z^2\bar{\zeta}^{|K|+1}+\ldots+z^{|K|-1}\bar{\zeta}^{2|K|-2})\overline{\Omega(\zeta)}]d\sigma_\zeta\Big\} \\ = Re\Big\{-\dfrac{1}{\pi}\displaystyle\iint_D [\dfrac{\zeta^{|K|}\Omega(\zeta)}{\zeta-z} + \dfrac{z\bar{\zeta}^{|K|}\overline{\Omega(\zeta)}}{1-\bar{\zeta}z}]d\sigma_\zeta\Big\} = 0 \ , \ z \in \Gamma_0 \ , \end{array}\right\} \tag{3.41}$$

one shows that $W(z) = \tilde{T}\Omega$ satisfies the boundary condition (3.33). □

Theorem 3.5 Let $\Omega(z) \in L_p(\overline{D})$, $p > 1$. Then $\tilde{T}\Omega$ possesses the following properties.

(1) $(\tilde{T}\Omega)_{\bar{z}} = \Omega(z)$, $\tilde{S}\Omega = (\tilde{T}\Omega)_z = \Pi\Omega + \sum_{k=0}^{N} \Pi_k\Omega + \Pi_*\Omega$,

where

$$\left.\begin{aligned} &\Pi\Omega = -\frac{1}{\pi}\iint_D \frac{\Omega(\zeta)}{(\zeta-z)^2}\, d\sigma_\zeta\,,\quad \Pi_0\Omega = \begin{cases} -\dfrac{1}{\pi}\displaystyle\iint_D \frac{(2K+1-2K\bar{\zeta}z)z^{2K}\overline{\Omega(\zeta)}d\sigma_\zeta}{(1-\bar{\zeta}z)^2}\,, & K \geq 0\,, \\ -\dfrac{1}{\pi}\displaystyle\iint_D \frac{\zeta^{2|K|}\overline{\Omega(\zeta)}d\sigma_\zeta}{(1-\bar{\zeta}z)^2}\,, & K < 0\,, \end{cases} \\ &\Pi_k\Omega = -\frac{1}{\pi}\iint_D \frac{e^{2i\theta_k}\overline{\Omega(\zeta)}\gamma_k^2 d\sigma_\zeta}{[\gamma_k^2-(\bar{\zeta}-\bar{z}_k)(z-z_k)]^2}\,,\quad k=1,\ldots,N\,,\quad \Pi_*\Omega = (T_*\Omega)_z\,. \end{aligned}\right\} \tag{3.42}$$

(2) $\tilde{S}$ is a bounded linear operator in $L_p(\overline{D})$, and satisfies

$$L_p[\tilde{S}\Omega,\overline{D}] \leq \lambda_p L_p[\Omega,\overline{D}]\,, \tag{3.43}$$

where λ_p is the smallest constant so that the above inequality holds, and

$$\lambda_2 \leq 1\,,\ \text{for}\ K < 0\,, \tag{3.44}$$

and for Problem C_0^{**}, $\lambda_2 = 1$ if $K = 0$.

(3) For a real constant $q_0 (0 \leq q_0 < 1)$, there exist two constants p_0, p_1 $(2 < p_0 < p_1 < \infty)$, so that

$$q_0 \lambda_{p_j} < 1\,,\ j=0,1\,,\ \text{when}\ K \leq 0\,. \tag{3.45}$$

(4) When $p > 2$ and $2 < p_0 \le p$, $\tilde{T}\Omega$ satisfies the estimate

$$C_\beta[\tilde{T}\Omega,\bar{D}] \le M_4 L_{p_0}(\Omega,\bar{D}) \ , \ M_4 = M_4(p_0,D) \ , \tag{3.46}$$

where $\beta = 1-2/p_0$.

Proof. (1) Noticing (3.34) - (3.37), it is clear that the formulae in (1) are all true.

(2) To prove (3.43), we are free to choose $\Omega(z) \in L_p(\bar{D})$, $p > 1$, and set $L_p[\Omega,\bar{D}] \neq 0$. Obviously, $L_p[\Omega^*,\bar{D}] = 1$, where $\Omega^*(z) = \Omega(z)/L_p[\Omega,\bar{D}]$. Similarly to § 9, Chapter 4 in [97]2), we can verify

$$L_p[\Pi\Omega^*,\bar{D}] \le \lambda_{1p} \ , \ L_p[\Pi_k\Omega^*,\bar{D}] \le \lambda_{2p} \ , \ k=0,1,\ldots,N \ , \ L_p[\Pi_*\Omega^*,\bar{D}] \le \lambda_{3p} \ , \tag{3.47}$$

where $\lambda_{jp}(j=1,2,3)$ are constants depending only on p and D . Combining the above formulae, we have $L_p[\tilde{S}\Omega^*,\bar{D}] \le \lambda_p < \infty$, so (3.43) holds.

Next we prove (3.44). Let $\Omega(z) \in D_\infty^0(D)$. From the Green formula, there are

$$\left.\begin{aligned}
||\tilde{S}\Omega||_{L_2}^2 &= [L_2(\tilde{S}\Omega,\bar{D})]^2 = \iint_D (\tilde{T}\Omega)_z\overline{(\tilde{T}\Omega)_z}d\sigma_z = \iint_D [\,\overline{\tilde{T}\Omega}\,(\tilde{T}\Omega)_z]_{\bar{z}}d\sigma_z \\
&- \iint_D \overline{\tilde{T}\Omega}\,\Omega_z d\sigma_z = \frac{1}{2i}\int_\Gamma \overline{\tilde{T}\Omega}\,(\tilde{T}\Omega)_z dz - \iint_D [\overline{\tilde{T}\Omega}\,\Omega]_z d\sigma_z \\
&+ \iint_D \bar{\Omega}\Omega d\sigma_z = Re\,\frac{1}{2i}\int_\Gamma \overline{\tilde{T}\Omega}\,(\tilde{T}\Omega)_z dz + \frac{1}{2i}\iint_\Gamma \overline{\tilde{T}\Omega}\,\Omega d\bar{z} + ||\Omega||_{L_2}^2 .
\end{aligned}\right\} \tag{3.48}$$

Because $\Omega(z)$ equals zero in the neighbourhood of Γ , it is easy to see that $\tilde{T}\Omega$ satisfies the boundary condition

$$Re[\overline{\Lambda(t)}\tilde{T}\Omega] = h(t) \ , \ t \in \Gamma \ . \tag{3.49}$$

Let

$$Q(z) = \sum_{m=-1}^{K} c_m t^m \ , \ c_m = c_m^+ + ic_m^- = H^+_{|K|+m} + iH^-_{|K|+m} \ ,$$

$$m=-1,\dots,K \ , \ c_K^+ = h_0 \ , \ c_K^- = 0 \ .$$

We denote $U(t) = \widetilde{T}\Omega - Q(t)$, for $t \in \Gamma_0$. From

$$Re[t^{|K|}U(t)] = 0 \ , \ \text{for} \ t \in \Gamma_0 \ , \tag{3.50}$$

and from (3.49) it follows that

$$\left.\begin{aligned} &Re[it^{|K|+1}U_t - it^{|K|-1}U_{\bar{t}} + i|K|t^{|K|}U(t)] = 0 \ , \ t \in \Gamma_0 \ , \\ &\text{i.e. } Re[it^{|K|+1}U_t] = Re[iKt^{|K|}U(t)] \ , \ t \in \Gamma_0 \ , \\ &\overline{\widetilde{T}\Omega} = -e^{-2i\theta_k}\widetilde{T}\Omega + 2h_k e^{-i\theta_k} \ , \ t \in \Gamma_k \ , \ k=1,\dots,N \ . \end{aligned}\right\} \tag{3.51}$$

Hence, we have

$$I = Re\frac{1}{2i}\int_{\Gamma}\overline{\widetilde{T}\Omega}\,(\widetilde{T}\Omega)_z dz = \sum_{k=1}^{N} Re[-\frac{1}{2i}\int_{\Gamma_k} e^{-2i\theta_k}\widetilde{T}\Omega(\widetilde{T}\Omega)_z dz$$

$$+\frac{1}{i}\int_{\Gamma_k} e^{-i\theta_k}h_k(\widetilde{T}\Omega)_z dz + Re\frac{1}{2i}\int_{\Gamma_0}[\overline{U(z)+Q(z)}][U_z+Q'(z)]dz$$

$$= \sum_{k=1}^{N} Re[-\frac{1}{4i}\int_{\Gamma_k} e^{-2i\theta_k}d(\widetilde{T}\Omega)^2 + \frac{1}{i}\int_{\Gamma_k} e^{-i\theta_k}h_k d\widetilde{T}\Omega] +$$

$$+ Re\frac{1}{2i}\int_{\Gamma_0}[\overline{U(z)}U_z + \overline{Q(z)}Q'(z) + \overline{U(z)}Q'(z) + \overline{Q(z)}U_z]dz$$

$$= Re\{\frac{1}{2}\int_{\Gamma_0}[\overline{iz^{|K|}U(z)}iz^{|K|+1}U_z + \overline{Q(z)}Q'(z)z]d\theta + \frac{1}{2i}\int_{\Gamma_0}[\overline{U(z)}Q'(z)dz + \overline{Q(z)}dU]\}$$

$$= Re\{\frac{1}{2}\int_{\Gamma_0}[\overline{iz^{|K|}U(z)}][Re\,iz^{|K|+1}U_z + i\,Im\,iz^{|K|+1}U_z]d\theta + \frac{1}{2}\int_{\Gamma_0}\overline{Q(z)}Q'(z)zd\theta\}$$

$$+ Re\int_{\Gamma_0}\overline{U(z)}Q'(z)zd\theta = \frac{K}{2}\int_{\Gamma_0}|Re\, iz^{|K|}U(z)|^2 d\theta - \frac{1}{2}\int_{\Gamma_0} Im\,\overline{iz^{|K|}U(z)}\, \times$$

$$\times\, Im[iz^{|K|+1}U_z]d\theta + \pi\sum_{m=-1}^{K} m|c_m|^2 + Re\int_{\Gamma_0}\overline{[iz^{|K|}U(z)]}[i\sum_{m=-1}^{K} mc_m z^{m+|K|}]d\theta$$

$$= \frac{K}{2}\int_{\Gamma_0}|Re\, iz^{|K|}U(z)|^2 d\theta + \frac{1}{2}\int_{\Gamma_0} Re[z^{|K|}U(z)]\, Im[iz^{|K|+1}U_z]d\theta$$

$$+ \pi\sum_{m=-1}^{K} m|c_m|^2 - \int_{\Gamma_0} Re[iz^{|K|}U(z)]\, Im[\sum_{m=-1}^{K} mc_m z^{m+|K|}]d\theta$$

$$= K\pi[(\frac{a_0}{2})^2 + \sum_{n=1}^{\infty}(\frac{a_n^2}{2} + \frac{b_n^2}{2})] + \pi\sum_{m=-1}^{K} m|c_m|^2$$

$$- \pi\sum_{m=-1}^{K} m[Re\, c_m\, b_{m+|K|} + Im\, c_m\, a_{m+|K|}]$$

$$\leq \pi\sum_{m=-1}^{K} m\,[(a_{m+|K|}^2 + b_{m+|K|}^2)/4 + |c_m|^2$$

$$- Re\, c_m\, b_{m+|K|} - Im\, c_m\, a_{m+|K|}] \leq 0\ , \qquad (3.52)$$

where a_n , $b_n (n=0,1,\ldots)$ are the coefficients of the Fourier series of $Re[iz^{|K|}U(z)]$ on Γ_0 . Due to (3.48) and (3.52), it is easily shown that

$$||\tilde{S}\Omega||_{L_2}^2 \leq ||\Omega||_{L_2}^2\ . \qquad (3.53)$$

From the dense property of $D_\infty^0(D)$ in $L_2(\overline{D})$, we see that (3.52) remains valid for any function $\Omega(z) \in L_2(\overline{D})$. Thus $\lambda_2 \leq 1$, if $K < 0$. In the case $K = 0$ and $\Omega(z) \in L_2(\overline{D})$, $\tilde{T}\Omega|_{z=a_0}$ may not be defined. Hence, we discuss Problem C_0^{**} with the condition

$$L_0\tilde{T}\Omega = \int_0^{2\pi}\tilde{T}\Omega d\theta =$$

$$iIm\int_{\Gamma_0}\{-\frac{1}{\pi}\iint_D[\frac{\Omega(\zeta)}{\zeta - z} + \frac{z\overline{\Omega(\zeta)}}{1-\overline{\zeta}z}]d\sigma_\zeta - \frac{1}{\pi}\iint_D[\sum_{j=1}^{N} g_j(z,\zeta)+g_*(z,\zeta)]\overline{\Omega(\zeta)}d\sigma_\zeta\}\frac{dz}{iz}\ .$$

Similarly to (3.52) we have

$$Re\,\frac{1}{2i}\int_{\Gamma}\overline{\tilde{T}\Omega}(\tilde{T}\Omega)_z dz =$$

$$Re\{-\frac{1}{2i}\int_{\Gamma_0}\tilde{T}\Omega(\tilde{T}\Omega)_z dz + \sum_{k=1}^{N}[\frac{-e^{-2i\theta_k}}{2i}\int_{\Gamma_k}\tilde{T}\Omega(\tilde{T}\Omega)_z dz + \frac{h_k e^{-i\theta_k}}{i}\int_{\Gamma_k}(\tilde{T}\Omega)_z dz]\} = 0 \ .$$

From the above formula and (3.48), $\lambda_2 = 1$ follows.

(3) On the basis of the Riesz theorem (see [97]2)) it can be seen that (3.45) holds.

(4) $2 < p_0 \le p < \infty$. From $\Omega(z) \in L_p(\overline{D})$, we can conclude $\Omega(z) \in L_{p_0}(\overline{D})$. According to Theorem 1.23 in [97]2), it is not difficult to verify

$$C_\beta[T\Omega^*,\overline{D}] \le M_5 \ , \ C_\beta[T_k\Omega^*,\overline{D}] \le M_6 \ , \ k=0,1,\ldots,N \ , \ C_\beta[T_*\Omega^*,\overline{D}] \le M_7 \ , \tag{3.54}$$

where $\beta = 1-2/p_0$, $\Omega^*(z) = \Omega(z)/L_{p_0}[\Omega,\overline{D}]$, $M_j = M_j(p_0,D)$, $j=5,6,7$. Hence we obtain (3.46). □

4. Properties of boundary value problems for first order complex equations

By virtue of the above results, we obtain easily the properties of the boundary value Problem B^* for (3.1).

Theorem 3.6 Problem B^* for (3.1) has a unique solution $w(z)$, and $w(z)$ possesses the integral representation

$$w(z) = \Phi(z) + \hat{T}\omega \ , \tag{3.55}$$

where $\Phi(z)$ is as in (3.25), and

$$\hat{T}\omega = -\frac{1}{\pi}\iint_D [G_1^*(z,\zeta)Re\ \omega(\zeta) + iG_2^*(z,\zeta)Im\ \omega(\zeta)]d\sigma_\zeta \ , \tag{3.56}$$

in which

$$\left.\begin{aligned} G_1^*(z,\zeta) &= \Delta(z)[G_1(z,\zeta)Re\ \Delta(\zeta) - G_2(z,\zeta)i\ Im\ \Delta(\zeta)]/|\Delta(\zeta)|^2 , \\ G_2^*(z,\zeta) &= \Delta(z)[-G_1(z,\zeta)i\ Im\ \Delta(\zeta) + G_2(z,\zeta)Re\ \Delta(\zeta)]/|\Delta(\zeta)|^2 , \end{aligned}\right\} \quad (3.57)$$

where $G_1(z,\zeta)$ and $G_2(z,\zeta)$ are as stated in (3.35), and $\Delta(\zeta)$ as in (3.5).

Proof. It is clear that $\hat{T}\omega$ is a solution of the complex equation (3.1) which satisfies the homogeneous boundary condition

$$Re[\overline{\lambda(t)}\hat{T}\omega] = h(t)e^{-ImS(t)}/|\Pi(t)| , \quad t \in \Gamma , \quad (3.58)$$

and $\Phi(z)$ is an analytic function satisfying the boundary condition of Problem B^* and the point conditions

$$Im[\overline{\lambda(a_j)}\Phi(a_j)] = b_j - Im[\overline{\lambda(t)}\hat{T}\omega]_{t=a_j} , \quad j \in \{j\} . \quad (3.59)$$

By Theorem 3.3, the above function $\Phi(z)$ exists and is unique. Hence the theorem is true. □

Theorem 3.7 The solution $w(z)$ of Problem B^* for the complex equation (3.1) satisfies the estimates

$$C_\beta[w,\overline{D}] \le M_8 , \quad L_{p_0}[|w_{\bar{z}}| + |w_z|,D_*] \le M_9 , \quad (3.60)$$

where $\beta = \min(\alpha, 1-2/p_0)$, $M_8 = M_8(p_0,\alpha,\ell,K,D)$, $M_9 = M_9(p_0,\alpha,\ell,K,D,D_*)$, D_* is a closed set in D. If the constant α satisfies the condition $1/2 < \alpha < 1$, then $w(z)$ satisfies the estimate

$$L_{p_0}[|w_{\bar{z}}| + |w_z|,\overline{D}] \le M_{10} = M_{10}(p_0,\alpha,\ell,K,D) . \quad (3.61)$$

Proof. Using Theorem 3.4 and Theorem 3.5 and noting that the function $\Delta(z)$ satisfies

$$C_\alpha[\Delta,\overline{D}] \le M_{11} = M_{11}(\alpha,\ell,K,D) \ , \tag{3.62}$$

we can obtain (3.60). If $1/2 < \alpha < 1$, we may select p_0 , so that $2 < p_0 < \min\,(p,\frac{1}{1-\alpha})$. From Theorem 1.11 in [97]2), (3.61) can be derived. □

§ 4. Some boundary value problems for simple second order equations

Now, we consider the simple equation

$$u_{z\bar{z}} = \rho(z) \ , \tag{4.1}$$

where $\rho(z)$ is real valued and $L_p[\rho,\overline{D}] \le \ell$, $2 < p < \infty$, and we give an integral representation and the properties of the oblique derivative problem for (4.1).

1. Some proper propositions of the oblique derivative problem for (4.1)

In this section, we discuss the irregular oblique derivative problem for (4.1).

Problem P^*. Find a continuously differentiable solution $u(z)$ on $\overline{D}$ to (4.1) satisfying the boundary condition

$$\left.\begin{aligned} &\frac{\partial u}{\partial \mu} = 2\tau(t) \ , \ t \in \Gamma \ , \ u(a_0) = b_0 \ , \\ &\text{i.e. } Re[\overline{\lambda(t)}u_t] = \tau(t) \ , \ t \in \Gamma \ , \ u(a_0) = b_0 \ , \end{aligned}\right\} \tag{4.2}$$

where $\vec{\mu}$, $\lambda(t)$, $\tau(t)$, a_0 , b_0 are as stated in (2.41) - (2.43). In order to give properties of solutions to Problem P^* for (4.1), we introduce proper modified propositions of the above problem, i.e. the boundary condition (4.2) and the equation (4.1) are replaced as follows.

$$\left.\begin{aligned} &Re[\overline{\lambda(t)}w(t)] = \tau(t) + h(t) \ , \ t \in \Gamma \ , \\ &Im[\overline{\lambda(a_j)}w(a_j)] = b_j \ , \ j \in \{j\} \ , \end{aligned}\right\} \tag{4.3}$$

$$\left.\begin{aligned} &Re[iu(t)] = H(t) \ , \ t \in \Gamma \ , \\ &Im[iu(a_0)] = b_0 \end{aligned}\right\} \tag{4.4}$$

and

$$w_{\bar{z}} = \rho(z) \ , \ u_{\bar{z}} = \overline{w(z)} \ , \ z \in D \ , \tag{4.5}$$

respectively, where $h(t)$, a_j , b_j are the same as in (1.2) and (1.3) and

$$H(t) = \begin{cases} 0 \ , \ t \in \Gamma_0 \ , \\ H_j \ , \ t \in \Gamma \ , \ j=1,\dots,N \ , \end{cases} \tag{4.6}$$

$H_j (j=1,\dots,N)$ are unknown real constants to be determined appropriately. The modified problem is called <u>Problem Q^*</u>. The second equations in (4.5) and (4.4) are replaced by

$$u(z) = 2Re \int_{a_0}^{z} [w(z) + \sum_{j=1}^{N} \frac{id_j}{z-z_j}]dz + b_0 \ , \tag{4.7}$$

where $d_j (j=1,\dots,N)$ are appropriate real constants so that the function determined by the integral in (4.7) is single-valued in D . The above problem is called <u>Problem Q^{**}</u>.

<u>Problem P^{**}.</u> If the index $K = \frac{1}{2\pi} \Delta_\Gamma \, arg \, \lambda(t) \geq N-1$ and $\vec{\mu} = \vec{\nu}$ on $\Gamma \setminus \Gamma_0$ in (4.2) satisfies the condition $cos(\nu,n) \geq 0$, and if $cos(\nu,n) = 0$, $t \in \Gamma_j$, we set

$$\int_{\Gamma_j} \tau(t)ds = 0 \ , \ u(a_j) = b_j \ , \ 1 \leq j \leq N \ , \tag{4.8}$$

where $a_j \in \Gamma_j$, b_j is a real constant, $|b_j| \leq \ell$. Without loss of generality, we may assume that $cos(\nu,n) = 0$ on $\Gamma_* = \Gamma_1 \cup \dots \cup \Gamma_{N_0}$, and it does not hold on $\Gamma_j (N_0 < j \leq N , \Gamma_{**} = \Gamma \setminus \Gamma_* \setminus \Gamma_0)$. In this case, the modified boundary condition is

$$\frac{\partial u}{\partial \mu} = 2[\tau(t)+h(t)]\ ,\ \text{i.e. } Re[\overline{\lambda(t)}u_t] = \tau(t) + h(t)\ ,\ t \in \Gamma\ , \tag{4.9}$$

where $h(t) = \begin{cases} h_0\ , & t \in \Gamma_0\ , \\ 0\ , & t \in \Gamma \setminus \Gamma_0, \end{cases}$ when $K = N-1$, and $h(t) = 0$ when $K \geq N$.

<u>Problem P^{***}.</u> If $\lambda(t)$ in (4.9) possesses the form

$$\overline{\lambda(t)} = \begin{cases} e^{-i\theta_0} t^{-K}\ , & t \in \Gamma_0\ , \\ e^{-i\theta_j}(t-z_j)\ , & t \in \Gamma_j\ ,\ j=1,\ldots,N\ , \end{cases} \tag{4.10}$$

the Problem P^{**} with the condition (4.10) is called <u>Problem P^{***}.</u> When $k = -1$, $\theta_j = \pm \frac{\pi}{2}$ and $\theta_j = 0$, $j=1,\ldots,N$, Problem P^{***} is the Dirichlet problem and the Neumann problem, respectively.

<u>2. Properties of solutions to Problem Q^* and Problem Q^{**} for (4.1)</u>

We first discuss the existence and uniqueness of solutions to Problem Q^* and Problem Q^{**} for (4.1), and then give their integral representations and estimates.

<u>Theorem 4.1</u> Problem Q^* and Problem Q^{**} for (4.1) have unique solutions $[w(z),u(z)]$. Moreover,

(1) if the index $K \geq N$, Problem P^* has N solvability conditions,

(2) if $0 \leq K < N$, Problem P^* has $2N-K$ solvability conditions,

(3) if $K < 0$, Problem P^* has $2N-K-1$ solvability conditions.

<u>Proof.</u> From Theorem 3.1, we see that Problem B^* for the first complex equation $w_{\bar{z}} = \rho(z)$ with the boundary condition (4.3) has a unique solution $w(z)$. By the same reason, we can find a unique solution $u(z)$ to the boundary value problem for $u_{\bar{z}} = \overline{w(z)}$ with the boundary condition (4.4). Hence Problem Q^* for (4.5) has a unique solution $[w(z),u(z)]$. Substituting the above function $w(z)$ into (4.7), we

obtain a unique single-valued function $u(z)$. This shows that Problem Q^{**} is uniquely solvable.

Substitute a solution $[w(z),u(z)]$ of Problem Q^{*} into (4.3) - (4.6). If $h(t) = 0$ and $H(t) = 0$, $t \in \Gamma$, then $[Im u(z)]_{z\bar{z}} = 0$, $z \in D$ and $Im u(t) = 0$, $t \in \Gamma$, and so $Im u(z) = 0$, $z \in \bar{D}$. Therefore, under the same conditions as in Theorem 4.1, Problem P^{*} for (4.1) is solvable. Similarly, substitute a solution $[w(z),u(z)]$ of Problem Q^{**} into (4.3), (4.7) and the first equation of (4.5). If $h(t) = 0$, $t \in \Gamma$ and $d_j = 0$ $(j=1,\ldots,N)$, then there is the same result. □

Theorem 4.2 The solution $[w(z),u(z)]$ of Problem Q^{**} can be expressed by the integral form (4.7), where $w(z)$ can be written in the following form

$$w(z) = \Phi(z) + \hat{T}\rho , \tag{4.11}$$

in which $\Phi(z)$ is an analytic function as stated in (3.25) with $r(t) = \tau(t)$ and the operator $\hat{T}$ is stated in (3.56), and $[w(z),u(z)]$ satisfies the estimates

$$C_\beta[w,\bar{D}] \le M_1 , \quad C_\beta^1[u,\bar{D}] \le M_2 , \tag{4.12}$$

where $M_j = M_j(p_0,\alpha,\ell,K,D)$, $j=1,2$, $\beta = 1-2/p_0$.

Proof. Because $w(z)$ is a solution of Problem B^{*} for $w_{\bar{z}} = \rho(z)$, $L_p[\rho,\bar{D}] \le \ell$, it follows from Theorem 3.6 and Theorem 3.7 that $w(z)$ is representable by (4.11) and satifies the first estimate in (4.12). Substituting $w(z)$ into (4.7), we can get bounds for the constants $d_j (j=1,\ldots,N)$. Hence from (4.7) the second estimate of (4.12) is derived. □

The solution to Problem Q^{*} for (4.1) possesses similar properties.

3. Properties of solutions to Problem P^{**} and Problem P^{***} for (4.1)

Theorem 4.3 Problem P^{**} for (4.1) is solvable.

Proof. Using a similar method as in the proof of Theorem 2.4, we can prove the theorem. In fact, by Theorem 3.6, we first find a solution $w_0(z)$ of Problem B^* for $w_{\bar{z}} = \rho(z)$, which satisfies the boundary condition

$$\left.\begin{aligned} &Re[\overline{\lambda(t)}w(t)] = \tau(t) + h(t) , \quad t \in \Gamma , \\ &Im[\overline{\lambda(a_j)}w(a_j)] = b_j , \; j \in \{j\} = \begin{cases} 1,\ldots,2K-N+1 , & K \geq N , \\ 1,\ldots,N , & K = N-1 . \end{cases} \end{aligned}\right\} \tag{4.13}$$

If

$$Re \int_{\Gamma_j} w_0(z)dz = 0 , \quad j = N_0+1,\ldots,N , \tag{4.14}$$

then the function

$$u(z) = 2Re \int_{a_0}^{z} w_0(z)dz + b_0 \tag{4.15}$$

is a single-valued solution of Problem P^{**} for (4.1). Otherwise, we can find $N - N_0$ solutions $w_{N_0+1}(z),\ldots,w_N(z)$ to Problem B^* for analytic functions with the boundary conditions

$$\left.\begin{aligned} &Re[\overline{\lambda(t)}w_k(t)] = h(t) , \quad t \in \Gamma , \\ &Im[\overline{\lambda(a_j)}w_k(a_j)] = \delta_{jk} = \begin{cases} 1 , & j = k , \; k = N_0+1,\ldots,N , \\ 0 , & j \neq k , \; j \in \{j\} . \end{cases} \end{aligned}\right\} \tag{4.16}$$

We can prove

$$J = \begin{vmatrix} I_{N_0+1,N_0+1} & \cdots & I_{N_0+1,N} \\ \vdots & & \vdots \\ I_{N,N_0+1} & \cdots & I_{N,N} \end{vmatrix} \neq 0 , \quad I_{j,k} = Re \int_{\Gamma_j} w_k(z)dz , \tag{4.17}$$

and seek a solution $c_{N_0+1},\dots,c_N$ of the system

$$c_{N_0+1}I_{j,N_0+1}+\dots+c_N I_{j,N} = -Re\int_{\Gamma_j} w_0(z)dz\ ,\ j=N_0+1,\dots,N\ . \tag{4.18}$$

Therefore, the function

$$u(z)=2Re\int_{a_0}^{z} w(z)dz+b_0\ ,\ w(z)=w_0(z)+\sum_{k=N_0+1}^{N} c_k w_k(z) \tag{4.19}$$

is a solution of Problem P^{**} for (4.1).

Under the condition (4.13) with $|b_j|\le \ell$, $j\in\{j\}$, we can show that the solution $u(z)$ of the above Problem P^{**} satisfies the second estimate in (4.12). □

Next, we give an integral representation for solutions to Problem P^{***} for (4.1) and its properties.

<u>Theorem 4.4</u> Let $u_0(z)$ be a solution of Problem P^{***} for harmonic functions. Then $u_0(z)$ admits the representation

$$\left.\begin{aligned}
&u_0(z)=v(z)+v_0(z),v(z)=\frac{2}{\pi}\int_{\Gamma} P(z,t)\tau(t)d\theta,P(z,t)=\sum_{m=0}^{N} s_m(z,t),z\in D,\\
&s_0(z,t)=\begin{cases}-Re[e^{i\theta_0}\{t^{K+1}\ell n(1-\bar tz)+\sum\limits_{m=1}^{K}(\bar tz)^m/m+z^{K+1}/2(K+1)\}],K\ge 0\ ,\ t\in\Gamma_0,\\ -Re[e^{i\theta_0}\ell n(1-\bar tz)],K=-1\ ,\ t\in\Gamma_0\ ,\\ 0\ ,\ K\le 1\ ,\ t\in\Gamma\setminus\Gamma_0\end{cases}\\
&s_m(z,t)=\begin{cases}-Re\ e^{i\theta_j}\ell n[\dfrac{(t-z)(-z_j)}{t(z-z_j)}]\ ,\ t\in\Gamma_m\ ,\\ 0\ ,\ t\in\Gamma\setminus\Gamma_m\ ,\end{cases}\qquad m=1,\dots,N\ ,
\end{aligned}\right\} \tag{4.20}$$

where $v_0(z)$ is a harmonic function in D, satisfying

$$\left.\begin{aligned}
&Re[\overline{\lambda(t)}v_{0t}] = -Re\,\Psi(t) + h(t)\ ,\ t\in\Gamma\ ,\ v_0(a_0) = b_0 - v(a_0)\ ,\\
&\Psi(t) = \overline{\lambda(t)}\ \{\sum_{\substack{m=0\\ m\neq j}}^{N}[\Phi_m(t)+\phi_m(t)] + \phi_j(t)\}\ ,\ t\in\Gamma_j\ ,\ j=0,1,\ldots,N\ ,\\
&\phi_0(z) = 0\ ,\ \phi_j(z) = \frac{e^{i\theta_j}}{2\pi i(z-z_j)}\int_{\Gamma_j}\frac{\tau(t)}{t-z_j}dt\ ,\ j=1,\ldots,N\ ,\\
&\Phi_0(z) = \begin{cases}\dfrac{e^{i\theta_0}z^K}{2\pi i}\displaystyle\int_{\Gamma_0}\frac{t+z}{t-z}\frac{\tau(t)}{t}dt\ ,\ K\geq 0\ ,\\[2ex] \dfrac{e^{i\theta_0}}{\pi i}\displaystyle\int_{\Gamma_0}\frac{\tau(t)}{(t-z)t}dt\ ,\ K=-1\ ,\end{cases}\\
&\Phi_j(z) = \frac{e^{i\theta_j}}{2\pi i(z-z_j)}\int_{\Gamma_j}\frac{t+z-2z_j}{t-z}\frac{\tau(t)}{t-z_j}dt\ ,\ j=1,\ldots,N\ .
\end{aligned}\right\}\tag{4.21}$$

Proof. When $K\geq 0$, it is easy to see that $u_0(a_0) = b_0$ and

$$\begin{aligned}
v_z &= \frac{1}{\pi}\int_{\Gamma_0} e^{i\theta_0}\tau(t)z^K[\frac{1}{1-\bar{t}z}-\frac{1}{2}]d\theta + \sum_{j=1}^{N}\frac{e^{i\theta_j}}{\pi}\int_{\Gamma_j}\frac{(t-z_j)\tau(t)}{(t-z)(z-z_j)}d\theta\\
&= \frac{e^{i\theta_0}z^K}{2\pi}\int_{\Gamma_0}\tau(t)\frac{t+z}{t-z}d\theta + \frac{1}{2\pi}\sum_{j=1}^{N}\frac{e^{i\theta_j}}{(z-z_j)}\int_{\Gamma_j}\tau(t)\,[\frac{t+z-2z_j}{t-z}+1]d\theta\\
&= \sum_{j=0}^{N}[\Phi_j(z)+\phi_j(z)]\ .
\end{aligned}$$

If $z\to t\in\Gamma_0$, $Re[\overline{\lambda(z)}\Phi_0(z)]\to\tau(t)$ and if $z\to t\in\Gamma_j$, $Re[\overline{\lambda(z)}\Phi_j(z)]\to\tau(t)$, $j=1,\ldots,N$. Hence $v_0(z)$ is a harmonic function, which satisfies the boundary condition (4.21).

If $K=-1$, we may similarly obtain the representation (4.20). □

Next, we introduce the double integral

$$u(z) = H\rho = \frac{2}{\pi}\iint_D G(z,\zeta)\rho(\zeta)d\sigma_\zeta = H_0\rho + \sum_{j=0}^{N} \widetilde{H}_j\rho + H_*\rho \;, \tag{4.22}$$

in which

$$\left.\begin{aligned}
&G(z,\zeta) = \ln\left|1-\frac{z}{\zeta}\right| + \sum_{j=0}^{N} g_j(z,\zeta) + g_*(z,\zeta) \;, \\
&H_0\rho = \frac{2}{\pi}\iint_D \ln\left|1-\frac{z}{\zeta}\right|\rho(\zeta)d\sigma_\zeta , \widetilde{H}_j\rho = \frac{2}{\pi}\iint_D g_j(z,\zeta)\rho(\zeta)d\sigma_\zeta \;, \; j=1,\dots,N \;, \\
&H_*\rho = \frac{2}{\pi}\iint_D g_*(z,\zeta)\rho(\zeta)d\sigma_\zeta , g_j(z,\zeta) = Re\left\{e^{2i\theta_j}\ln\frac{-z_j[\gamma_j^2-(\overline{\zeta}-\overline{z_j})(z-z_j)]}{[\gamma_j^2+z_j(\overline{\zeta}-\overline{z_j})](z-z_j)}\right\} \\
&\qquad\qquad j=1,\dots,N \;, \\
&g_0(z,\zeta) = \begin{cases} Re\; e^{2i\theta_0}\,\overline{\zeta}^{-2K-2}[\ln(1-\overline{\zeta}z) + \sum_{m=1}^{2k+1} (\overline{\zeta}z)^m/m] \;, & \text{if } K \ge 0 \;, \\ Re\; e^{2i\theta_0}\,\overline{\zeta}^{-2K-2}\ln(1-\overline{\zeta}z) \;, & \text{if } K < 0 \;, \end{cases}
\end{aligned}\right\} \tag{4.23}$$

and $g_*(z,\zeta)$ is harmonic function in $z \in D$ satisfying the boundary condition

$$\left.\begin{aligned}
&Re[\overline{\lambda(z)}g_z] = Re\;\sigma(z,\zeta) + h(z) \;, \; z \in \Gamma \;, \; \zeta \in D \;, \; g(0,\zeta) = 0 \;, \\
&Im[\overline{\lambda(a_j)}g_z(a_j,\zeta)] = Im\;\sigma(a_j,\zeta) \;, \; j \in \{j\} \;, \\
&\sigma(z,\zeta) = \begin{cases} e^{-i\theta_0}\overline{z}^K \sum_{m=1}^{N} g_{mz}(z,\zeta) \;, & z \in \Gamma_0 \;, \\ e^{-i\theta_j}(z-z_j) \sum_{\substack{m=0 \\ m\neq j}}^{N} g_{mz}(z,\zeta) \;, & z \in \Gamma_j \;, \; j=1,\dots,N \;, \end{cases}
\end{aligned}\right\} \tag{4.24}$$

where the existence of $g(z,\zeta)$ is known from the proof of Theorem 4.4.

<u>Theorem 4.5</u> Let $\rho(z) \in L_p(\overline{D})$, $p > 1$. Then $H\rho$ possesses the following properties.

(1) $\widetilde{T}\rho = (H\rho)_z = T\rho + \sum_{j=0}^{N} \widetilde{T}_j\rho + \widetilde{T}_*\rho$, $S\rho = (H\rho)_{zz} = \Pi\rho + \sum_{j=0}^{N} \widetilde{\Pi}_j\rho + \Pi_*\rho$,

where

$$
\left.\begin{aligned}
&T\rho = -\frac{1}{\pi}\iint_D \frac{\rho(\zeta)}{\zeta - z}\,d\sigma_\zeta\,, \quad \Pi\rho = -\frac{1}{\pi}\iint_D \frac{\rho(\zeta)}{(\zeta-z)^2}\,d\sigma_\zeta\,,\\
&\widetilde{T}_0\rho = \begin{cases} -\dfrac{1}{\pi}\displaystyle\iint_D \frac{e^{2i\theta_0} z^{2K+1}\rho(\zeta)}{1-\overline{\zeta}z}\,d\sigma_\zeta\,, & K \ge 0\,,\\[2ex] -\dfrac{1}{\pi}\displaystyle\iint_D \frac{e^{2i\theta_0}\overline{\zeta}^{-2K-1}\rho(\zeta)}{1-\overline{\zeta}z}\,d\sigma_\zeta\,, & K < 0\,,\end{cases}\\
&\widetilde{\Pi}_0\rho = \begin{cases} -\dfrac{1}{\pi}\displaystyle\iint_D \frac{e^{2i\theta_0}(2K+1-2K\overline{\zeta}z)\,z^{2K}\rho(\zeta)}{(1-\overline{\zeta}z)^2}\,d\sigma_\zeta\,, & K \ge 0\\[2ex] -\dfrac{1}{\pi}\displaystyle\iint_D \frac{e^{2i\theta_0}\overline{\zeta}^{-2K}\rho(\zeta)}{(1-\overline{\zeta}z)^2}\,d\sigma_\zeta\,, & K < 0\,,\end{cases}\\
&\widetilde{T}_j\rho = -\frac{1}{\pi}\iint_D \frac{e^{2i\theta_j}\gamma_j^2\rho(\zeta)\,d\sigma_\zeta}{\gamma_j^2(z-z_j)-(\overline{\zeta}-\overline{z_j})(z-z_j)^2}\,,\\
&\widetilde{\Pi}_j\rho = -\frac{1}{\pi}\iint_D \frac{e^{2i\theta_j}\gamma_j^2[\gamma_j^2-2(\overline{\zeta}-\overline{z_j})(z-z_j)]\rho(\zeta)}{[\gamma_j^2(z-z_j)-(\overline{\zeta}-\overline{z_j})(z-z_j)^2]^2}\,d\sigma_\zeta\,, \quad j=1,\dots,N\,,\\
&\widetilde{T}_*\rho = (H_*\rho)_z\,, \quad \Pi_*\rho = (H_*\rho)_{zz}\,.
\end{aligned}\right\} \qquad (4.25)
$$

(2) Here S given by $S\rho = (H\rho)_{zz}$ is a bounded linear operator from $L_p(\overline{D})$ into itself satisfying

$$L_p[S\rho,\overline{D}] \le \Lambda_p L_p[\rho,\overline{D}]\,, \qquad (4.26)$$

where Λ_p is the smallest constant so that the above inequality holds, and

$$\Lambda_p \le 1 \text{ , for } N = 0 \text{ and } K = -1 \text{ .} \tag{4.27}$$

(3) For a real constant $q_0 (0 \le q_0 < 1)$, there exist two constants p_0, p_1 $(2 < p_0 < p_1 < \infty)$, so that

$$q_0 \Lambda_{p_j} < 1 (j=0,1) \text{ , for } N = 0 \text{ and } K = -1 \text{ .} \tag{4.28}$$

(4) When $p > 2$ and $2 < p_0 < p$, $H\rho$ satisfies

$$C^1_\beta[H\rho,\overline{D}] \le M_3 L_p(\rho,\overline{D}) \text{ , } M_3 = M_3(p_0,D) \text{ ,} \tag{4.29}$$

where $\beta = 1 - 2/p_0$.

<u>Proof.</u> Using the method in the proof of Theorem 3.5, we can derive the formulae (4.25), (4.26), (4.28), and (4.29). It remains to estimate the constant Λ_p . For any real measurable function $\rho(z) \in D^0_\infty(D)$, we have

$$\begin{aligned} \|S\rho\|^2_{L_2} &= \{L_2[S\rho,\overline{D}]\}^2 = \iint_D (H\rho)_{\overline{z}\overline{z}} \, (H\rho)_{zz} d\sigma_z = \\ &= Re \frac{1}{2i} \int_\Gamma (H\rho)_{\overline{z}} \, (H\rho)_{zz} dz + \|\rho\|^2_{L_2} \text{ .} \end{aligned} \tag{4.30}$$

From

$$\begin{aligned} &Re[e^{-i\theta_0} z(H\rho)_z] = h_0 \text{ , } z \in \Gamma_0 \text{ , } K=-1 \text{ ,} \\ &Re[e^{-i\theta_j}(z-z_j)(H\rho)_z] = 0 \text{ , } z \in \Gamma_j \text{ , } j=1,\dots,N \text{ ,} \end{aligned} \tag{4.31}$$

it follows

$$\left.\begin{aligned} (H\rho)_{\overline{z}} &= -e^{-2i\theta_0} z^2 (H\rho)_z + 2h_0 \, e^{-i\theta_0} z \text{ , } z \in \Gamma_0 \text{ ,} \\ (H\rho)_{\overline{z}} &= -e^{-2i\theta_j}(z-z_j)^2 (H\rho)_z/\gamma_j^2 \text{ , } z \in \Gamma_j \text{ , } j=1,\dots,N \text{ .} \end{aligned}\right\} \tag{4.32}$$

Thus

$$I = Re\frac{1}{2i}\int_{\Gamma} (H\rho)_{\bar{z}}(H\rho)_{zz}dz = -\sum_{j=0}^{N} Re\left\{\frac{1}{2i}\int_{\Gamma_j} e^{-2i\theta_j}\frac{(z-z_j)^2}{\gamma_j^2}(H\rho)_z(H\rho)_{zz}dz\right\} \qquad (4.33)$$

$$+ Re\left\{\frac{1}{i}\int_{\Gamma_0} e^{-i\theta_0}h_0 z(H\rho)_{zz}dz\right\} = \sum_{j=0}^{N} Re\left\{-\frac{1}{4\gamma_j^2 i}\int_{\Gamma_j} e^{-2i\theta_j}(z-z_j)^2 d[(H\rho)_z]^2\right\}$$

$$+ Re\left\{\frac{1}{i}\int_{\Gamma_0} e^{-i\theta_0}h_0 z d(H\rho)_z\right\} = \sum_{j=0}^{N} Re\,\frac{1}{2\gamma_j^2 i}\int_{\Gamma_j} e^{-2i\theta_j}(z-z_j)[(H\rho)_z]^2 dz$$

$$- Re\left\{\frac{h_0}{i}\int_{\Gamma_0} e^{-i\theta_0}(H\rho)_z dz\right\} = \sum_{j=0}^{N} Re\left\{-\frac{1}{2\gamma_j^2 i}\int_{\Gamma_j}\overline{(z-z_j)}(H\rho)_{\bar{z}}(H\rho)_z dz\right\} - 2\pi h_0^2$$

$$= -\frac{1}{2}\int_0^{2\pi}|(H\rho)_z|^2 d\theta + \sum_{j=1}^{N}\frac{1}{2\gamma_j^2}\int_0^{2\pi}|(z-z_j)(H\rho)_z|^2 d\theta - 2\pi h_0^2 ,$$

where $z_0 = 0$, $\gamma_0 = 1$, $\theta = \arg(z-z_j)$ for $z \in \Gamma_j$, $j=1,\ldots,N$.

If D is the unit disc $|z| < 1$, then $N = 0$. Hence

$$I = -\frac{1}{2}\int_0^{2\pi}|(H\rho)_z|^2 d\theta - 2\pi h_0^2 \le 0 , \qquad (4.34)$$

and so

$$\|(H\rho)_{zz}\|_{L_2}^2 \le \|\rho\|_{L_2}^2 . \qquad (4.35)$$

Similarly to the proof of Theorem 3.5, we obtain (4.27). □

Finally, from Theorem 4.4 and Theorem 4.5, we conclude

Theorem 4.6 Let $u(z)$ be a solution of Problem P^{***} for (4.1). Then $u(z)$ can be expressed in the form

$$u(z) = u_0(z) + H\rho , \qquad (4.36)$$

where $u_0(z)$ and $H\rho$ are stated in (4.20) and (4.22), respectively.

The general boundary condition

$$\frac{\partial u}{\partial \mu} + 2\sigma(t)\, u(t) = 2[\tau(t)+h(t)] \;, \; t \in \Gamma \;, \qquad (4.37)$$

can be written in the form

$$\frac{\partial u}{\partial \mu} = 2[R(t)+h(t)] \;, \; R(t) = \tau(t) - \sigma(t)\, u(t) \;, \; t \in \Gamma \;.$$

Hence, the solution $u(z)$ of the boundary value problem with the boundary condition (4.37) for (4.1) can be expressed in a form as in (4.36) (cf. [103],24),25).

II Boundary value problems for elliptic complex equations of first order

In this chapter, we first transform linear and nonlinear uniformly elliptic systems of two first order equations into a complex form, and discuss properties of solutions for complex equations of first order. Afterwards, we consider various boundary value problems for elliptic complex equations, i.e. the Haseman boundary value problem, the Riemann-Hilbert boundary value problem and the general compound boundary value problem in a multiply connected domain. In addition, we give some applications of the boundary value problems to quasiconformal mappings. The method for handling the above problems is based on the representation theorems and a priori estimates of solutions for the complex equation and the method of continuity, the Schauder fixed point theorem and the Leray-Schauder theorem.

§ 1. Reductions of elliptic systems of two first order equations

1. Reduction of linear elliptic systems of two first order equations

Let us consider a system of two linear equations of first order

$$\begin{aligned} a_{11}u_x+a_{12}u_y+b_{11}v_x+b_{12}v_y &= a_1u+b_1v+c_1 , \\ a_{21}u_x+a_{22}u_y+b_{21}v_x+b_{22}v_y &= a_2u+b_2v+c_2 , \end{aligned} \tag{1.1}$$

where the coefficients a_{jk} , b_{jk} , a_j , b_j , c_j $(j,k=1,2)$ are real functions of $(x,y) \in D$, D is a bounded domain in $\mathbb{R}^2$. The system (1.1) is called elliptic, if the quadratic form in ξ and η

$$K(\xi,\eta) = \begin{vmatrix} a_{11}\xi+a_{12}\eta & b_{11}\xi+b_{12}\eta \\ a_{21}\xi+a_{22}\eta & b_{21}\xi+b_{22}\eta \end{vmatrix} = \tag{1.2}$$

$$= K_1\xi^2+(K_2+K_3)\xi\eta+K_4\eta^2$$

is positive definite in D , where

$$K_1 = \begin{vmatrix} a_{11} & b_{11} \\ a_{21} & b_{21} \end{vmatrix} , \quad K_2 = \begin{vmatrix} a_{11} & b_{12} \\ a_{21} & b_{22} \end{vmatrix} ,$$

$$K_3 = \begin{vmatrix} a_{12} & b_{11} \\ a_{22} & b_{12} \end{vmatrix} , \quad K_4 = \begin{vmatrix} a_{12} & b_{12} \\ a_{22} & b_{22} \end{vmatrix} .$$

This is equivalent to

$$J = 4K_1K_4-(K_2+K_3)^2 > 0 , \ K_1 > 0 , \ \text{i.e.} \tag{1.3}$$

$$J = 4K_5K_6-(K_2-K_3)^2 > 0 , \ K_5 = \begin{vmatrix} a_{11} & a_{12} \\ a_{21} & a_{22} \end{vmatrix} , \quad K_6 = \begin{vmatrix} b_{11} & b_{12} \\ b_{21} & b_{22} \end{vmatrix} .$$

From $J > 0$, it follows that

$$K_1K_6 > 0 \ \text{or} \ K_1K_6 < 0 , \ \text{i.e.} \ K_6 \neq 0 .$$

Hence, the elliptic system (1.1) can be transformed into the form

$$\begin{aligned} v_y &= au_x+bu_y+a_0u+b_0v+f_0 , \\ -v_x &= du_x+cu_y+c_0u+d_0v+g_0 , \end{aligned} \tag{1.4}$$

where $a = K_1/K_6$, $b = K_3/K_6$, $c = K_4/K_6$, $d = K_2/K_6$, and the ellipticity condition (1.3) is reduced to

$$\Delta = \frac{1}{4} JK_6^{-2} = ac - \frac{1}{4}(b+d)^2 > 0 , \ a > 0 . \tag{1.5}$$

If the coefficients a_{jk} , b_{jk} $(j,k=1,2)$ are bounded in the domain D and

$$J \geq J_0 > 0 \ , \ K_1 > 0 \ , \tag{1.6}$$

in which J_0 is a positive constant, then the system (1.1) is called uniformly elliptic, and the uniform ellipticity condition for (1.4) is

$$\Delta = \frac{1}{4} JK_6^{-2} \geq \Delta_0 > 0 \ , \ K_1 > 0 \ , \tag{1.7}$$

where Δ_0 is a positive constant. It is easily seen that a, b, c, d are bounded in D .

We introduce complex notations

$$z = x+iy \ , \ \bar{z} = x-iy \ , \ w = u+iv \ , \ \bar{w} = u-iv$$

and

$$\begin{aligned}
w_z &= \frac{1}{2}(w_x - iw_Y) = \frac{1}{2}[(u_x+v_y)+i(v_x-u_y)] \ , \\
w_{\bar{z}} &= \frac{1}{2}(w_x + iw_y) = \frac{1}{2}[(u_x-v_y)+i(v_x+u_y)] \ , \\
u_x &= \frac{1}{2}(w_z+\bar{w}_{\bar{z}}+w_{\bar{z}}+\bar{w}_z) \ , \ u_y = \frac{i}{2}(w_z-\bar{w}_{\bar{z}}-w_{\bar{z}}+\bar{w}_z) \ , \\
v_x &= \frac{i}{2}(-w_z+\bar{w}_{\bar{z}}-w_{\bar{z}}+\bar{w}_{\bar{z}}) \ , \ v_y = \frac{1}{2}(w_z+\bar{w}_{\bar{z}}-w_{\bar{z}}-\bar{w}_z) \ .
\end{aligned} \tag{1.8}$$

The system (1.4) can be transformed into the complex form

$$(q_1+1)w_{\bar{z}}+q_2\bar{w}_z = -q_2 w_z-(q_1-1)\bar{w}_{\bar{z}}+r_1 w+r_2\bar{w}+r_3 \ , \tag{1.9}$$

which can be written into the following form

$$w_{\bar{z}} = Q_1(z)w_z+Q_2(z)\bar{w}_{\bar{z}}+A_1(z)w+A_2(z)\bar{w}+A_3(z) \ , \tag{1.10}$$

where

$$Q_1(z) = \frac{-2q_2}{|q_1+1|^2-|q_2|^2} \ , \quad Q_2(z) = \frac{|q_2|^2-(q_1-1)(\bar{q}_1+1)}{|q_1+1|^2-|q_2|^2} \ ,$$

$$q_1(z) = \frac{1}{2}[a+c+i(d-b)] \ , \ q_2(z) = \frac{1}{2}[a-c+i(d+b)] \ .$$

On account of

$$|q_1+1|^2-|q_2|^2 = \frac{1}{4}[(2+a+c)^2+(d-b)^2] - \frac{1}{4}[(a-c)^2+(d+b)^2] =$$

$$= 1+a+c+\left(\frac{d-b}{2}\right)^2 +\Delta \geq 1+\Delta ,$$

the ellipticity condition (1.5) can be reduced to the complex form

$$|Q_1(z)|+|Q_2(z)| = \frac{\sqrt{(a+c)^2-4\Delta}+\sqrt{(1+\sigma)^2-4\Delta}}{1+a+c+\sigma} < 1 \tag{1.11}$$

and the uniform ellipticity condition (1.7) can be reduced to

$$|Q_1(z)|+|Q_2(z)| \leq q_0 < 1 , \tag{1.12}$$

in which $\sigma = \frac{1}{4}(d-b)^2+\Delta$ and q_0 is a nonnegative constant.

If the coefficients of (1.4) satisfy the conditions

$$b = d , \; a_0 = b_0 = c_0 = d_0 = f_0 = g_0 = 0 , \text{ and } \Delta = 1 ,$$

then (1.4) is the Beltrami system of equations, the complex form of which is

$$w_{\bar{z}} = Q_1(z)w_z . \tag{1.13}$$

When $a = c = p$, $b = -d = q$, $a_0 = b_0 = c_0 = d_0 = f_0 = g_0 = 0$ in (1.4), the complex equation can be written in the form

$$w_z = Q_2(z)\bar{w}_{\bar{z}} . \tag{1.14}$$

The solutions $w(z)$ of (1.14) have been called (p,q)-analytic functions by G.N. Polozhy [83] and pseudo-analytic functions of the second kind by L. Bers [10]1), respectively. If the coefficients $Q_1(z) = Q_2(z) = 0$, $A_3(z) = 0$ in (1.10), then we obtain the complex equation

$$w_{\bar{z}} = A_1(z)w + A_2(z)\bar{w} \ . \tag{1.15}$$

Its solutions $w(z)$ in D have been called pseudo-analytic functions of the first kind and generalized analytic functions by L. Bers [10]1) and I.N. Vekua [97]2), respectively. If the coefficients of (1.10) satisfy the conditions

$$Q_j(z) \in W^1_p(D) \ , \ j=1,2 \ ,$$

the uniformly elliptic complex equation (1.10) can be transformed into a complex equation similar to the type of (1.15) (cf. [67]1), [6]5)).

For this purpose define

$$Q = \frac{1+|Q_1|^2-|Q_2|^2-\sqrt{\nabla}}{2\bar{Q}_1} = \frac{2Q_1}{1+|Q_1|^2-|Q_2|^2+\sqrt{\nabla}} \ , \tag{1.16}$$

and

$$P = \frac{1-|Q_1|^2+|Q_2|^2-\sqrt{\nabla}}{2\bar{Q}_2} = \frac{2Q_2}{1-|Q_1|^2+|Q_2|^2+\sqrt{\nabla}} \tag{1.17}$$

where

$$\nabla = (1-|Q_1|^2-|Q_2|^2)^2 - 4|Q_1Q_2|^2 = (1-(|Q_1|+|Q_2|)^2)(1-(|Q_1|-|Q_2|)^2) \ .$$

It is easy to see that $\nabla > 0$. In fact

$$0 < (1-(|Q_1|+|Q_2|)^2)^2 \le \nabla$$

from which we obtain

$$|P| \le |Q_1| + |Q_2| \ , \ |Q| \le |Q_1| + |Q_2| \ .$$

Moreover, a simple computation shows

$$P = \frac{Q_2}{|Q_2|^2 - |Q_1|^2 + Q_1 Q^{-1}} .$$

This together with the indentities

$$|Q_1|^2(1-|Q|^2) = \sqrt{\nabla}\,\bar{Q}_1 Q \ , \ Q_2(1-|P|^2) = P\sqrt{\nabla} \ ,$$

and

$$(|Q_2|^2-|Q_1|^2)(1-|Q|^2) + \sqrt{\nabla} = |1-\bar{Q}_1 Q|^2 - |Q_2 Q|^2$$

leads to an alternative form

$$P = \frac{Q_2(1-|Q|^2)}{|1-\bar{Q}_1 Q|^2 - |Q_2 Q|^2} . \qquad (1.18)$$

Lemma 1.1 Let $w \in W_p^1(D)$ fulfil the equation

$$w_{\bar{z}} = Q_1 w_z + Q_2 \bar{w}_{\bar{z}} + A \qquad (1.19)$$

and let ζ be a complete homeomorphism of

$$\zeta_{\bar{z}} = Q\zeta_z \qquad (1.20)$$

with Q given by (1.16). Then by changing the variable from z to ζ and the corresponding unknown from w to ω according to

$$\omega = w - P\bar{w} \qquad (1.21)$$

with P given by (1.17), equation (1.19) can be transformed into the canonical form

$$\omega_{\bar{\zeta}} = A_1\omega + A_2\bar{\omega} + A_3 \qquad (1.22)$$

with

$$A_1 = \frac{\bar{P}P_{\bar{\zeta}}}{1-|P|^2}, \quad A_2 = -\frac{P_{\bar{\zeta}}}{1-|P|^2},$$

$$A_3 = \frac{(1-\bar{Q}_1 Q)A + Q_2 Q\bar{A}}{[|1-\bar{Q}_1 Q|^2 - |Q_2 Q|^2]\overline{\zeta_z}} .$$

Remark. The inverse transformation transforms (1.22) back into (1.19) accordingly.

Proof. In view of (1.20) we note that

$$w_z = \zeta_z w_\zeta + \bar{Q}\overline{\zeta_z} w_{\bar{\zeta}} , \quad w_{\bar{z}} = Q\zeta_z w_\zeta + \overline{\zeta_z} w_{\bar{\zeta}} .$$

Substituting these into (1.19) and its conjugate equation, we obtain two equations in four unknowns $w_{\bar{\zeta}}$, $\bar{w}_{\bar{\zeta}}$, $\bar{w}_\zeta$, and w_ζ as in (1.9). However, because of the choice of Q in (1.16), here it is not difficult to see that both $\bar{w}_\zeta$ and w_ζ can be eliminated. Thus, we arrive at the equation

$$[|1-Q_1\bar{Q}|^2 - |Q_2 Q|^2]\overline{\zeta_z} w_{\bar{\zeta}} + [(\bar{Q}-\overline{Q_1})Q_2 Q - (1-\overline{Q_1 Q})Q_2]\overline{\zeta_z}\bar{w}_{\bar{\zeta}} = (1-\bar{Q}_1 Q)A + Q_2 Q\bar{A} ,$$

from which formula (1.22) follows immediately from (1.21) and the definition of P in (1.18). □

Remark. If Q_1 , $Q_2 \in W^1_p(D)$ with $p > 2$ then $Q \in W^1_p(D)$. Identify Q with 0 outside the finite domain D . Then, see [2]1), [63], there exists a complete homeomorphism of (1.20) with $\zeta(0) = 0, \zeta(\infty) = \infty$ of the Riemann spere $\hat{\mathbb{C}}$ and a constant M depending only on p and on the uniform ellipticity constant q_0 from (1.12) such that

$$\exp\{-ML_p(Q_z\bar{D})\} \le |\zeta_z(z)| \le \exp\{ML_p(Q_z,\bar{D})\} .$$

This shows that the Jacobian J of ζ defined by

$$J = |\zeta_z|^2 - |\zeta_{\bar{z}}|^2$$

is bounded below away from zero

$$(1-q_0^2)\ \exp\{-2ML_p(Q_z,\bar{D})\} \le J\ .$$

2. Reduction of the quasilinear elliptic system of two first order equations

If the coefficients of (1.1) are functions of x, y, u, v, then system (1.1) is called quasilinear. In the domain D, the definition for ellipticity as well as for uniform ellipticity of (1.1) are the same as that given for the linear system (1.1). In this case, the complex form of (1.1) becomes

$$w_{\bar{z}} = Q_1(z,w)w_z + Q_2(z,w)\bar{w}_{\bar{z}} + A_1(z,w)w + A_2(z,w)\bar{w} + A_3(z,w) \tag{1.23}$$

and its ellipticity condition and uniform ellipticity condition can be written as

$$|Q_1(z,w)| + |Q_2(z,w)| < 1\ , \tag{1.24}$$

$$|Q_1(z,w)| + |Q_2(z,w)| \le q_0 < 1 \tag{1.25}$$

in D, respectively.

We denote the right-hand side of (1.23) by $F(z,w,w_z)$. Obviously, the system (1.23) can be written in the form

$$\begin{aligned} &w_{\bar{z}} = F(z,w,w_z)\ , \\ &F(z,w,0) = A_1(z,w)w + A_2(z,w)\bar{w} + A_3(z,w) \end{aligned} \tag{1.26}$$

and from (1.24) and (1.25), we may derive

$$|F(z,w,U_1)-F(z,w,U_2)| < |U_1-U_2|\ , \tag{1.27}$$

$$|F(z,w,U_1)-F(z,w,U_2)| \le q_0|U_1-U_2|, \tag{1.28}$$

for $z \in \mathbb{C}$ and U_1, $U_2 \in \mathbb{C}$.

3. Reduction of the nonlinear elliptic system of two first order equations

Let $\Phi_j(x,y,\zeta_1,\zeta_2,\ldots,\zeta_6)(j=1,2)$ be continuous real functions in $(x,y) \in D$ and the real variables $\zeta_1,\zeta_2,\ldots,\zeta_6$ and possess continuous partial derivatives with respect to ζ_3,ζ_4,ζ_5 and ζ_6. The system of first order equations

$$\Phi_j(x,y,u,v,u_x,u_y,v_x,v_y) = 0 \ , \ j=1,2 \qquad (1.29)$$

is called elliptic, if the following quadratic form is positive definite in D

$$K(\xi,\eta) = \begin{vmatrix} \Phi_{1u_x}\xi+\Phi_{1u_y}\eta & \Phi_{1v_x}\xi+\Phi_{1v_y}\eta \\ \Phi_{2u_x}\xi+\Phi_{2u_y}\eta & \Phi_{2v_x}\xi+\Phi_{2v_y}\eta \end{vmatrix} = \qquad (1.30)$$

$$= K_1\xi^2 + (K_2+K_3)\xi\eta + K_4\eta^2 > 0 \quad \text{in } D \text{ for } \xi^2+\eta^2 > 0 \ ,$$

where $K_1 = \dfrac{D(\Phi_1,\Phi_2)}{D(u_x,v_x)}$, $K_2 = \dfrac{D(\Phi_1,\Phi_2)}{D(u_x,u_y)}$, $K_3 = \dfrac{D(\Phi_1,\Phi_2)}{D(u_y,v_x)}$,

$K_4 = \dfrac{D(\Phi_1,\Phi_2)}{D(u_y,v_y)}$, $K_5 = \dfrac{D(\Phi_1,\Phi_2)}{D(u_x,u_y)}$, $K_6 = \dfrac{D(\Phi_1,\Phi_2)}{D(v_x,v_y)}$,

i.e.

$$I = 4K_1K_4 - (K_2+K_3)^2 > 0 \ , \ K_1 > 0 \quad \text{in } D \ . \qquad (1.31)$$

If Φ_{ju_x} , Φ_{ju_y} , Φ_{jv_x} , Φ_{jv_y} $(j=1,2)$ are bounded in D and there exists a positive constant Δ_0 , so that

$$I \ge \Delta_0 > 0 \ , \ K_1 > 0 \ , \qquad (1.32)$$

then the nonlinear system (1.29) is called uniformly elliptic.
Denoting $\Phi = \Phi_1 + i\Phi_2$, we have

$$\Phi_{u_x} = \frac{1}{2}[\Phi_{w_z} + \Phi_{\overline{w}_{\overline{z}}} + \Phi_{w_{\overline{z}}} + \Phi_{\overline{w}_z}] ,$$

$$\Phi_{u_y} = \frac{i}{2}[-\Phi_{w_z} + \Phi_{\overline{w}_{\overline{z}}} + \Phi_{w_{\overline{z}}} - \Phi_{\overline{w}_z}] ,$$

$$\Phi_{v_x} = \frac{i}{2}[\Phi_{w_z} - \Phi_{\overline{w}_{\overline{z}}} + \Phi_{w_{\overline{z}}} - \Phi_{\overline{w}_z}] , \tag{1.33}$$

$$\Phi_{v_y} = \frac{1}{2}[\Phi_{w_z} + \Phi_{\overline{w}_{\overline{z}}} - \Phi_{w_{\overline{z}}} - \Phi_{\overline{w}_z}] ,$$

and

$$I = \frac{1}{4}[|\Phi_{w_{\overline{z}}}|^2 + |\Phi_{w_z}|^2 - |\Phi_{\overline{w}_z}|^2 - |\Phi_{\overline{w}_{\overline{z}}}|^2]^2 - |\Phi_{w_{\overline{z}}}|^2 |\Phi_{w_z}|^2$$
$$- |\Phi_{\overline{w}_z}|^2 |\Phi_{\overline{w}_{\overline{z}}}|^2 + \Phi_{w_{\overline{z}}} \Phi_{\overline{w}_z} \overline{\Phi}_{\overline{w}_{\overline{z}}} \Phi_{w_z} + \overline{\Phi}_{\overline{w}_z} \overline{\Phi}_{w_{\overline{z}}} \Phi_{w_z} \Phi_{\overline{w}_{\overline{z}}} . \tag{1.34}$$

Theorem 1.1 (1) If the nonlinear system (1.29) is uniformly elliptic in D , then it is solvable with respect to $w_{\overline{z}}$ or w_z and

$$w_{\overline{z}} = F(z,w,w_z) \;,\; \text{or} \quad w_z = F(z,w,w_{\overline{z}}) \tag{1.35}$$

respectively, and the uniform ellipticity condition (1.32) is equivalent to

$$I_F = \frac{1}{4}(1+|F_U|^2 - |F_{\overline{U}}|^2)^2 - |F_U|^2 \geq d > 0 , \tag{1.36}$$

where d is a positive constant.

(2) Conversely, if the complex equation (1.35) is uniformly elliptic and

$$J_{w_{\overline{z}}} = \frac{D(\Phi,\overline{\Phi})}{D(w_{\overline{z}},\overline{w}_z)} \geq d_0 > 0 \quad \text{or}$$

$$J_{w_z} = \frac{D(\Phi,\overline{\Phi})}{D(w_z,\overline{w}_{\overline{z}})} \geq d_0 > 0 \quad \text{in } D ,$$

then a solution $w(z)$ of (1.35) in D is also a solution of the nonlinear equation $\Phi = 0$.

<u>Proof.</u> (1) Through an immediate computation we can conclude

$$J_{w_{\bar z}} = \frac{D(\Phi,\bar\Phi)}{D(w_{\bar z},\bar w_z)} = |\Phi_{w_{\bar z}}|^2 - |\Phi_{\bar w_z}|^2 = K_1 + K_4 + K_5 + K_6 , \qquad (1.37)$$

$$J_{w_z} = \frac{D(\Phi,\bar\Phi)}{D(w_z,\bar w_{\bar z})} = |\Phi_{w_z}|^2 - |\Phi_{\bar w_{\bar z}}|^2 = K_1 + K_4 - K_5 - K_6 , \qquad (1.38)$$

$$I = 4K_1 - (K_2+K_3)^2 = 4K_5K_6 - (K_2-K_3)^2 \geq \Delta_0 > 0 . \qquad (1.39)$$

It follows from (1.31) and (1.39) that $K_1K_4 > 0$ and $K_5K_6 > 0$. Hence, $K_1K_5 > 0$ or $K_1K_5 < 0$. From (1.37) or (1.38) we therefore obtain $J_{w_{\bar z}} \neq 0$ or $J_{w_z} \neq 0$, and so (1.35).

If $K_1K_5 > 0$, we substitute $F(z,w,w_z)$ instead of $w_{\bar z}$ into $\Phi = 0$, and differentiate the identity with respect to w_z , $\bar w_{\bar z}$ and obtain

$$\begin{aligned} &\Phi_{w_{\bar z}} F_{w_z} + \Phi_{\bar w_z} \bar F_{w_z} + \Phi_{w_z} = 0 , \\ &\Phi_{w_{\bar z}} F_{\bar w_{\bar z}} + \Phi_{\bar w_z} \bar F_{\bar w_{\bar z}} + \Phi_{\bar w_{\bar z}} = 0 . \end{aligned} \qquad (1.40)$$

Because of $J_{w_{\bar z}} = |\Phi_{w_{\bar z}}|^2 - |\Phi_{\bar w_z}|^2 \neq 0$, we can solve (1.40) for F_{w_z} and get

$$F_{w_z} = (\Phi_{\bar w_z} \bar\Phi_{w_z} - \Phi_{w_z} \bar\Phi_{\bar w_z})/J_{w_{\bar z}} . \qquad (1.41)$$

From the boundedness of $J_{w_{\bar z}}$ and

$$I = J^2_{w_{\bar z}} [\tfrac{1}{4}(1+|F_{w_z}|^2-|F_{\bar w_{\bar z}}|^2)^2-|F_{w_z}|^2] \geq \Delta_0 > 0 , \qquad (1.42)$$

(1.36) is derived. Similarly we can conclude (1.36) for the case $K_1K_5 < 0$.

(2) According to the condition $|J_{w_{\bar{z}}}| \geq d_0$ or $|J_{w_z}| \geq d_0 > 0$, and (1.34), we see by (1.29) that the uniform ellipticity condition for

$$\Phi = w_{\bar{z}} - F(z,w,w_z) \quad \text{or} \quad \Phi = w_z - F(z,w,w_{\bar{z}}) \ ,$$

is that in (1.36), and that a solution $w(z)$ of (1.35) is also a solution of (1.29), i.e. $\Phi = 0$. □

Besides, form (1.36), we can obtain the discriminant of the uniform ellipticity condition for the complex equation (1.35) as

$$|F_U| + |F_{\bar{U}}| \leq q_0 < 1 \ , \tag{1.43}$$

where q_0 is a nonnegative constant.

Furthermore we can derive

$$|F(z,w,U_1)-F(z,w,U_2)| \leq q_0|U_1-U_2| \ . \tag{1.44}$$

In fact, if we set $U = U_2 + t(U_1-U_2)$, $0 \leq t \leq 1$, we easily see that

$$F(z,w,U_1) - F(z,w,U_2) = \int_{U_2}^{U_1} dF(z,w,U) =$$

$$= \int_0^1 dF[z,w,U_2+t(U_1-U_2)] = \int_0^1 F_U dt(U_1-U_2)$$

$$+ \int_0^1 F_{\bar{U}} dt(\bar{U}_1-\bar{U}_2) = Q_3(U_1-U_2) + Q_4(\bar{U}_1-\bar{U}_2) \ .$$

Hence, (1.44) is true. If we take $U_1 = U$ and $U_2 = 0$ and denote

$$Q_1(z,w,U) = Q_3(z,w,U_1,U_2) \ , \ Q_2(z,w,U) = Q_4(z,w,U_1,U_2) \ ,$$

then the complex equation $w_{\bar{z}} = F(z,w,w_z)$ can be written in the form

$$w_{\bar z} = Q_1(z,w,w_z)w_z + Q_2(z,w,w_z)\bar w_{\bar z} + F(z,w,0) \ ,$$
$$F(z,w,0) = A_1(z,w)w + A_2(z,w)\bar w + A_3(z,w) \ . \tag{1.45}$$

The equation $w_z = F(z,w,w_{\bar z})$ can be written into a similar form. Further, if we denote

$$Q(z,w,w_z) = \begin{cases} Q_1(z,w,w_z) + Q_2(z,w,w_z)\dfrac{\bar w_{\bar z}}{w_z} \ , \ w_z \neq 0 \ , \\ 0 \ , \ w_z = 0 \ , \end{cases} \tag{1.46}$$

the complex equation (1.45) can be written as

$$w_{\bar z} = F(z,w,w_z) \ ,$$
$$F(z,w,w_z) = Q(z,w,w_z)w_z + A_1(z,w)w + A_2(z,w)\bar w + A_3(z,w) \ . \tag{1.47}$$

4. Conditions for uniformly elliptic complex equations of first order

In the following we always suppose that the domain D is an $(N+1)$-connected circular domain in the unit disc $\mathbb{D} = \{|z|<1\}$ with the boundary

$$\Gamma = \sum_{j=0}^{N} \Gamma_j \ , \ \Gamma_j = \{|z-z_j|=\gamma_j\}(j=1,\ldots,N) \ , \ \Gamma_0 = \Gamma_{N+1} = \{|z|=1\}$$

and $0 \in \mathbb{D}$, until further notice. This can be realized through a conformal mapping. Actually, let $\zeta = \zeta(z)$ be a conformal mapping from a general $(N+1)$-connected domain G with the boundary $L \in C^1_\mu(0<\mu\le 1)$ onto the circular domain D and $z = z(\zeta)$ its inverse mapping. Then the complex equation (1.45) is transformed into the following form

$$w_{\bar\zeta} = Q_1[z(\zeta),w,\frac{w_\zeta}{z'(\zeta)}]\frac{\overline{z'(\zeta)}}{z'(\zeta)}w_\zeta + Q_2[z(\zeta),w,\frac{w_\zeta}{z'(\zeta)}]\bar w_{\bar\zeta}$$
$$+A_1[z(\zeta),w]\overline{z'(\zeta)}w + A_2[z(\zeta),w]\overline{z'(\zeta)}\bar w + A_3[z(\zeta),w]\overline{z'(\zeta)} \tag{1.48}$$

Due to $L \in C^1_\mu$, we can derive that $\zeta(z) \in C^1_\mu(\bar G)$ and $z(\zeta) \in C^1_\mu(\bar D)$ (cf. [97]2). It is not difficult to see that the complex equations (1.45) and (1.48) belong to the same type.

In this chapter, we assume that the complex equation (1.45) or (1.47) satisfies the following conditions.

Condition C

(1) $Q_j(z,w,U)(j=1,2)$, $A_j(z,w)(j=1,2,3)$ are measurable in $z \in D$ for all continuous functions $w(z)$ and all measurable functions $U(z)$ in D and satisfy

$$L_p[A_j(z,w(z)),\bar{D}] \le k_0 < \infty \ , \ j = 1,2,3 \ , \tag{1.49}$$

where $p(2<p<\infty)$, $k_0(0\le k_0<\infty)$ are real constants and $A_j = 0$, for $z \notin D$, $j = 1,2,3$.

(2) $Q_j(z,w,U)(j=1,2)$, $A_j(z,w)(j=1,2,3)$ are continuous in $w \in \mathbb{C}$ for almost every point $z \in D$ and $U \in \mathbb{C}$.

(3) The equation (1.45) or (1.47) satisfies the uniform ellipticity condition (1.44), i.e.

$$|F(z,w,U_1)-F(z,w,U_2)| \le q_0|U_1-U_2| \ , \tag{1.50}$$

for almost every point $z \in D$ and $w,U_1,U_2 \in \mathbb{C}$.

Condition C^*

When the domain D is the whole z-plance $\mathbb{C}$, the above conditions (1), (2) and (3) hold on $D = \mathbb{C}$, up to condition (1.49) which is replaced by

$$\begin{aligned} L_{p,2}[A_j(z,w(z)),\mathbb{C}] &= L_p[A_j(z,w(z)),\bar{\mathbb{D}}] + \\ &+ L_p[z^{-2}A_j(z^{-1},w(z^{-1})),\bar{\mathbb{D}}] \le k_0 \ , \ j = 1,2,3 \ , \end{aligned} \tag{1.51}$$

where $p(2<p<\infty)$ and k_0 are real constants.

A function $w(z) \in W^1_{p_0}(D)$, $2 < p_0 \le p$ which is continuous in D and satisfies (1.45) or (1.47) will be called a solution. If $w(z)$ only is continuous up to isolated singularites in D it will be considered as a generalized solution.

§ 2. Properties of solutions to the elliptic complex equation of first order

In this section, we shall give some representation theorems and compactness principles of solutions for the uniformly elliptic complex equation of first order.

1. A representation theorem for solutions to complex equations of first order

We first prove a Lemma.

Lemma 2.1 Let $w(z)$ be a homeomorphic solution for the complex Beltrami equation

$$w_{\bar{z}} = Q(z)w_z \ , \ |Q(z)| \le q_0 < 1 \ , \ z \in D \tag{2.1}$$

with the conditions $w(0) = 0$, $w(1) = 1$, mapping the circular domain D onto a circular domain G in $|w| < 1$ with the boundary

$$L = \sum_{j=0}^{N} L_j \ , \ L_j = \{|w-w_j|=\rho_j\} \ , \ j = 1,\ldots,N \ ,$$

and $L_0 = L_{N+1} = \{|w|=1\}$, $0 \in G$. Then $w(z)$ and its inverse function $z(w)$ satisfy the estimates

$$C_\beta[w(z),\bar{D}] \le M_1 \ , \ L_{p_0}[|w_{\bar{z}}|+|w_z|,\bar{D}] \le M_2 \ , \tag{2.2}$$

$$C_\beta[z(w),\bar{G}] \le M_3 \ , \ L_{p_0}[|z_{\bar{w}}|+|z_w|,\bar{G}] \le M_4 \ , \tag{2.3}$$

where $\beta = 1-2/p_0$, $2 < p < \infty$, $M_j = M_j(q_0,p_0,D)$, $j = 1,2,3,4$.

Proof. First of all, we continuously extend $w(z)$ across the boundary Γ of D by

$$w^*(z) = \begin{cases} w(z) , & z \in \bar{D} , \\ \dfrac{\rho_j^2}{\overline{w\left(\frac{\gamma_j^2}{\bar{z}-\bar{z}_j}+z_j\right)-w_j}} + w_j , & z \in D_j , \ j = 0,1,\ldots,N , \end{cases} \tag{2.4}$$

$$Q^*(z) = \begin{cases} Q(z) , & z \in D , \\ \overline{Q\left(\frac{\gamma_j^2}{\bar{z}-\bar{z}_j}+z_j\right)} \left(\frac{z-z_j}{\bar{z}-\bar{z}_j}\right)^2 , & z \in D_j , \ j = 0,1,\ldots,N , \end{cases} \tag{2.5}$$

where

$$D_0 = \{1<|z|<\frac{1}{1-d}\} , \quad D_j = \{\frac{\gamma_j^2}{\gamma_j+d} < |z-z_j|<\gamma_j\} , \quad j = 1,\ldots,N$$

and $2d$ is the least distance between $\Gamma_j (j=1,\ldots,N)$ and $z = 0$ and $w_0 = z_0 = 0$, $\gamma_0 = 1$.

It is easy to see that $|Q^*(z)| \le q_0 < 1$, $z \in D^* = \bar{D} \cup \bigcup_{j=0}^{N} D_j$. We can prove that $w^*(z)$ is a homeomorphic solution for the complex Beltrami equation

$$w^*_{\bar{z}} = Q^*(z) w^*_z , \tag{2.6}$$

which maps the circular domain D^* onto a domain G^* in the w-plane.

We denote

$$D^*_\ell = \bar{D} \cup \{1<|z|\le\frac{1}{1-\frac{d}{\ell}}\} \cup \bigcup_{j=1}^{N} \{\frac{\gamma_j^2}{\gamma_j+\frac{d}{\ell}}\le|z-z_j|<\gamma_j\} , \quad 2 \le \ell < \infty$$

and prove the estimate

$$C[w^*,D^*_\ell] \le M_5 = M_5(q_0,p_0,D^*_\ell) . \tag{2.7}$$

In fact, by Theorem 2.16, Chapter 2 in [97]2), $w^*(z)$ can be expressed as

$$w^*(z) = \Phi[\chi(z)] \ , \ \chi(z) = z + T\omega - T_0\omega \ , \tag{2.8}$$

where $T_0\omega = T\omega\big|_{z=0}$, $w(z) \in L_{p_0}(\bar{D}^*)$, $\chi(z)$ is a complete homeomorphism on the z-plane, $\chi(z)$ and its inverse function $z(\chi)$ satisfies the estimates

$$C_\beta[\chi(z)-z,\mathbb{C}] \le M_6 \ , \ C_\beta[z(\chi)-\chi,\mathbb{C}] \le M_7 \ , \tag{2.9}$$

$$L_{p_0}[|\chi_{\bar{z}}|+|\chi_z|,D_R] \le M_8 \ , \ L_{p_0}[|z_{\bar{\chi}}|+|z_\chi|,D_R] \le M_9 \ , \tag{2.10}$$

in which $D_R = \{|z|<R\}(R<\infty)$,

$$M_j = M_j(q_0,p_0) \ , \ j = 6,7 \ , \ M_j = M_j(q_0,p_0,R) \ , \ j = 8,9 \ .$$

Hence, the domain $H = \chi(D^*)$ contains the disc $H_0 = \{|\chi-\chi(1)| \le M_{10}\}$, $M_{10} = M_{10}(q_0,p_0,d) > 0$ and the schlicht function $\Phi(\chi)$ maps H_0 onto a domain G_0 which includes the point $w = 1$. Let b_2 be a point on the closed Jordan curve $\Phi[\{|\chi-\chi(1)| = M_{10}\}]$, whose distance from $b_1 = 1$ is the smallest, and $w = \Phi(\chi)$ maps a curve γ onto the line segment $\overline{b_1b_2}$. It is not difficult to see that

$$\begin{aligned} 2 \ge |b_2-1| &= \left|\int_\gamma \Phi'(\chi)d\chi\right| = \int_\gamma |\Phi'(\chi)||d\chi| \ge \\ &\ge |\Phi'(a)|M_{10} \ , \ \text{i.e.} \quad |\Phi'(a)| \le \frac{2}{M_{10}} \ , \ \text{and} \quad |\Phi(a)| < 3 \ , \end{aligned} \tag{2.11}$$

where $a \in \gamma \cap H_0$. According to the deformation theorem (cf. Theorem 5, § 4, Chapter 2, in [45]), we have

$$|\Phi(\chi)| \le |\Phi(a)| + M_{11}|\Phi'(a)| \le M_{12} \quad \text{on} \quad \chi(D_\ell^*) \ , \tag{2.12}$$

where $M_j = M_j(q_0,p_0,D^*,D_\ell^*)$, $j = 11,12$. From (2.8), (2.9), and (2.12), (2.7) follows. Using the Cauchy formula for analytic functions, we obtain

$$C^1[\Phi(\chi),\chi(D_4^*)] \le M_{13} = M_{13}(q_0,p_0,D^*,D_4^*) \ . \tag{2.13}$$

Furthermor, we conclude (2.2).

To prove (2.3), using the reductio ad absurdum, we first show

$$h \ge M_{14} = M_{14}(q_0,p_0,d) > 0 \ , \tag{2.14}$$

where h is the minimum of $\rho_j(j=1,\ldots,N)$ and the least distance between $L_j(j=1,\ldots,N)$ and $w = 0$. Moreover, we note that $z = z(w)$ is a homeomorphic solution of the complex Betrami equation

$$z_{\bar{w}} + \tilde{Q}(w)z_w = 0 \ , \ \tilde{Q}(w) = Q[z(w)] \frac{\bar{z}_{\bar{w}}}{z_w} \tag{2.15}$$

in the (N+1)-connected circular domain, which satisfies the condition $z(0) = 0$, $z(1) = 1$. Similarly to (2.2), we can obtain (2.3). □

<u>Theorem 2.2</u> Let the complex equation (1.45) satisfy Condition C, and $w(z)$ be a generalized solution to (1.45) in the (N+1)-connected circular domain. Then $w(z)$ admits the representation

$$w(z) = \{\Phi[\zeta(z)]+\psi(z)\}e^{\phi(z)} \ , \tag{2.16}$$

where $\psi(z) = Tf - T_1f$, $T_1f = Tf\big|_{z=1}$, $\phi(z) = Tg - T_1g$, $\zeta(z) = \Psi[\chi(z)]$, $\chi(z) = z + Th - T_0h$, $T_0h = Th\big|_{z=0}$. The functions $f(z)$, $g(z)$ and $h(z)$ satisfy the estimates

$$L_{p_0}[f,\bar{D}] \le k_1 \ , \ L_{p_0}[g,\bar{D}] \le k_1 \ , \ L_{p_0}[h,\bar{D}] \le k_1 \ , \tag{2.17}$$

in which $k_1 = k_1(q_0,p_0,k_0,D)$, and $\chi(z)$ is a complete homeomorphism of $\mathbb{C}$, $\psi(\chi)$ is a schlicht function mapping $\chi(D)$ onto a circular domain G similar to the domain D and $\Psi[\chi(0)] = 0$, $\Psi[\chi(1)] = 1$, $\Phi(\zeta)$ is an analytic function in $\zeta(D)$ except for some isolated singularities and $\psi(z)$, $\phi(z)$, $\zeta(z)$ and its inverse function $z(\zeta)$ satisfy the following estimates

$$C_\beta[\psi,\mathbb{C}] \le M_{15} \ , \ L_{p_0}[|\psi_{\bar z}|+|\psi_z|,\mathbb{C}] \le M_{16} \ , \tag{2.18}$$

$$C_\beta[\phi,\mathbb{C}] \le M_{17} \ , \ L_{p_0}[|\phi_{\bar z}|+|\phi_z|,\mathbb{C}] \le M_{18} \ , \tag{2.19}$$

$$\begin{aligned} &C_\beta[\zeta,\bar D] \le M_{19} \ , \ L_{p_0}[|\zeta_{\bar z}|+|\zeta_z|,\bar D] \le M_{20} \ , \\ &C_\beta[z,\bar G] \le M_{21} \ , \ L_{p_0}[|z_{\bar\zeta}|+|z_\zeta|,\bar G] \le M_{22} \ , \end{aligned} \tag{2.20}$$

where $\beta = 1-2/p_0$, $2 < p_0 < p$, $M_j = M_j(q_0,p_0,k_0,D) \ge 0$, $J = 15,\ldots,22$.

Proof. We first discuss the integral equation

$$g = Q\Pi g + A \ , \ \Pi g = -\frac{1}{\pi}\iint_D \frac{g(\zeta)}{(\zeta-z)^2}\, d\sigma_\zeta \ , \tag{2.21}$$

$$Q = \begin{cases} Q_1 + Q_2\bar w_{\bar z}/w_z \ , & \text{for } w_z \neq 0 \ , \\ 0 \ , & \text{for } w_z = 0 \text{ or } z \notin D \ , \end{cases}$$

$$A = \begin{cases} A_1 + A_2\bar w/w \ , & \text{for } w \neq 0 \ , \\ 0 \ , & \text{for } w = 0 \text{ or } z \notin D \ . \end{cases}$$

Noting $q_0\|\Pi\|_{L_{p_0}} < 1$ and according to the principle of contraction, we can obtain a solution $g(z)$ to (2.21) satisfying the second estimate in (2.17), so that $\phi = Tg - T_1g$ satisfies (2.19).

Next, from the integral equation

$$f = Q\Pi f + A_3e^{-\phi} \ , \tag{2.22}$$

we similarly may find a solution $f(z)$ satisfying the first estimate in (2.17), and so $\psi = Tf - T_1f$ satisfies (2.18).

Finally, setting

$$W(z) = w(z)e^{-\phi(z)} - \psi(z) \ , \tag{2.23}$$

it is not difficult to verify that $W(z)$ is a solution to the complex Beltrami equation

$$W_{\bar{z}} = QW_z \tag{2.24}$$

in D , and that $W(z)$ can be expressed as

$$W(z) = \Phi\{\Psi[\chi(z)]\} = \Phi[\zeta(z)] \ , \tag{2.25}$$

where $\chi(z) = z + Th - T_0h$ is a complete homeomorphism, in which $h(z)$ satisfies the third estimate in (2.17), and by Lemma 2.1, the homeomorphic solution $\zeta(z) = \Psi[\chi(z)]$ of the Beltrami equation (2.24) and its inverse function $z(\zeta)$ satisfy (2.20). □

<u>Corollary 2.3</u> Let $w_1(z) = T\omega - T_1\omega(\omega \in L_{p_0}(\bar{D}))$ be a solution of (1.38) on the whole plane $\mathbb{C}$. Then $w_1(z)$ can be expressed as

$$w_1(z) = \psi(z)e^{\phi(z)} = T\omega - T_1\omega \tag{2.26}$$

where ψ and ϕ are as stated in Theorm 2.2, and w_1 satisfies the estimate

$$C_\beta[w_1,\mathbb{C}] \le M_{23} \ , \ L_{p_0}[|w_{1\bar{z}}|+|w_{1z}|,\mathbb{C}] \le M_{24} \tag{2.27}$$

in which $\beta = 1-2/p_0$ and $M_j = M_j(q_0,p_0,D)$, $j = 23,24$.

<u>Proof.</u> By Theorem 2.2, $w_1(z)$ admits the representation

$$w_1(z) = \{\Phi[\chi(z)]+\psi(z)\}e^{\phi(z)} \ , \tag{2.28}$$

where $\phi(z) = Tg - T_1g$, $\psi(z) = Tf - T_1f$, $\chi(z) = z + Th - T_0h$ satisfy (2.17)-(2.19) and $\phi(1) = \psi(1) = 0$, $\phi(\infty) = -T_1g$,

$\psi(\infty) = -T_1 f$ and $\Phi(\chi)$ is an entire function on the χ-plane. Because $\chi(z) \to \infty$ as $z \to \infty$ it follows $\Phi(\chi) \to w_1(\infty)e^{-\phi(\infty)} - \psi(\infty) = -T_1 \omega e^{T_1 g} + T_1 f$. Thus $\Phi(\chi)$ = constant, i.e. $\Phi(\chi) = w_1(1) = 0$. Therefore, w_1 possesses the representation (2.26), and satisfies the estimate (2.27). □

2. Another representation for solutions to the complex equation of first order

We now shall prove another representation for the equation (1.45) or (1.47) which is different to (2.16).

Theorem 2.4 Under the hypothesis of Theorem 2.2, the solution $w(z)$ of the complex equation (1.45) can be expressed as

$$w(z) = \Phi[\zeta(z)]e^{\phi(z)} + \psi(z) \tag{2.29}$$

where $\psi(z) = Tf - T_1 f$, $\phi(z) = Tg - T_1 g$, $\zeta(z) = \Phi[\chi(z)]$, $\chi(z) = z + Th - T_0 h$, $\chi(z)$ is a complete homeomorphism of the Betrami equation (2.24) on the whole-plane, $\Psi(\chi)$ and $\Phi(\zeta)$ are as stated in Theorem 2.2, and $f(z)$, $g(z)$, $h(z)$, $\Phi(z)$, $\psi(z)$, $\zeta(z)$ and its inverse function $z(\zeta)$ satisfy the estimates (2.17)-(2.20).

Proof. The result of this theorem can be obtained by replacing the integral equations (2.22), (2.21), and (2.24) by the integral equations

$$f(z) = Q\Pi f + A_1(Tf - T_1 f) + A_2(\overline{Tf - T_1 f}) + A_3 , \tag{2.30}$$

$$g(z) = Q\Pi g + A , \tag{2.31}$$

$$W_{\bar{z}} = QW_z , \tag{2.32}$$

respectively, where Q and A are as stated in (2.21). In order to solve the integral equation (2.30), we ought to use Theorem 3.4 of the next paragraph or the Fredholm theorem for integral equations. The first estimates in (2.17) and (2.18) are derived by the estimate (2.27). □

Corollary 2.5 Let D be a bounded domain, and let the complex equation (1.45) satisfy Condition C in D. Then the generalized solution $w(z)$ to (1.45) in D can be expressed by

$$w(z) = \Phi[\chi(z)]e^{\phi(z)} + \psi(z) , \tag{2.33}$$

where $\Phi(z)$, $\psi(z)$ and $\chi(z)$ are as stated in Theorem 2.2, $\chi(z)$ and its inverse function $z(\chi)$ satisfy the estimates (2.9) and (2.10), and $\Phi(\chi)$ is an analytic function except for some isolated singular points in $\chi(D)$.

As Theorem 2.4 we can prove Corollary 2.5.

3. The compactness principle for sequences of solutions to complex equations

Let us introduce the functions

$$\sigma_n(z) = \begin{cases} 1, & \text{for } z \in D_n , \\ 0, & \text{for } z \notin D_n , \end{cases} \tag{2.34}$$

where D_n is a point set in D whose distance from Γ is not less than $1/n$, n is a positive integer, and consider the complex equation

$$w_{\bar{z}} = F_n(z,w,w_z) , \quad F_n(z,w,w_z) = \sigma_n(z)F(z,w,w_z) , \tag{2.35}$$

where $F(z,w,w_z)$ is as stated in (1.45), $Q^{(n)} = \sigma_n Q$, $A_j^{(n)} = \sigma_n A_j$, $j = 1,2,3$.

Theorem 2.6 Let the complex equation (1.45) satisfy Condition C and $w_n(z)$ be a solution to the equation (2.35) in the bounded domain D. If the solutions $w_n(z)$ $(n=1,2,\ldots)$ are uniformly bounded on any closed set D^* in D, $n = 1,2,\ldots$, then from $\{w_n(z)\}$ we can select a subsequence $\{w_{n_k}(z)\}$ which uniformly converges to a solution $w(z)$ of (1.47) on D^*.

Proof. By Corollary 2.5, the solution $w_n(z)$ to (2.35) can be expressed as

$$w_n(z) = \Phi_n[\chi_n(z)]e^{\phi_n(z)} + \psi_n(z) , \tag{2.36}$$

where

$$\psi_n = Tf_n - T_1 f_n , \quad \phi_n = Tg_n - T_1 g_n ,$$
$$\chi_n(z) = z + Th_n(z) - T_0 h_n$$

satisfy the estimates (2.17) - (2.19), (2.9) and (2.10), and $f_n(z)$, $g_n(z)$, and $h_n(z)$ are the solutions of the following integral equations

$$f_n = F_n(z,w_n,\Pi f_n) - F_n(z,w_n,0) + A_1^{(n)}(z,w_n)\psi_n + A_2^{(n)}(z,w_n)\bar{\psi}_n + A_3^{(n)}(z,w_n) , \tag{2.37}$$

$$W_n g_n = F_n(z,w_n,W_n\Pi g_n+\Pi f_n) - F_n(z,w_n,\Pi f_n) + A_1^{(n)}(z,w_n)W_n + A_2^{(n)}(z,w_n)\bar{W}_n , \tag{2.38}$$

$$\Phi_n'(\chi)h_n e^{\phi_n} = F_n(z,w_n,\Phi_n'(\chi)(1+\Pi h_n)e^{\phi_n} + W_n\Pi g_n+\Pi f_n) - F_n(z,w_n,W_n\Pi g_n+\Pi f_n) , \tag{2.39}$$

respectively, where $W_n(z) = w_n(z) - \psi_n(z)$. Because $\{w_n\}$ is uniformly bounded on any closed set D^* in D, we can choose subsequences of $\{\psi_n\}$, $\{\phi_n\}$, $\{\chi_n\}$, $\{W_n\}$, and $\{w_n\}$, which uniformly converge to $\psi_0(z)$, $\phi_0(z)$, $\chi_0(z)$, $W_0(z)$, and $w_0(z)$ on D, respectively, and subsequences of $\{\Phi_n\}$, $\{\Phi_n'\}$ which uniformly converge to $\Phi_0(\chi), \Phi_0'(\chi)$ on any closed set of $\chi_0(D)$, respectively. For the sake of convenience, in the future we shall denote subsequences of $\{\psi_n\},\ldots,\{w_n\}$ again by $\{\psi_n\},\ldots,\{w_n\}$. Noting the uniform ellipticity condition (1.50), we have

$$|F(z,w_0,\Pi f) - F(z,w_0,0)| \le q_0|\Pi f| , \tag{2.40}$$

and by the principle of contraction, the integral equation

$$f_0 = F(z,w_0,\Pi f_0) - F(z,w_0,0) + A_1(z,w_0)\psi_0 + A_2(z,w_0)\overline{\psi_0} + A_3(z,w_0) \tag{2.41}$$

has a unique solution $f_0 \in L_{p_0}(\bar{D})$. Let

$$f_n - f_0 = F_n(z,w_n,\Pi f_n) - F_n(z,w_n,\Pi f_0) + C_n(z) , \tag{2.42}$$

where
$$\begin{aligned} C_n(z) &= [F_n(z,w_n,\Pi f_n)-F_n(z,w_n,0)] - [F(z,w_0,\Pi f_0) \\ &\quad - F(z,w_0,0)] + [A_1^{(n)}(z,w_n)\psi_n - A_1(z,w_0)\psi_0] \\ &\quad + [A_2^{(n)}(z,w_n)\overline{\psi_n} - A_2(z,w_0)\overline{\psi_0}] + [A_3^{(n)}(z,w_n)-A_3(z,w_0)] \\ &= [Q_n(z,w_n,\Pi f_0)-Q(z,w_0,\Pi f_0)]\Pi f_0 + [A_1^{(n)}(z,w_n)\psi_n \\ &\quad - A_1(z,w_0)\psi_0] + [A_2^{(n)}(z,w_n)\overline{\psi_n} - A_2(z,w_0)\overline{\psi_0}] \\ &\quad + [A_3^{(n)}(z,w_n)-A_3(z,w_0)] . \end{aligned}$$

It is clear that when $n \to \infty$, $\{C_n(z)\}$ converges to 0 for almost every point $z \in D$. Hence, for two arbitrary sufficiently small positive constants ε_1 and ε_2 there exist a subset D_* in D , a sufficiently large positive integer N , so that meas $D_* < \varepsilon_1$ and $|C_n| < \varepsilon_2$ on $\bar{D} \setminus D_*$ for $n > N$. By the Hölder inequality and the Minkowski inequality, we have

$$\begin{aligned} L_{p_0}[C_n,\bar{D}] &\le L_{p_0}[C_n,D_*] + L_{p_0}[C_n,\bar{D}\setminus D_*] \le \\ &\le L_{p_1}[C_n,D_*]L_{p_2}[1,D_*] + \varepsilon_2 L_{p_0}[1,\bar{D}\setminus D_*] \le \\ &\le [2L_{p_1}(\Pi f_0,D_*)+2(M_{23}+1)k_2]\varepsilon_1^{1/p_2} + \varepsilon_2\pi^{1/p_0} = \varepsilon , \end{aligned} \tag{2.43}$$

in which $L_{p_1}[A_j, D_*] \le k_2 < \infty$, $j = 1,2,3$, $p_2 = p_0 p_1/(p_1 - p_0)$ $(2 < p_0 < p_1 \le p_2 < \infty)$, $n > N$, and M_{23} is a constant. This shows $L_{p_0}[C_n, \bar{D}] \to 0$, if $n \to \infty$. From (2.42) it follows that

$$L_{p_0}[f_n - f_0, \bar{D}] \le L_{p_0}[C_n, \bar{D}]/(1 - q_0 \Lambda_{p_0}) , \tag{2.44}$$

and $L_{p_0}[f_n - f_0, \bar{D}] \to 0$, if $n \to \infty$.

Consequently, $\psi_0 = Tf_0 - T_1 f_0$.

Similarly, we can find a unique solution $g \in L_{p_0}(\bar{D})$ of the integral equation

$$\begin{aligned} W_0 g_0 &= F(z, w_0, W_0 \Pi g_0 + \Pi f_0) - F(z, w_0, \Pi f_0) \\ &\quad + A_1(z, w_0) W_0 + A_2(z, w_0) \bar{W}_0 , \end{aligned} \tag{2.45}$$

and prove that $L_{p_0}[g_n - g_0, \bar{D}] \to 0$ if $n \to \infty$.

Hence $\phi_0 = Tg_0 - T_1 g_0$.

Besides, from the integral equation

$$\begin{aligned} \Phi_0'(\chi) h e^{\phi_0} &= F(z, w_0, \Phi_0'(\chi)(1 + \Pi h) e^{\phi_0} + W_0 \Pi g_0 + \Pi f_0) \\ &\quad - F(z, w_0, W_0 \Pi g_0 + \Pi f_0) , \end{aligned} \tag{2.46}$$

we can obtain a unique $h \in L_{p_0}(\bar{D})$ and prove that $L_{p_0}(h_n - h_0, \bar{D}) \to 0$ as $n \to \infty$.

If $w \equiv 0$ in D or $\Phi'(\chi) \equiv 0$ in $\chi(D)$, we take $g(z) \equiv 0$ and $h(z) \equiv 0$, respectively.

Hence $\chi_0(z) = z + Th_0 - T_0 h$.

Finally, taking

$$w_0(z) = \Phi_0[\chi_0(z)]e^{\phi_0(z)} + \psi_0(z) , \; w_{0\bar{z}} = \Phi_0(\chi)h_0e^{\phi_0} + W_0g_0 + f_0 , \; w_{0z} = \Phi_0'(\chi)(1+\Pi h_0)e^{\phi_0} + W_0\Pi g_0 + \Pi f_0 \tag{2.47}$$

into account, we see that a subsequence of $\{w_n\}$ uniformly converges to $w_0(z)$ on any closed set D^* in D . □

By using the same method as in the proof of Theorem 2.6, we can get the next result.

Theorem 2.7 Let (1.45) satisfy Condition C and $\{w_n\}$ be a sequence of solutions to (1.45) in a bounded domain D , where $w_n (n=1,2,\ldots)$ are uniformly bounded on any closed set D^* in D . Then we can choose a subsequence of $\{w_n\}$, which uniformly converges on D^* to a solution of (1.45).

§ 3. The Haseman boundary value problem for elliptic complex equations of first order

Before considering the Haseman boundary value problem, it is necessary to give some existence theorems for solutions to the uniformly elliptic complex equation (1.47) on the whole plane $\mathbb{C}$.

1. An existence theorem for solutions to the nonlinear complex equation

Lemma 3.1 Let the complex equation (1.47) satisfy Condition C^*, and

$$w(z) = \Phi(z) + T\omega - T_1\omega \tag{3.1}$$

be a bounded continuous solution to (1.47) on $E(E = \mathbb{C} \setminus E_0)$, where E_0 is a set of bounded curves, $w(z) \in L_{p,2}(\mathbb{C})$ and $\Phi(z)$ is analytic in E and $\Phi'(z) \in L_{p,2}(\mathbb{C})$. We suppose that the left and right nontangential limits of $w(z)$ on E_0 exist and are continuous.

Then $w(z)$ and $\psi(z) = w(z) - \Phi(z)$ satisfy the estimates

$$C[w,\mathbb{C}] \le M_1 \ , \ L_{p_0,2}[|w_{\bar{z}}| + |w_z| \ , \mathbb{C}] \le M_2 \ , \tag{3.2}$$

$$C_\beta[\ ,\bar{\mathbb{C}}] \le M_3 \ , \ L_{p_0,2}[|\psi_{\bar{z}}| + |\psi_z| \ , \mathbb{C}] \le M_4 \ , \tag{3.3}$$

where $\beta = 1-2/p_0$, $M_j = M_j(q_0,p_0,k_0,E_0,\Phi)$, $j = 1,\ldots,4$.

Proof. Substituting $w(z) = \Phi(z) + T\omega - T_1\omega$ in the complex equation (1.47), we obtain

$$\begin{aligned} &\psi_{\bar{z}} = F(z,w,w_z) \ , \\ &F(z,w,w_z) = Q(z,w,w_z)\psi_z + A_1(z,w)\psi + A_2(z,w)\bar{\psi} + A \ , \\ &A = Q(z,w,w_z)\Phi' + A_1(z,w)\Phi + A_2(z,w)\bar{\Phi} + A_3(z,w) \ . \end{aligned} \tag{3.4}$$

From Condition C*, it is easy to see that

$$\begin{aligned} &|Q(z,w,w_z)| \le q_0 < 1 \ , \\ &L_{p_0}[A_j(z,w),\bar{\mathbb{C}}] \le k_3 \ , \ j=1,2 \ , \ L_{p_0}[A,\bar{\mathbb{C}}] \le k_3 \ , \end{aligned} \tag{3.5}$$

where $k_3 = k_3(q_0,p_0,k_0,E_0,\Phi)$. According to the method used in the proof of Corollary 2.3, we can obtain the estimate (3.3), and then (3.2) holds. □

Lemma 3.2 Let k_4 be a nonnegative constant. If the condition $L_{p,2}[A_3(z,w),\mathbb{C}] \le k_0$ in (1.47) is replaced by

$$L_{p_0,2}[A_3(z,w),\mathbb{C}] \le k_4 \ , \tag{3.6}$$

then the solution $w = T\omega - T_1\omega \ (\omega \in L_{p_0,2}(\mathbb{C}))$ of (1.47) satisfies the estimate

$$C_\beta[w,\bar{\mathbb{C}}] \le M_5 k_4 \ , \ L_{p_0,2}[|w_{\bar{z}}| + |w_z|,\mathbb{C}] \le M_6 k_4 \ , \tag{3.7}$$

where $M_j = M_j(q_0, p_0, k_0, E_0, \Phi)$, $j = 5,6$.

Proof. When $k_4 > 0$, we set $W(z) = w(z)/k_4$, and $W(z)$ is a solution of the following equation

$$\begin{aligned} W_{\bar{z}} &= Q(z,w,w_z)W_z + A_1(z,w)W + \\ &\quad + A_2(z,w)\bar{W} + A_3(z,w)/k_4 . \end{aligned} \tag{3.8}$$

It is evident that $L_{p_0,2}[A_3(z,w)/k_4,\mathbb{C}] \le 1$
By Lemma 3.1, we conclude

$$C_\beta[W,\bar{\mathbb{C}}] \le M_5 , \; L_{p_0,2}[|W_{\bar{z}}| + |W_z|,\mathbb{C}] \le M_6 \tag{3.9}$$

and so (3.7) is true.

Obviously, when $k_4 = 0$, (3.7) holds. □

Next we discuss the nonlinear complex equation

$$\begin{aligned} w_{\bar{z}} &= f(z,w,w_z) + A_3(z) , \\ f(z,w,w_z) &= Q(z,w_z)w_z + A_1(z)w + A_2(z)\bar{w} \end{aligned} \tag{3.10}$$

and assume that (3.10) satisfies Condition C*.

Theorem 3.3 Suppose that $\Phi(z)$ is as stated in Lemma 3.1 and (3.10) satisfies Condition C*. Then (3.10) has a solution $w(z)$ as follows:

$$w(z) = \Phi(z) + T\omega - T_1\omega = \Phi + \psi , \tag{3.11}$$

where $\omega \in L_{p_0,2}(\mathbb{C})$.

Proof. In order to use the method of continuity, we introduce the integral equation with parameter $t(0 \le t \le 1)$

$$\omega(z) - tf(z,\Phi+\psi,\Phi'+\Pi\omega) = A(z)\ ,\ A \in L_{p_0,2}(\mathbb{C})\ , \tag{3.12}$$

which is equivalent to the complex equation

$$w_{\bar{z}} - tf(z,w,w_z) = A(z)\ , \tag{3.13}$$

where $A \in L_{p_0,2}(\mathbb{C})$, $\Pi\omega = -\frac{1}{\pi}\iint\limits_{\mathbb{C}} \frac{\omega(\zeta)}{(\zeta-z)^2}\, d\sigma_\zeta$.

Let T be a point set in the interval $[0,1]$ so that, for every $t \in T$, (3.12) possesses a solution $\omega \in L_{p_0,2}(\mathbb{C})$ for any measurable function $A \in L_{p_0,2}(\mathbb{C})$. It is clear that when $t = 0$, (3.12) has the solution $\omega(z) = A(z)$. Hence T is non-empty. If we can prove that T is both open and closed in $[0,1]$, then we can deduce that T is $[0,1]$. In particular, when $t = 1$ and $A(z) = A_3(z)$, (3.12) has a solution $\omega \in L_{p_0,2}(\mathbb{C})$. So (3.13) with $t = 1$ and $A(z) = A_3(z)$, i.e. (3.10) possesses a solution w of the type (3.11).

To prove that T is an open set in $0 \le t \le 1$, let $t_0 \in T$. We rewrite the equation (3.12) in the following form

$$\omega(z) - t_0 f(z,\Phi+\psi,\Phi'+\Pi\omega) = (t-t_0)f(z,\Phi+\psi,\Phi'+\Pi\omega) + A(z)\ . \tag{3.14}$$

Since for $t = t_0 \in T$, (3.14) with any $A \in L_{p_0,2}(\mathbb{C})$ is solvable. (3.12) for $t = t_0$ possesses a solution $\omega \in L_{p_0,2}(\mathbb{C})$.
We are free to choose a function ω_0 in $L_{p_0,2}(\mathbb{C})$ $(2 < p_0 < p)$ and in particular $\omega_0 = 0$. Let $\omega_0 = 0$ replace $\omega(z)$ in the right-hand side of (3.14). With $\psi = \psi_0 = T\omega_0 - T_0\omega_0$ it is obvious that $(t-t_0)f(z,\Phi+\psi_0,\Phi'+\Pi\omega_0) + A(z) \in L_{p_0,2}(\mathbb{C})$, and hence (3.14), with a right-hand side as indicated, has a solution ω_1 . According to Lemma 3.2, we see that $\omega_1 \in L_{p_0,2}(\mathbb{C})$. Thus, by successive iteration, we obtain a sequence of solutions ω_n , $n = 1,2,\ldots,$ which satisfy

$$\omega_n(z) - t_0 f(z,\Phi+\psi_n,\Phi'+\Pi\omega_n) = \\ = (t-t_0)f(z,\Phi+\psi_{n-1},\Phi'+\Pi\omega_{n-1}) + A(z) \quad , \ n = 1,2,\ldots, \tag{3.15}$$

where $\psi_n(z) = T\omega_n - T_1\omega_n$. From (3.15), we have

$$\omega_{n+1}-\omega_n-t_0[f(z,\Phi+\psi_{n+1},\Phi'+\Pi\omega_{n+1})-f(z,\Phi+\psi_n,\Phi'+\Pi\omega_n)] \\ =(t-t_0)[f(z,\Phi+\psi_n,\Phi'+\Pi\omega_n)-f(z,\Phi+\psi_{n-1},\Phi'+\Pi\omega_{n-1})], n = 1,2,\ldots . \tag{3.16}$$

By (3.7) in Lemma 3.2, choosing $\delta = 1/2\ M_6(q_0\lambda_{p_0} + k_0M_5)$ and denoting $\|\omega\| = L_{p_0,2}(|\omega|,\mathbb{C})$ we obtain for $|t-t_0| < \delta$

$$\|\omega_{n+1}-\omega_n\| \le M_6k = |t-t_0|M_6(q_0\lambda_{p_0}+2k_0M_5)\|\omega_n-\omega_{n-1}\| < \frac{1}{2}\|\omega_n-\omega_{n-1}\| . \tag{3.17}$$

Hence

$$\|\omega_{n+1}-\omega_n\| < \frac{1}{2^n}\|\omega_1-\omega_0\|$$

so that

$$\|\omega_n-\omega_m\| < \frac{1}{2^N}\sum_{j=1}^{\infty}\frac{1}{2^j}\|\omega_1-\omega_0\| < \frac{1}{2^N}\|\omega_1-\omega_0\|$$

for $n,m > N+1$, where N is a positive integer. This shows that $\|\omega_n-\omega_m\| \to 0$, for $n,m \to \infty$. By the completeness of the Banach space $L_{p_0,2}(\mathbb{C})$, there exists a function $\omega_*(z) \in L_{p_0,2}(\mathbb{C})$, so that $\|\omega_n-\omega_*\| \to 0$ for $n \to \infty$. Thus we can select subsequences of $\{\omega_n(z)\}$ and $\{\Pi\omega_n\}$ which converge to $\omega_*(z)$ and $\Pi\omega_*$ for almost every point $z \in \mathbb{C}$, respectively. $\omega_*(z)$ is a solution of (3.12) for $|t-t_0| < \delta$ and $0 \le t \le 1$. Consequently, $w_*(z) = \Phi(z) + T\omega_* - T_1\omega_*$ is a solution of (3.13). This proves that T is an open set in $0 \le t \le 1$.

Next, we prove that T is a closed set in $[0,1]$. Let $t_n \in T$, $n = 1,2,\ldots,$ and $t_n \to t_0$, for $n \to \infty$. We shall verify that when $t = t_0$, (3.12) is solvable. From $t_n \in T$, $n = 1,2,\ldots$ and (3.12) corresponding to t_n , we conclude

$$\omega_n-\omega_m-t_n[f(z,\Phi+\psi_n,\Phi'+\Pi\omega_n)-f(z,\Phi+\psi_n,\Phi'+\Pi\omega_m)] + c_{n,m} ,$$
$$c_{n,m} = t_n f(z,\Phi+\psi_n,\Phi'+\Pi\omega_m)-t_m f(z,\Phi+\psi_m,\Phi'+\Pi\omega_m) , \tag{3.18}$$

which shows

$$\|\omega_n-\omega_m\| \le \frac{1}{1-t_n q_0 \lambda_{p_0}} \|c_{n,m}\| . \tag{3.19}$$

According to the method in (2.43), we can derive $\|c_{n,m}(z)\| \to 0$ and $\|\omega_n-\omega_m\| \to 0$ for $n,m \to \infty$. Hence, there exists $\omega_0(z) \in L_{p_0,2}(\mathbb{C})$, so that $\|\omega_n-\omega_0\| \to 0$ for $n \to \infty$. Thus, $w_0(z) = \Phi(z) + T\omega_0 - T_1\omega_0$ is a solution of (3.13) with $t = t_0$. This completes the proof. □

Theorem 3.4 Suppose that $\Phi(z)$ is as stated in Lemma 3.1 and the nonlinear complex equation (1.47) satisfies Condition C*. Then (1.47) has a solution $w(z)$ of the form (3.11).

Proof. We only discuss the following equation

$$\psi_{\bar{z}} = F(z,\Phi+\psi,\Phi'+\psi_z) , \psi(z) = T\omega - T_1\omega \tag{3.20}$$

and find a solution of (3.20) in the form

$$\psi(z) = T\omega - T_1\omega , \omega(z) \in L_{p_0,2}(\mathbb{C}) . \tag{3.21}$$

In order to apply the Schauder fixed point theorem let us introduce a closed and convex set B in the Banach space $C(\mathbb{C})$, the elements of which are the continuous functions $\psi(z)$ on $\mathbb{C}$ satisfying the condition

$$C[\psi,\mathbb{C}] \le M_3 , \tag{3.22}$$

where M_3 is the nonnegative constant from (3.3). We select an arbitrary function $\Psi(z) \in B$ and insert it into the coefficients of (1.47). It is clear that the complex equation

$$\psi_{\bar{z}} = Q(z,\Phi+\Psi,\Phi'+\psi_z)(\Phi'+\psi_z)+A_1(z,\Phi+\Psi)(\Phi+\psi)$$
$$+ A_2(z,\Phi+\Psi)(\overline{\Phi}+\overline{\psi}) + A_3(z,\Phi+\Psi) \tag{3.23}$$

satisfies Condition C* as (3.13) with $t = 1$ and $A = A_3(z,\Phi+\Psi)$. By Lemma 3.2 and the proof of Theorem 3.3, the complex equation (3.23) has a solution $\psi(z)$ of the type (3.21), and $\psi(z)$ satisfies the first estimate of (3.3). We denote the mapping from $\Psi(z)$ onto $\psi(z)$ given by (3.23) by $\psi = S(\Psi)$. It remains to show that $S(\Psi)$ is a continuous mapping in B. We choose an arbitrary sequence $\Psi_n(z) \in B$, $n = 0,1,2,\ldots$, so that $C[\Psi_n-\Psi_0,\mathbb{C}] \to 0$ for $n \to \infty$. Set $\psi_n = S(\Psi_n) = T\omega_n - T_1\omega_n$, $n = 0,1,\ldots$.
From the equation corresponding to $\psi_n = S(\Psi_n)$ and $\psi_0 = S(\Psi_0)$, we obtain

$$\begin{aligned}
\omega_n-\omega_0 &= [F(z,\Phi+\Psi_n,\Phi'+\Pi\omega_n) - F(z,\Phi+\Psi_n,\Phi'+\Pi\omega_0)] \\
&\quad + A_1(z,\Phi+\Psi_n)(\psi_n-\psi_0)+A_2(z,\Phi+\Psi_n)(\overline{\psi_n}-\overline{\psi_0})+c_n(z) , \\
c_n &= [Q(z,\Phi+\Psi_n,\Phi'+\Pi\omega_0)-Q(z,\Phi+\Psi_0,\Phi'+\Pi\omega_0)]\Pi\omega_0 + \\
&\quad + [A_1(z,\Phi+\Psi_n)-A_1(z,\Phi+\Psi_0)](\Phi+\psi_0)+[A_2(z,\Phi+\Psi_n) \\
&\quad - A_2(z,\Phi+\Psi_0)](\overline{\Phi}+\overline{\psi_0})+A_3(z,\Phi+\Psi_n)-A_3(z,\Phi+\Psi_0) .
\end{aligned} \tag{3.24}$$

Similar to (3.18), we can prove that $\|\omega_n-\omega_0\| \to 0$ and $C[\psi_n-\psi_0,\mathbb{C}] \to 0$ for $n \to \infty$. This shows that S is a continuous operator. Hence, by the Schauder fixed point theorem, the mapping $\psi = S(\Psi)$ possesses a fixed point $\psi(z) = T\omega - T_1\omega$. Consequently, $w(z) = \Phi(z) + \psi(z)$ is a solution of the nonlinear complex equation (1.47). □

By using the above method, we can prove the corresponding result on a bounded domain D.

Corollary 3.5 Let $\Phi(z)$ be an analytic function in a bounded domain D with the condition $\Phi'(z) \in L_{p_0}(\overline{D})$ $(2 < p_0 < p)$ and the equation (1.47) satisfy Condition C in D. Then (1.47) has a solution $w(z)$ possessing the form

$$w(z) = \Phi(z) + T\omega - T_1\omega , \tag{3.25}$$

where

$$\omega(z) \in L_{p_0}(\bar{D}) , \quad T\omega = -\frac{1}{\pi}\iint_D \frac{\omega(\zeta)}{\zeta - z} d\sigma_\zeta , \quad T_1\omega = T\omega\Big|_{z=1} .$$

2. Formulation and reduction of the Haseman boundary value problem

We denote by D^+ an $(N+1)$-connected bounded domain with the boundary

$$\Gamma = \bigcup_{j=0}^{N} \Gamma_j \in C^1_\mu (0 < \mu < 1) ,$$

where $\Gamma_1, \ldots, \Gamma_N$ are situated inside Γ_0, and by $D^- = \bigcup_{j=0}^{N} D_j^-$ the complement of $\overline{D^+}$ in the z-plane $\mathbb{C}$, where D_j^- is the domain surrounded by $\Gamma_j (j=1,\ldots,N)$ and D_0^- is an unbounded domain with the boundary Γ_0. We will discuss the complex equation (1.47) and assume that (1.47) satisfies Condition C^* in $\mathbb{C}$.

Problem H. The so-called Haseman boundary value problem for the equation (1.47) is to find a sectionally bounded continuous solution $w^\pm$ of (1.47) in $D^\pm$ satisfying the boundary condition

$$w^+[\alpha(t)] = G(t)w^-(t) + g(t) , \quad t \in \Gamma , \tag{3.26}$$

where $\alpha(t)$ topologically maps Γ_j onto itself and is a positive shift on Γ_j, $j = 0,1,\ldots,N$, and $G(t)$, $g(t)$ satisfy

$$\begin{aligned} &C_\nu[G,\Gamma] \le \ell , \ |G(t)| \ge \ell^{-1} , \ C_\nu[g,\Gamma] \le \ell , \ \frac{1}{2} < \nu < 1 , \\ &C^1_\mu[\alpha,\Gamma] \le \ell , \ |\alpha'(t)| \ge \ell^{-1} > 0 , \ 0 < \mu < 1 , \end{aligned} \tag{3.27}$$

in which μ,ν and ℓ are positive constants.

$$K = \frac{1}{2\pi} \Delta_\Gamma \, \mathit{arg} \, G(t)$$

is called the index of Problem H. In particular, when $\alpha(t) = t$, the problem is the Riemann boundary value Problem R.

In order to discuss the solvability of Problem H, we introduce the modified boundary value problem H*, i.e. when $K < 0$, we permit that the solution $w^{\pm}(z)$ of (1.47) has a pole of order $\leq |K| - 1$ at $z = \infty$. Similarly, the corresponding modified problem R is called Problem R*.

Using the conformal glue theorem (cf. [69]), Problem H* for (1.47) can be transformed into Problem R* for another nonlinear complex equation.

Theorem 3.6 Let (1.47) satisfy Condition C*. A necessary and sufficient condition for

$$w(z) = \begin{cases} w^{+}(z), z \in D^{+} \\ w^{-}(z), z \in D^{-} \end{cases}$$

being a solution of Problem H* for (1.47) is that $w^*(\zeta) = w[z(\zeta)]$ is a solution of Problem R* with a boundary condition of the type

$$w^{*+}(\zeta) = G^{*}(\zeta)w^{*-}(\zeta) + g^{*}(\zeta) \ , \ \zeta \in \Gamma^{*} = \zeta[\Gamma] \tag{3.28}$$

for the nonlinear complex equation

$$w^{*}_{\bar{\zeta}} = \overline{z'(\zeta)}\ F[z(\zeta), w^{*}, w^{*}_{\zeta}/z'(\zeta)] \ , \ \zeta \in D^{*\pm} = \zeta(D^{\pm}) \tag{3.29}$$

where

$$\zeta(z) = \begin{cases} \zeta^{+}(z), z \in D^{+} \\ \zeta^{-}(z), z \in \overline{D^{-}} \end{cases}$$

is the conformal glue function satisfying the boundary condition

$$\zeta^{+}[\alpha(t)] = \zeta^{-}(t) \ , \ t \in \Gamma \ ,$$

and $\zeta(0) = 0$, $\zeta(1) = 1$, $\zeta(\infty) = \infty$, $z(\zeta)$ is the inverse function of $\zeta(z)$, $G^*(\zeta)$ and $g^*(\zeta)$ on $\Gamma^* = \zeta[\Gamma]$ satisfy conditions similar to those for $G(t)$, $g(t)$ on Γ .

We can verify that the complex equation (3.29) satisfies a Condition C^*, where the constants k_0 , ℓ in (1.51) and (3.27) are replaced by other constants k_0' , ℓ' . Hence, we may only discuss the solvability of Problem R* and Problem R for (3.29) or (1.47).

3. Solvability of the Haseman boundary value problem

We first prove an equivalence theorem for Problem R* of the complex equation (1.47), and then derive a solvability result for Problem R and Problem H of (1.47).

Theorem 3.7 A necessary and sufficient condition for

$$w(z) = \begin{cases} w^+(z), z \in D^+ \\ w^-(z), z \in D^- \end{cases}$$

being a solution of Problem R^* for (1.47) with Condition C^* is that

$$V(z) = [w(z) - \phi(z)]/X(z)$$

is a continuous solution to the equation

$$V_{\bar{z}} = F[z,\phi(z)+X(z)V(z),\phi'(z)+X'(z)V(z)+X(z)V_z]/X(z) \tag{3.30}$$

on $\mathbb{C}$ and $V(\infty) = 0$, if $K < 0$, where $\phi(z)$ and $X(z)$ are analytic functions in $D^{\pm}\setminus\{\infty\}$ satisfying the boundary conditions

$$\begin{aligned} \phi^+(t) &= G(t)\phi^-(t)+g(t), t \in \Gamma , \\ X^+(t) &= G(t)X^-(t), t \in \Gamma , \end{aligned} \tag{3.31}$$

which can be expressed as

$$\phi(z) = X(z)\psi(z), \psi(z) = \frac{1}{2\pi i}\int_\Gamma \frac{g(t)dt}{X^+(t)(t-z)} + P(z) , \tag{3.32}$$

$$X(z) = \begin{cases} X^+(z) = e^{\Gamma(z)}/\Pi(z), z \in D^+ , \\ X^-(z) = z^{-K} e^{\Gamma(z)}, z \in D^- , \end{cases} \qquad P(z) = \begin{cases} c_0 + \ldots + c_K z^K, K \geq 0 , \\ 0 , K < 0 , \end{cases}$$

$$\Gamma(z) = \frac{1}{2\pi i}\int_\Gamma \frac{\ell n[t^{-K} G(t)\Pi(t)]}{t-z} ,$$

$$\Pi(z) = \prod_{j=1}^{N} (z-s_j)^{K_j}, s_j \in D_j, j=1,\ldots,N ,$$

$K_j = \frac{1}{2\pi}\Delta_{\Gamma_j} \arg G(t)$, $j = 1,\ldots,N$, and $c_j (j=0,1,\ldots,K, K \geq 0)$ are complex constants.

<u>Proof.</u> Let

$$w(z) = \begin{cases} w^+(z) , z \in D^+ \\ w^-(z) , z \in D^- \end{cases}$$

be a solution of Problem R^* for (1.47). Noting that $\phi(z)$ and $X(z)$ satisfy the boundary conditions (3.31), we see that

$$w(z) = \phi(z) + X(z)V(z) \tag{3.33}$$

satisfies the following boundary condition

$$\begin{aligned} w^+(t) &= \phi^+(t) + X^+(t)V^+(t) \\ &= G(t)[\phi^-(t) + X^-(t)V^+(t)] + g(t) \\ &= G(t)w^-(t) + g(t) = G(t)[\phi^-(t) + X^-(t)V^-(t)] \\ &\quad + g(t) , t \in \Gamma . \end{aligned} \tag{3.34}$$

From $X^-(t) \neq 0$, $t \in \Gamma$, it follows that $V^+(t) = V^-(t)$, $t \in \Gamma$, and so

$$V(z) = \begin{cases} V^+(z) \ , \ z \in D^+ \\ V^-(z) \ , \ z \in D^- \end{cases}$$

is a continuous solution of (3.30) on $\mathbb{C}$ as stated in this theorem.

Conversely, it is clear that the second part of the result is true. □

<u>Theorem 3.8</u> Let the nonlinear complex equation (1.47) satisfy Condition C*, and the coefficients Q and A_3 satisfy

$$L_{p,2}[Q/z,\mathbb{C}] \le \ell < \infty \ , \ L_{p,2}[A_3 z^K,\mathbb{C}] \le \ell \ (K>0) \ . \tag{3.35}$$

Then Problem R* and Problem H* for (1.47) are solvable and if $K \ge 0$, the general solutions of Problem R and Problem H contain $2K + 2$ arbitrary real constants; if $K < 0$ Problem R and Problem H have at least $-2K-2$ solvability conditions arising from

$$\lim_{z\to\infty} [\psi(z) + V(z)]z^j = 0, j = -1,\ldots,K+1 \ .$$

<u>Proof.</u> From Condition C* for (1.47) and (3.35), it is not difficult to see that the nonlinear equation (3.30) satisfies a similar Condition C*. By Theorem 3.4, the equation (3.30) possesses a solution $V(z) = T\omega$, $\omega(z) \in L_{p_0,2}(\mathbb{C})$. Hence from Theorem 3.7, Problem R* for equation (1.47) has a solution of the type (3.33). If $K \ge 0$, the solution of Problem R* for (1.47) is a solution of Problem R, and the general solutions contains $K + 1$ arbitrary complex constants $c_0,\ldots,c_K$, i.e. $2K + 2$ arbitrary real constants. If $K < 0$ and $\psi(z) + V(z)$ has a zero-point of order $-K$ at $z = \infty$, the solution of Problem R* is bounded on $D^{\pm}$, and so is a solution of Problem R. Therefore, Problem R for (1.47) has at least $-2K-2$ solvability conditions. From Theorem 3.6, the corresponding results for solvability of Problem H^* and Problem H for (1.47) are derived

(cf. [106]). □

§ 4. The Riemann-Hilbert problem for the elliptic complex equation of first order

In Chapter 1, we have considered the solvability of the Riemann-Hilbert boundary value problem for simple complex equations of first order. Now, we will discuss the estimates of solutions and the solvability of the Riemann-Hilbert problem for nonlinear elliptic complex equations of first order, in particular, we shall give an expression for the general solutions of the Riemann-Hilbert problem for linear elliptic complex equations, which contain the total number of linearly independent solutions to the homogeneous problem and the total number of solvability conditions for the nonhomogeneous problem.

1. Formulation and uniqueness of the solution of the Riemann-Hilbert problem

Let D be an $(N+1)$-connected domain as stated in § 1 of Chapter 1.

Problem A. The so-called Riemann-Hilbert problem for (1.45) or (1.47) is to find a solution $w(z)$ in the $(N+1)$-connected domain D, continuous on $\bar{D}$ and satisfying the boundary condition

$$Re[\overline{\lambda(t)}w(t)] = r(t) \ , \ t \in \Gamma \ , \tag{4.1}$$

where $|\lambda(t)| = 1$, $C_\alpha[\lambda,\Gamma] \le \ell$, $C_\alpha[r,\Gamma] \le \ell$ and $\alpha(0 < \alpha < 1)$, $\ell(0 < \ell < \infty)$ are constants.

Problem A with $r(t) = 0(t \in \Gamma)$ and $A_3 = 0(z \in D)$ is called Problem A_0. The corresponding modified Riemann-Hilbert Problem B for (1.45) is to find a solution $w(z)$, continuous on $\bar{D}$ and satisfying the boundary conditions

$$Re[\overline{\lambda(t)}w(t)] = r(t) + h(t) \ , \ t \in \Gamma \ , \tag{4.2}$$

$$Im[\overline{\lambda(a_j)}w(a_j)] = b_j \ , \ j \in \{j\} = \begin{cases} 1,\dots,2K-N+1, K \ge N \ , \\ N-K+1,\dots,N+1, 0 \le K < N \ , \end{cases} \tag{4.3}$$

where

$$h(t) = \begin{cases} 0 \ , & t \in \Gamma \ , \ K \geq N \ , \\ h_j \ , & t \in \Gamma_j \ , \ j = 1,\ldots,N-K \ , \\ 0 \ , & t \in \Gamma_j \ , \ j = N-K+1,\ldots,N+1 \ , \end{cases} \Big\} \ 0 \leq K < N \ , \qquad (4.4)$$

$$\begin{cases} h_j \ , & t \in \Gamma_j \ , \ j = 1,\ldots,N \ , \\ h_0 + \sum_{j=1}^{|K|-1} Re[(H_m^+ + iH_m^-)t^m] \ , & t \in \Gamma_0 \ , \end{cases} \Big\} \ K < 0 \ ,$$

in which $h_j(j=1,\ldots,N+1)$, $H_m^{\pm}(m=1,\ldots,|K|-1)$ are unknown real constants to be determined appropriately, and
$a_j \in \Gamma_j(j=1,\ldots,N)$, $a_j \in \Gamma_0(j=N+1,\ldots,2K-N+1)$ are distinct points, $b_j(j=1,\ldots,2K-N+1)$ are real constants. Problem B with $r(t) = 0$ and $A_3(z,w) = 0$ is called Problem B_0.

In order to get uniqueness for the solution of Problem B for (1.45) or (1.47), we have to add a condition. For any continuous functions $w_1(z)$, $w_2(z)$ in $\bar{D}$ and any measurable function $V(z)$ in D ,

$$F[z,w_1,V] - F[z,w_2,V] = A(z,w_1,w_2,V)(w_1 - w_2) \ , \qquad (4.5)$$
$$L_p[A(z,w_1,w_2,V) \ , \ \bar{D}] \leq k_0 < \infty \ , \ 2 < p < \infty \ .$$

Theorem 4.1 Let the nonlinear complex equation (1.45) or (1.47) satisfy Condition C and (4.5). Then the solution of Problem B for (1.45) or (1.47) is unique.

Proof. Let $w_1(z)$ and $w_2(z)$ be two solutions of Problem B for (1.45) or (1.47). It is clear that $w(z) = w_1(z) - w_2(z)$ is a solution to the complex equation

$$w_{\bar{z}} = Q\, w_z + Aw\ , \tag{4.6}$$

$$Q = \begin{cases} [F(z,w_1,w_{1z})-F(z,w_1,w_{2z})]/(w_{1z}-w_{2z}),\ \text{for}\ \ w_{1z} \neq w_{2z}\ , \\ 0\ ,\ \text{for}\ \ w_{1z} = w_{2z}\ , \end{cases}$$

$$A = \begin{cases} [F(z,w_1,w_{2z})-F(z,w_2,w_{2z})]/(w_1-w_2),\ \text{for}\ \ w_1(z) \neq w_2(z)\ , \\ 0\ ,\ \text{for}\ \ w_1(z) = w_2(z)\ , \end{cases}$$

where $|Q| \le q_0 < 1$, $L_p[A,\bar{D}] \le k_0 < \infty$, $2 < p < \infty$.

According to Theorem 2.2, the solution $w(z)$ of (4.6) can be expressed as

$$w(z) = \Phi[\zeta(z)]e^{\phi(z)}\ , \tag{4.7}$$

in which $\phi(z)$, $\zeta(z)$ and its inverse function $z(\zeta)$ satisfy the estimates in (2.19), (2.20) and $\Phi(\zeta)$ is an analytic function in $G = \zeta(D)$, continuous on $\zeta(\bar{D})$ satisfying the boundary conditions

$$Re[\overline{\Lambda(\zeta)}\Phi(\zeta)] = h[z(\zeta)]\ ,\ \zeta \in L = \zeta(\Gamma)\ , \tag{4.8}$$

$$Im[\overline{\Lambda(\tilde{a}_j)}\Phi(\tilde{a}_j)] = 0\ ,\ j \in \{j\} = \begin{cases} 1,\ldots,2K-N+1, K \ge N\ , \\ N-K+1,\ldots,N+1, 0 \le K < N\ , \end{cases} \tag{4.9}$$

where $\overline{\Lambda(\zeta)} = \overline{\lambda[z(\zeta)]}\ e^{\phi[z(\zeta)]}$, $\tilde{a}_j = \zeta(a_j)$, $j \in \{j\}$.

Obviously, $\frac{1}{2\pi}\Delta_L\, arg\, \Lambda(\zeta) = \frac{1}{2\pi}\Delta_L\, arg\, \lambda[z(\zeta)]\overline{e^{\phi(z(\zeta))}} =$

$$= \frac{1}{2\pi}\Delta_\Gamma\, arg\, \lambda(z) = K\ .$$

If $K \ge N$, from Theorem 1.2 in Chapter 1, it is clear that $\Phi(\zeta) = 0$ for $\zeta \in G$, and then $w(z) = 0$, $w_1(z) = w_2(z)$ for $z \in D$.

If $K < 0$, in analogy to the proof of Theorem 3.1 in Chapter 1, $h[z(\zeta)]$ satisfies the conditions

$$\int_L [\overline{\Lambda(\zeta)}]^{-1} \Psi_n(\zeta)\, h[z(\zeta)]\zeta'(s)\, ds = 0 \,, \tag{4.10}$$

in which $\Psi_n(\zeta)$ $(n=1,\ldots,N-2K-1)$ are linearly independent solutions to the corresponding conjugate homogenuous problem B_0' with the boundary condition

$$Re\{[\overline{\Lambda(\zeta)}]^{-1} \Psi_n(\zeta)\zeta'(s)\} = 0 \,, \ \zeta \in L \,, \ n = 1,\ldots,N-2K-1 \,.$$

Hence,

$$\sum_{j=0}^{N} h_j \int_{L_j} Im\{[\overline{\Lambda(\zeta)}]^{-1} \Psi_n(\zeta)\zeta'(s)\}ds \, + \tag{4.11}$$

$$+ \sum_{m=1}^{-K-1} \int_{L_0} Re\{(H_m^+ + iH_m^-)[z(\zeta)]^m\} Im\{[\overline{\Lambda(\zeta)}]^{-1} \Psi_n(\zeta)\zeta'(s)\}ds = 0 \,,$$
$$n = 1,\ldots,N-2K-1 \,.$$

If $h_j (j=0,1,\ldots,N)$, $H_m^{\pm}(m=1,\ldots,-K-1)$ are not all equal to zero, then the coefficients determinant of the algebraic system (4.11) certainly equals zero. Therefore, we can find real constants $c_1,\ldots,c_{N-2K-1}$, which are not all equal to zero, so that

$$\int_{L_j} Im\{[\overline{\Lambda(\zeta)}]^{-1} \Psi(\zeta)\zeta'(s)\}ds = 0 \,, \ j = 0,1,\ldots,N \,,$$

$$\int_{L_0} Im\left\{[\overline{\Lambda(\zeta)}]^{-1} \Psi(\zeta)\zeta'(s) \begin{Bmatrix} cos(m\ arg\ z(\zeta)) \\ sin(m\ arg\ z(\zeta)) \end{Bmatrix}\right\} ds = 0 \,, \tag{4.12}$$
$$m = 1,\ldots,-K-1 \,,$$

where $\Psi(\zeta) = \sum_{n=1}^{N-2K-1} c_n \Psi_n(\zeta)$ is a solution of Problem B_0' and $\Psi(\zeta) \not\equiv 0$ in $G = \zeta(D)$. From the first formula in (4.12), there exist points $a_j^* \in L_j$ $(j=1,\ldots,N)$, so that

$$[\Lambda(\zeta)]^{-1}\ \Psi(\zeta)\zeta'(s)\Big|_{\zeta=a_j^*}=0\ ,\ \text{i.e.}\ \Psi(a_j^*)=0\ ,\ j=1,\ldots,N\ . \tag{4.13}$$

In addition, let $U(z)$ be a harmonic function in $\mathbb{D}$ which satisfies the boundary condition

$$U(z) = Im\{[\overline{\Lambda(\zeta)}]^{-1}\ \Phi(\zeta)\zeta'(s)\}\Big|_{\zeta=\zeta(z)}\ ,\ \text{for}\ \ |z| = 1\ .$$

Let $s = s(\theta)$ denote the corresponding relation between s and θ in $\zeta = e^{is} = \zeta(e^{i\theta})$. Then we have

$$U(z) = \frac{1}{2\pi}\int_0^{2\pi} U(t)\ Re\ \frac{t+z}{t-z}\ ds(\theta) \tag{4.14}$$

$$= Re\ \frac{z^{|K|}}{\pi}\int_0^{2\pi}\frac{U(t)ds(\theta)}{t^{|K|-1}(t-z)}\quad \text{for}\ \ |z| < 1\ .$$

From the above formula and the second formula in (4.12), similarly to (3.19) of Chapter 1, we see that there are points $a_j^* \in L_0(j=N+1,\ldots,N-2K)$, so that

$$[\overline{\Lambda(\zeta)}]^{-1}\ \Psi(\zeta)\zeta'(s)\Big|_{\zeta=a_j^*} = 0\ , \tag{4.15}$$

i.e. $\Psi(a_j^*) = 0$, $j = N+1,\ldots,N-2K$.

Thus, we get the absurd inequality

$$2N - 2K \leq 2N_G + N_L = 2(N-K-1) = 2N - 2K - 2\ . \tag{4.16}$$

This contradiction proves that

$$h_j = 0\ (j=0,1,\ldots,N)\ ,\ H_m^{\pm} = 0\ (m=1,\ldots,-K-1) \tag{4.17}$$

in (4.11), and so $\Phi(\zeta) = 0$. Consequently, $w(z) = 0$ in D . Thus $w_1(z) = w_2(z)$ in D .

For $0 \leq K < N$, we discuss the boundary value problem B with the boundary conditions

$$Re[\overline{\Lambda^*(\zeta)\Psi(\zeta)}] = h_j \ , \ \zeta \in L_j \ , \ j = 1,\dots,N-K \ ,$$
$$Im[\overline{\Lambda^*(\tilde{a}_j)\Psi(\tilde{a}_j)}] = 0 \ , \ j = N-K+1,\dots,N+1 \ , \qquad (4.18)$$

where $\Lambda^*(\zeta) = \Lambda(\zeta)\bar{\zeta}^{-K-N}$, $\Psi(\zeta) = \zeta^{N-K}\Phi(\zeta)$, $\frac{1}{2\pi}\Delta_L \, arg\, \Lambda^*(\zeta) = N$. Suppose that $\Phi(\zeta) \not\equiv 0$, for $\zeta \in G$. By (1.11) in Chapter 1 we obtain the absurd inequality

$$2(N-K) + 2(K+1) = 2(N+1) \leq 2N_G + N_L = 2N \ , \qquad (4.19)$$

where N_G and N_L are the total number of zero points of Ψ in G and on L, respectively. This contradiction proves that $w_1(z) = w_2(z)$ for $z \in D$. □

2. A priori estimates of solutions for the Riemann-Hilbert problem

Theorem 4.2 Suppose that (1.45) satisfies Condition C. Then the solution $w(z)$ of Problem B for (1.45) satisfies the estimates

$$C_{\alpha_1}[w,\bar{D}] \leq M_1 \ , \ L_{p_0}[|w_{\bar{z}}|+|w_z|,D_m] \leq M_2 \ , \qquad (4.20)$$

where $\alpha_1 = \alpha\beta^2$, $\beta = 1-2/p_0$, $2 < p_0 < p < \infty$, $D_m = \{z | z \in D, d(z,\Gamma) \geq 1/m\}$, $d(z,\Gamma)$ denotes the distance between z and Γ, m is a positive integer, $M_1 = M_1(q_0,p_0,k_0,\alpha,\ell,K,D)$, $M_2 = M_2(q_0,p_0,k_0,\alpha,\ell,K,D,m)$.

Proof. Let $w(z)$ be a solution of Problem B for (1.45). By Theorem 2.4, $w(z)$ can be expressed as

$$w(z) = \Phi[\zeta(z)]e^{\phi(z)} + \psi(z) \ , \qquad (4.21)$$

where $\phi(z)$, $\psi(z)$, $\zeta(z)$ and its inverse function $z(\zeta)$ satisfy the estimates (2.18)-(2.20), and $\Phi(\zeta)$ is an analytic function in the circular domain $G = \zeta(D)$, satisfying the boundary condition

$$Re[\overline{\Lambda(\zeta)}\Phi(\zeta)] = R(\zeta) + h[z(\zeta)] \ , \ \zeta \in L = \zeta(\Gamma) \ , \tag{4.22}$$

$$Im[\overline{\Lambda(\tilde{a}_j)}\Phi(\tilde{a}_j)] = B_j \ , \ j \in \{j\} = \begin{cases} 1,\ldots,2K-N+1, K\geq N \ , \\ N-K+1,\ldots,N+1, 0\leq K<N \ , \end{cases} \tag{4.23}$$

in which

$$\overline{\Lambda(\zeta)} = \overline{\lambda[z(\zeta)]}e^{\phi[z(\zeta)]} \ , \ R(\zeta) = r[z(\zeta)] - Re\{\overline{\lambda[z(\zeta)]}\psi[z(\zeta)]\} \ ,$$

$$\tilde{a}_j = \zeta(a_j) \ , \ B_j = b_j - Im[\overline{\lambda(a_j)}\psi(a_j)] \ , \ j \in \{j\} \ ,$$

and $\Lambda(\zeta)$, $R(\zeta)$ and B_j satisfy the conditions

$$C_{\alpha\beta}[\Lambda,L] \leq M_3 \ , \ C_{\alpha\beta}[R,L] \leq M_4 \ , \ |B_j| \leq M_5 \ , \ j \in \{j\} \ , \tag{4.24}$$

where $M_j = M_j(q_0,p_0,k_0,\alpha,\ell,K,D)$, $j = 3,4,5$.

On the basis of Theorem 1.4 in Chapter 1, it is not difficult to see that $\Phi(\zeta)$ satisfies the following estimates

$$C_{\alpha\beta}[\Phi,\bar{G}] \leq M_6 \ , \ C[\Phi',G_m] \leq M_7 \ , \tag{4.25}$$

where

$$G_m = \zeta(D_m) \ , \ M_6 = M_6(q_0,p_0,k_0,\alpha,\ell,K,D) \ , \ M_7 = M_7(q_0,p_0,k_0,\alpha,\ell,K,D,m) \ .$$

From (2.18)-(2.20), (4.21), and (4.25), (4.20) follows. □

Assuming $1/2 < \alpha < 1$, in the following we will derive a stronger estimate of the solution $w(z)$ to the above problem B.

Theorem 4.3 Let the nonlinear complex equation (1.45) or (1.47) satisfy Condition C and $1/2 < \alpha < 1$. Then the solution $w(z)$ of Problem B for (1.45) or (1.47) satisfies

$$C_\beta[w,\bar{D}] \le M_8 \ , \ L_{p_0}[|w_{\bar{z}}|+|w_z|,\bar{D}] \le M_9 \ , \tag{4.26}$$

where $2 < p_0 < p$, $M_j = M_j(q_0,p_0,\alpha,\ell,K,D)$, $j = 8,9$.

<u>Proof.</u> Here we need standardized boundary conditions corresponding to Problem B. In the case of analytic functions the standardized boundary condition is the boundary condition of Problem C, which is discussed in § 1 of Chapter 1. By using a method similar to that used there, we find an analytic function $S(z)$ in D , which satisfies the boundary condition of the modified Dirichlet problem, as stated in (1.7) of Chapter 1. Let

$$W(z) = e^{-iS(z)}\Pi(z)w(z) = w(z)/Y(z) \ , \ z \in D \ . \tag{4.27}$$

Then the nonlinear complex equation (1.45) and the boundary conditions (4.2), (4.3) can be transformed into

$$W_{\bar{z}} = F(z,Y(z)W(z),Y'(z)W(z)+Y(z)W_z)/Y(z) \ , \tag{4.28}$$

$$\begin{aligned} &Re[\overline{\Lambda(t)}W(t)] = R(t) + H(t) \ , \ t \in \Gamma \ , \\ &Im[\overline{\Lambda(a_j)}W(a_j)] = B_j \ , \ j \in \{j\} \ , \end{aligned} \tag{4.29}$$

where $\Lambda(t)$, $R(t)$, $H(t)$ and $B_j(j \in \{j\})$ are the same as those in (3.6) and (3.7) of Chapter 1. Taking the integral expression for $S(z)$ (see (3.25) in Chapter 1) and $1/2 < \alpha < 1$ into account, we can derive that $Y(z)$ satisfies the estimate

$$C_\alpha[Y,\bar{D}] \le M_{10} \ , \ L_{p_0}[Y',\bar{D}] \le M_{11} \ , \tag{4.30}$$

where

$$2 < p_0 < min(\frac{1}{1-\alpha}, p) \ , \ M_j = M_j(p_0,\alpha,\ell,K,D) \ , \ j = 10,11 \ .$$

Hence the nonlinear complex equation (4.28) satisfies Condition C, where the constants p, k_0 and ℓ are replaced by other constants

p_0, k_0' and ℓ', respectively, where $k_0' = k_0'(q_0, p_0, k_0, \alpha, \ell, K, D)$, $\ell' = \ell'(q_0, p_0, k_0, \alpha, \ell, K, D)$. Using (2.18)-(2.20), (4.21), (4.25), and (4.30), we can obtain the estimate

$$L_{p_0}[|w_{\bar{z}}|+|w_z|, D_m] \le M_{12} = M_{12}(q_0, p_0, k_0, \alpha, \ell, K, D, m) , \tag{4.31}$$

where D_m is stated in connection with (4.20). In the following, we will estimate the L_{p_0}-norm of $|w_{\bar{z}}| + |w_z|$ in the neighbourhood of Γ. To this end we introduce a continuously differentiable function $\tau_0(z)$ in $\tilde{\mathbb{D}}_0 = \{|z| \le 1\}$ i.e.

$$\begin{aligned} &\tau_0(z) = 0 , \text{ for } |z| < 1-d , \\ &\tau_0(z) = 1 , \text{ for } z \in D_0 = \{1-\frac{d}{2} \le |z| \le 1\} , \\ &0 \le \tau_0(z) \le 1 , \text{ for } 1-d \le |z| \le 1-\frac{d}{2} \end{aligned} \tag{4.32}$$

and continuously differentiable functions $\tau_j(z)$ in $\tilde{D}_j = \{|z-z_j| \ge \gamma_j\}$ $(j=1,\dots,N)$

$$\left.\begin{aligned} &\tau_j(z) = 0 , \text{ for } |z-z_j| \ge \gamma_j + d , \\ &\tau_j(z) = 1 , \text{ for } z \in D_j = \{\gamma_j \le |z-z_j| \le \gamma_j + \frac{d}{2}\} , \\ &0 \le \tau_j(z) \le 1 , \text{ for } \gamma_j + \frac{d}{2} \le |z-z_j| \le \gamma_j + d , \end{aligned}\right\} j = 1,\dots,N . \tag{4.33}$$

Multiply (4.28) by $\tau_0(z)$ and denote $W_0(z) = \tau_0(z)W(z)$. Then $W_0(z)$ is a solution of the complex equation

$$\left.\begin{aligned} &W_{0\bar{z}} = Q(z,w,w_z)W_{0z} + F_0(z,W,W_z) , \\ &F_0(z,W,W_z) = \tau_{0\bar{z}}W - Q\tau_{0z}W + \tau_0(z)F(z,w,0)/Y(z) \end{aligned}\right\} \tag{4.34}$$

in $\tilde{\mathbb{D}}_0$, and satisfies the boundary condition

$$Re[\overline{\Lambda(t)}W_0(t)] = R(t) + H(t) , \ t \in \Gamma_0 = \{|t|=1\} . \tag{4.35}$$

From (4.26) the estimate

$$L_{p_0}[F_0(z,W,W_z),\overline{\tilde{\mathbb{D}}_0}] \le M_{13}(q_0,p_0,k_0,\alpha,\ell,K,D) \ , \tag{4.36}$$

is easily seen. Because $W_0(z) = 0$, for $|z| < 1-d$ the function $W_*(z) = z^{-K}W_0(z)$ is a solution of Problem D for the complex equation

$$W_{*\bar{z}} = Q(z,w,w_z)W_{*z} + Kz^{-1}QW_* + z^{-K}F_0(z,w,w_z) \tag{4.37}$$

in $\tilde{\mathbb{D}}_0$, and $Im\, W_*(a_j) = B_0'$, $W_*(z) = W_0(z) = 0$ for $|z| < 1-d$. From Lemma 1.3 and Theorem 3.6 of Chapter 1, the function $W_*(z)$ can be expressed as

$$W_*(z) = \Phi(z) + P_0\omega - P_0\omega\big|_{z=a_0} \ , \tag{4.38}$$

where

$$\Phi(z) = \frac{1}{2\pi i}\int_{\Gamma_0} [R(t)+H(t)]\,\frac{t+z}{t-z}\,\frac{dt}{t} + iB_0' \ ,$$

$$P_0\omega = -\frac{1}{\pi}\iint_{\tilde{\mathbb{D}}_0} \Big[\frac{\omega(\zeta)}{\zeta-z} + \frac{z\omega(\zeta)}{1-\bar{\zeta}z}\Big]d\sigma_\zeta \ ,$$

B_0' is a real constant, and the analytic function $\Phi(z)$ satisfies the estimate (1.16) of Chapter 1, and according to the uniqueness of the solution of Problem D for (4.37) (see Theorem 4.1 and [103]25), $W_*(z)$ satisfies the estimate

$$L_{p_0}[|W_{*\bar{z}}|+|W_{*z}|,\overline{\tilde{\mathbb{D}}_0}] \le M_{14} = M_{14}(q_0,p_0,k_0,\alpha,\ell,K,D) \ . \tag{4.39}$$

From $W_*(z) = z^{-K}W_0(z) = z^{-K}W(z)$ in D_0 , it follows that $w(z) = Y(z)W(z)$ satisfies the estimate

$$L_{p_0}[|w_{\bar{z}}|+|w_z|,\bar{D}_0] \le M_{15} = M_{15}(q_0,p_0,k_0,\alpha,\ell,K,D) \ . \tag{4.40}$$

Next, multiply (4.28) by $\tau_j(z)$ and denote $W_j(z) = \tau_j(z)W(z)$, $j = 1,\ldots,N$. Then $W_j(z)$ is a solution of the following complex equation

$$
\begin{aligned}
&W_{j\bar{z}} = Q(z,w,w_z)W_{jz} + F_j(z,W,W_z) , \\
&F_j(z,W,W_z) = \tau_{j\bar{z}}W - Q\tau_{jz}\overline{W} + \tau_j F(z,w,0)/Y(z)
\end{aligned}
\tag{4.41}
$$

in $\tilde{D}_j$, and satisfies the boundary condition

$$
Re[\overline{\Lambda(t)}W_j(t)] = R(t) + H(t) , \quad t \in \Gamma_j \tag{4.42}
$$

and $W_j(z) = 0$, for $|z-z_j| > \gamma_j + d$, $j = 1,\ldots,N$.
Denoting by $z = z_j(\zeta)$ the inverse function of $\zeta = \zeta_j(z) = \gamma_j/(z-z_j)$ the function $W_j^*(\zeta) = W_j[z_j(\zeta)]$ is a solution of the complex equation

$$
W^*_{j\bar{\zeta}} = Q(z,w,w_z)\,\frac{\overline{z'(\zeta)}}{z'(\zeta)}\,W_{j\zeta} + \overline{z'(\zeta)}\,F_j(z,W,W_z) \tag{4.43}
$$

in $\tilde{\mathbb{D}}_0$, and satisfies the boundary condition

$$
Re[\overline{\Lambda[z_j(\zeta)]}W_j^*(\zeta)] = R[z_j(\zeta)]+H[z_j(\zeta)] , \quad \text{for} \quad |\zeta| = 1 , \tag{4.44}
$$

$j = 1,\ldots,N$. As before, we can derive the estimates

$$
L_{p_0}[|W^*_{j\bar{\zeta}}|+|W^*_{j\zeta}|,\overline{\tilde{\mathbb{D}}_0}] \le M_{16} , \quad L_{p_0}[|w_{\bar{z}}|+|w_z|,\overline{D_j}] \le M_{17} , \quad j = 1,\ldots,N , \tag{4.45}
$$

where $M_j = M_j(q_0,p_0,k_0,\alpha,\ell,K,D)$, $j = 16,17$. Combining (4.31), (4.40) and (4.45), we easily see that the second formula of (4.26) is true.

Finally, let $\omega(z) = w_{\bar{z}}$. The function

$$
T\omega = -\frac{1}{\pi}\iint_D \frac{\omega(\zeta)}{\zeta-z}\,d\sigma_\zeta
$$

satisfies the estimate

$$C_\beta[T\omega,\bar{D}] \leq M_{18} L_{p_0}[\omega,\bar{D}] \ , \ M_{18} = M_{18}(p_0,D) \ . \tag{4.46}$$

The solution $w(z)$ can be expressed as

$$w(z) = \Phi(z) + T\omega = \Phi(z) + \Psi(z) \tag{4.47}$$

(see [97]2), where $\Phi(z)$ is an analytic function satisfying the boundary condition

$$Re[\overline{\Lambda(t)}\Phi(t)] = r(t) - Re[\overline{\lambda(t)}T\omega] + h(t) \ , \ t \in \Gamma \ . \tag{4.48}$$

Applying Theorem 1.4 of Chapter 1, the function $\Phi(z)$ satisfies the estimate

$$C_\beta[\Phi,\bar{D}] \leq M_{19} = M_{19}(q_0,p_0,k_0,\alpha,\ell,K,D) \ . \tag{4.49}$$

From (4.46) and (4.49) the first estimate in (4.26) follows. □

By the above theorem, we immediately obtain the following result.

Corollary 4.4 Under the hypothesis of Theorem 4.3, the solution $w(z)$ of Problem B for (1.45) or (1.47) can be expressed in the form (4.47), where $\Phi(z)$, $\Psi(z)$ satisfy the estimates

$$C_\beta[\Phi,\bar{D}] \leq M_{20} \ , \ C_\beta[\Psi,\bar{D}] \leq M_{20} \ , \tag{4.50}$$

$$L_{p_0}[\Phi',\bar{D}] \leq M_{21} \ , \ L_{p_0}[|\Psi_{\bar{z}}|+|\Psi_z|,\bar{D}] \leq M_{21} \ , \tag{4.51}$$

where $M_j = M_j(q_0,p_0,k_0,\alpha,\ell,K,D)$, $j = 20,21$.

3. Solvability of the Riemann-Hilbert problem for nonlinear complex equations

To apply the Leray-Schauder theorem, we first discuss the nonlinear complex equation

$$w_{\bar{z}} = F_m(z,w,w_z) \ , \ F_m(z,w,w_z) = \sigma_m(z)F(z,w,w_z) \ , \tag{4.52}$$

where $\sigma_m(z)$ is defined as in (2.34).

Lemma 4.5 Under the hypothesis of Theorem 4.2 the solution $w(z)$ of Problem B for (4.52) can be expressed in the form (4.47), where $\Phi(z)$ and $\Psi(z)$ satisfy the estimates

$$C_{\alpha_1}[\Phi,\bar{D}] \le M_{22} \ , \ L_{p_0}[\Phi',D_m] \le M_{23} \ , \ C_\beta[\Psi,\bar{D}] \le M_{24} \ , \tag{4.53}$$

$$L_{p_0}[\omega,\bar{D}] = L_{p_0}[\omega,D_m] < M_{25} \ , \ L_{p_0}[\Pi\omega,\bar{D}] < M_{26} \ , \tag{4.54}$$

where $\alpha_1 = \alpha\beta^2, \Pi\omega = -\frac{1}{\pi}\iint_D \frac{\omega(\zeta)}{(\zeta-z)^2} d\sigma_\zeta \ , \ M_j = M_j(q_0,p_0,k_0,\alpha,\ell,K,D,m)$, $j = 22,\ldots,26$.

Proof. Obviously, the expression (4.47) for the solution $w(z)$ to (4.52) remains valid. From the second estimate in (4.20), the estimates in (4.54) follow, and so the third estimate in (4.53) is true. By the first estimate in (4.20) and the third estimate in (4.53), we obtain the first and second estimates in (4.53). □

Theorem 4.6 Let the nonlinear complex equation (1.45) or (1.47) satisfy Condition C. Then Problem B for (4.52) has a solution.

Proof. According to the representation (4.47), we only need to find a solution $\Psi(z) = T\omega$ $(\omega(z) \in L_{p_0}(D_m))$ of

$$\Psi_{\bar{z}} = F_m[z,\Phi(z)+\Psi(z),\Phi'(z)+\Psi_z] \ , \tag{4.55}$$

so that $w(z) = \Phi(z) + \Psi(z)$ satisfies the boundary conditions (4.2)-(4.4). It is equivalent to finding a solution $\omega(z) \in L_{p_0}(D_m)$ for the integral equation

$$\omega(z) = F_m(z,w(z),\Phi'(z)+\Pi\omega) \ , \tag{4.56}$$

so that $w(z) = \Phi(z) + T\omega$ satisfies (4.2)-(4.4) on Γ .

Thus, we introduce the Banach space $B = L_{p_0}(D_m)$ and denote by B_M the set of measurable functions $\omega(z)$ satisfying the inequality

$$L_{p_0}[\omega, D_m] < M_{25} , \tag{4.57}$$

where M_{25} is the constant in (4.54). It is evident that B_M is a bounded open set in B . We are free to choose $\omega(z) \in B_M$ and set

$$\Psi(z) = -\frac{1}{\pi}\iint_{D_m} \frac{\omega(\zeta)}{\zeta - z}\, d\sigma_\zeta .$$

Afterwards we find an analytic function $\Phi(z)$ in D so that $w(z) = \Phi(z) + \Psi(z)$ satisfies the boundary conditions (2.2)-(2.4). Substitute $w(z)$, $\Phi'(z)$ into the integral equation with the parameter k

$$\Omega(z) = k\, F_m(z, w(z), \Phi'(z) + \Pi\Omega) , \quad 0 \le k \le 1 , \tag{4.58}$$

where

$$\Pi\Omega = -\frac{1}{\pi}\iint_{D} \frac{\Omega(\zeta)}{(\zeta - z)^2}\, d\sigma_\zeta .$$

By Condition C and the principle of contraction, the integral equation (4.58) has unique solution. Denote by $\Omega = S(\omega,k)\,(0 \le k \le 1)$ this mapping from $\omega(z)$ onto $\Omega(z)$. Moreover, if $\omega(z)$ is a solution of the integral equation

$$\omega(z) = k\, F_m(z, \Phi(z) + T\omega, \Phi'(z) + \Pi\omega) , \quad 0 \le k \le 1 , \tag{4.59}$$

then by the first estimate of (4.54), $\omega(z)$ satisfies the inequality (4.57), i.e. $\omega(z) \in B_M$. Set $B_0 = B_M \times [0,1]$. Now we verify that the mapping $\Omega = S(\omega,k)$ satisfies the three conditions of the Leray-Schauder theorem:

1. For every $k \in [0,1]$, $\Omega = S(\omega,k)$ continuously maps the Banach space B into itself, and is completely continuous on B_M . Besides, for $\omega(z) \in \overline{B_M}$, $S(\omega,k)$ is uniformly continuous with respect to $k \in [0,1]$.

In fact, let us arbitrarily select $\omega_n(z) \in \overline{B_M}$, $n = 1,2,\ldots$. For $\Omega_n = S(\omega_n,k)$ and $\Omega_\ell = S(\omega_\ell,k)$, we have

$$\Omega_n - \Omega_\ell = k[F_m(z,w_n,\Phi_n'+\Pi\Omega_n)-F_m(z,w_n,\Phi_n'+\Pi\Omega_\ell)+c_{n\ell}] , \tag{4.60}$$

where

$$w_j(z) = \Phi_j(z)+T\omega_j, j = n,\ell, c_{n\ell} = F_m(z,w_n,\Phi_n'+\Pi\Omega_\ell) - F_m(z,w_\ell,\Phi_\ell'+\Pi\Omega_\ell) .$$

In view of (4.53), we can choose subsequences of $\{w_n(z)\}$ and $\{\Phi_n'(z)\}$ which uniformly converge to $\omega_0(z)$ and $\Phi_0'(z)$ on D_m , respectively. Relabeling these subsequences as $\{w_n(z)\}$, $\{\Phi_n'(z)\}$ and using the same method as in (2.43), we can prove $L_{p_0}[c_{n\ell},\overline{D}] = L_{p_0}[c_{n\ell},D_m] \to 0$ for $n,\ell \to \infty$. Due to

$$L_{p_0}[\Omega_n-\Omega_\ell,D_m] \le L_{p_0}[c_{n\ell},D_m]/(1-q_0\lambda_{p_0}) ,$$

it is easy to show that $L_{p_0}[\Omega_n-\Omega_\ell,D_m] \to 0$ as $n,\ell \to \infty$. Because of the completeness of the Banach space B , there exists a measurable function $\omega_0(z) \in B$, so that $L_{p_0}[\omega_n-\omega_0,D_m] \to 0$ as $n \to \infty$. This shows the complete continuity of $\Omega = S(\omega,k)(0 \le k \le 1)$ on $\overline{B_M}$. By a similar method, we can prove also that $\Omega = S(\omega,k)(0 \le k \le 1)$ continuously maps $\overline{B_M}$ into B , and $S(\omega,k)$ is uniformly continuous with respect to $k \in [0,1]$ for $\omega(z) \in \overline{B_M}$.

2. For $k = 0$ from (4.59) and (4.47) it is evident that $\Omega = S(\omega,0) = \Phi(z) \in B_M$.

3. From (4.54) we see that $\omega = S(\omega,k)(0 \le k \le 1)$ does not have a solution $w(z)$ on the boundary $\partial B_M = \overline{B_M} \setminus B_M$.

Hence, by the Leray-Schauder theorem (cf. [64]), we know that the integral equation (4.59) with $k = 1$ (i.e. (4.56)) has a solution $\omega(z) \in B_M$. So $\Psi(z) = T\omega$ is a solution of (4.55). □

Theorem 4.7 Under the hypothesis of Theorem 4.2, Problem B for (1.45) or (1.47) is solvable.

Proof. On the basis of Theorem 4.2 and Theorem 4.6, it is obvious that Problem B for the complex equations (4.52) has a solution $w_m(z) = \Phi_m(z) + T\omega$ and this solution satisfies the estimate (4.20) $(m=1,2,\dots)$. Hence, we can choose a subsequence of $\{w_m(z)\}$, which uniformly converge to $w_0(z)$ satisfying (4.20). By Theorem 2.6 $w_0(z)$ is a solution of (1.45) or (1.47) and is therefore a solution of Problem B for (1.45) or (1.47). □

Afterwards, according to the method in the proof of Theorem 1.9 in Chapter 1, we can derive the following result about the solvability of Problem A for (1.45) or (1.47).

Theorem 4.8 Let (1.45) or (1.47) satisfy Condition C. Problem A for (1.45) or (1.47) possesses the following results about solvability:

(1) If the index $K \geq N$, Problem A is solvable.

(2) If $0 \leq K < N$, the total number of solvability conditions for Problem A does not exceed $n - K$.

(3) If $K < 0$, Problem A has $N-2k-1$ solvability conditions (cf. [103]3)).

4. The Riemann-Hilbert problem for linear equations

Finally, we discuss the linaer case of the complex equation (1.45), i.e.

$$w_{\bar{z}} = F(z,w,w_z), \tag{4.61}$$
$$F(z,w,w_z) = Q_1(z)w_z + Q_2(z)\bar{w}_{\bar{z}} + A_1(z)w + A_2(z)\bar{w} + A_3(z),$$

and assume that (4.61) satisfies Condition C, i.e.

$$|Q_1(z)|+|Q_2(z)| \le q_0 < 1 , \quad L_p[A_j,\bar{D}] \le k_0 , \quad j = 1,2,3, \tag{4.62}$$

where $q_0, p(2 < p < \infty)$, k_0 are nonnegative constants.

Theorem 4.9 Let the linear complex equation (4.61) satisfy Condition C.

(1) If $K \ge N$, the general solution of Problem A for (4.61) can be expressed as

$$w(z) = w_0(z) + \sum_{j=1}^{J} c_j w_j(z) , \tag{4.63}$$

where $J = 2K - N+1$, $w_0(z)$ is a special solution to Problem A for (4.61), and $w_j(z)(j=1,\dots,2K-N+1)$ are linear independent solutions to the homogeneous Problem A_0.

(2) If $0 \le K < N$, Problem A for (4.61) has $N - K - S$ solvability conditions $(0 \le S \le \min(N-K,K+1))$. If these conditions are satisfied, then the general solution $w(z)$ can be expressed by (4.63) with $J = K+1$.

Proof. (1) On account of Theorem 4.7, Problem A for (4.61) has a special solution $w_0(z)$ with conditions

$$Im[\overline{\lambda(a_j)}w(a_j)] = 0 , \quad j = 1,\dots,N - 2K - 1 , \tag{4.64}$$

and Problem A for the homogeneous complex equation

$$w_{\bar{z}} = Q_1(z)w_z + Q_2(z)\bar{w}_{\bar{z}} + A_1(z)w + A_2(z)\bar{w} \tag{4.65}$$

has $2K - N+1$ linear independent solutions $w_k(z)(k=1,\dots,2K-N+1)$ with the conditions

$$Im[\overline{\lambda(a_j)}w_k(a_j)] = \delta_{jk} = \begin{cases} 1, j = k , \\ 0, j \ne k , \end{cases} \quad j,k = 1,\dots,2K - N+1 .$$

Let $w(z)$ be any solution of Problem A for (4.61). Then $W(z) = w(z) - w_0(z)$ is a solution of (4.65). Set

$$c_j = Im[\overline{\lambda(a_j)}W(a_j)] \ , \ j = 1,\ldots,2K - N+1 \ . \tag{4.66}$$

By the uniqueness of the solution of Problem B for (4.65), we have

$$W(z) = \sum_{j=1}^{2K-N+1} c_j w_j(z) \ . \tag{4.67}$$

Hence, $w(z)$ can be expressed as (4.63). Conversely, selecting arbitrary real constants $c_1,\ldots,c_{2K-N+1}$, it is easy to see that

$$w(z) = w_0(z) + \sum_{j=1}^{2K-N+1} c_j w_j(z)$$

is a solution of Problem A for (4.61). Thus we have (4.63).

(2) We denote by Problem B* the boundary value problem for (4.61) and (4.2). Similarly to (1), the general solution $w(z)$ can be expressed as (4.63) with $J = K + 1$. Set

$$h_k' = r(t) - Re[\overline{\lambda(t)}w_0(t)] \ , \ t \in \Gamma_k \ , \ k = 1,\ldots,N-K \ , \tag{4.68}$$

and denote by S the rank of the coefficient matrix for the system of albegraic equations

$$\sum_{j=1}^{K+1} c_j \ Re[\overline{\lambda(t)}w_j(t)] = h_k' = r(t) - Re[\overline{\lambda(t)}w_0(t)] \ , \ t \in \Gamma_k \ , \quad k = 1,\ldots,N-K \ . \tag{4.69}$$

Thus, from (4.69) we can determine S constants out of the $K + 1$ arbitrary constants $c_1,\ldots,c_{K+1}$ and there are S identities in (4.69) linearly independent. Substituting

$$w(z) = w_0(z) + \sum_{j=1}^{K+1} c_j w_j(z)$$

into (4.2) we obtain

$$Re[\overline{\lambda(t)}w(t)] = r(t) + h_k \, , \, t \in \Gamma_k \, , \, k = 1,\ldots,N\text{-}K \, , \tag{4.70}$$

where S constants in $\{h_1,\ldots,h_{N-K}\}$ are equal to zero. If the other constants are equal to zero, too, then the solution $w(z)$ of Problem B* is a solution of Problem A. This completes the proof. □

§ 5. Applications of boundary value problems to nonlinear quasiconformal mappings

It is not difficult to show that the complex equation (1.45) or (1.47) may not possess homeomorphic solutions. For instance, the special complex equation

$$w_{\bar z} = A_1(z)w \, , \; A_1(z) = 16(z-2z^2\bar z) \, , \; z \in \mathbb{D} = \{|z|<1\} \, , \tag{5.1}$$

has no homeomorphic solution $w(z)$ so that $w(z)$ maps $\overline{\mathbb{D}}$ onto itself and $w(0) = 0$, $w(1) = 1$. Otherwise, let $w(z)$ be a homeomorphic solution which maps $\overline{\mathbb{D}}$ onto itself with $w(0) = 0$, $w(1) = 1$. It is clear that $w(z)$ can be expressed in the form

$$w(z) = \Phi(z) \, e^{16(z\bar z - z^2\bar z^2)} \, , \; z \in \mathbb{D} \, , \tag{5.2}$$

where $\Phi(z)$ is an analytic function in $\mathbb{D}$, $\Phi(0) = 0$, $\Phi(1) = 1$ and $|\Phi(z)| \to 1$ for $|z| \to 1$. Thus $\Phi(z) = z$ and $w(1/2) = 1/2\, e^3 > 1$. From the schlichtness property of $w(z)$ in $\mathbb{D}$ this is impossible.

We will consider the nonlinear complex equation

$$\begin{aligned} & w_{\bar z} = Q_1(z,w,w_z)w_z + Q_2(z,w,w_z)\bar w_{\bar z} \\ & \text{or} \quad w_{\bar z} = Q(z,w,w_z)w_z \end{aligned} \tag{5.3}$$

and call a homeomorphic solution of (5.3) a nonlinear quasiconformal mapping. In the particular case, $Q_j = Q_j(z)$ or $Q_j = Q_j(z,w)(j=1,2)$, the existence of quasiconformal mappings for (5.3), which map a simply connected domain or a multiply connected domain onto a canonical domain were studied by L. Bers and L. Nirenberg [12]1), B. Bojarski [15]1), S. Pater [82] and Li Zhong [66]1). In the case of a simply connected domain, M.A. Lavrent'ev [60]3), B. Bojarski and

T. Iwaniec [16] proved the fundamental theorem of the existence of a quasiconformal mapping for (5.3). The existence of such a quasiconformal mapping in the case of a multiply connected domain was proved independently by Fang Ai-nong [32], Wen Guo-chun [103]2), V.N. Monakhov [75]2), and T. Iwaniec [56]2) using different methods.

To prove the existence of quasiconformal mappings for (5.3), we will use the result on solvability for Problem D for (5.3).

1. Existence theorems for nonlinear quasiconformal mappings

We first give some estimates of the quasiconformal mapping from D onto G for (5.3).

Theorem 5.1 Let (5.3) satisfy Condition C and $w(z)$ be a quasiconformal mapping which maps an (N+1)-connected circular domain D onto a rectilinear slit domain G and satisfies the conditions: $w(a) = 0$, $a \in D$, $w(0) = \infty$ and $|w(1)| = 1$. Then $w(z)$ satisfies the estimates:

$$C_\beta[w\zeta,\bar{D}] \le M_1 , \quad L_{p_0}[|w_{\bar{z}}|+|w_z|,D_*] \le M_2 , \tag{5.4}$$

where $\zeta(z)$ is a homeomorphic solution of the Betrami equation $\zeta_{\bar{z}} = q(z)\zeta_z$ $(|q(z)| \le q_0 < 1)$ which maps D onto a circular domain H similar to D and $\zeta(0) = 0$, $\zeta(1) = 1$, $D_* = \bar{D}\setminus\{|z| \le \varepsilon\}$, ε is a sufficiently small positive number, $M_1 = M_1(q_0,p_0,D)$, $M_2 = M_2(q_0,p_0,D,\varepsilon)$.

Remark A so-called rectilinear slit domain G means a domain whose boundary consits of N+1 rectilinear slits $L_j(j=0,1,\ldots,N)$ with oblique angles $\theta_j(j=0,1,\ldots,N)$, respectively.

Proof. We substitute $w(z)$ into (5.3), and denote $q(z) = Q[z,w(z),w_z]$. Obviously, $|q(z)| \le q_0 < 1$. From Theorem 2.2, $w(z)$ admits the representation

$$w(z) = \Phi[\zeta(z)] , \tag{5.5}$$

where $\zeta(z)$ is as stated above, $\Phi(\zeta)$ is an analytic function in $\zeta(D)\setminus\{0\}$ with a pole of order 1 at $\zeta = 0$ and satisfies the boundary condition

$$Re\left\{[e^{-i(\theta_j+\frac{\pi}{2})}\zeta^{-1}][\zeta\Phi(\zeta)]\right\} = h_j \ , \ j = 0,1,\ldots,N \tag{5.6}$$

and $\Phi[\zeta(a)] = 0$, $|\Phi(1)| = 1$. According to the proof of Theorem 1.4 in Chapter 1 and using the reductio ad absurdum, we can obtain

$$|h_j| \le M_3 = M_3(q_0,p_0,D,\varepsilon) \ , \ j = 0,1,\ldots,N \ . \tag{5.7}$$

Thus (5.4) is true. □

<u>Theorem 5.2</u> Let (5.3) satisfy Condition C. Then there exists a quasiconformal mapping $w(z)$ for (5.3) which maps the (N+1)-connected domain D onto a rectilinear slit domain G as stated in Theorem 5.1.

<u>Proof.</u> We first assume that the coefficient $Q(z,w,w_z)$ of (5.3) vanishes indentically in $|z| \le \varepsilon$, where $\varepsilon_n \to 0$, when $n \to \infty$, and write this equation in the form

$$w_{\bar{z}} = F_n(z,w,w_z) \ , \ F_n(z,w,w_z) = Q_n(z,w,w_z)w_z \ . \tag{5.8}$$

Similarly to the proof of Theorem 4.6, we can find a solution

$$w_n(z) = \Phi_n(z) + \Psi_n(z) = \Phi_n(z) + T\omega_n \tag{5.9}$$

satisfying the boundary condition

$$Re\left[e^{-i(\theta_j+\frac{\pi}{2})}w_n(t)\right] = h_j \ , \ t \in \Gamma_j \ , \ j = 0,1,\ldots,N \ , \tag{5.10}$$

where $\omega_n(z) \in L_{p_0}(\bar{D})$, $2 < p_0 < p$, $\Phi_n(z)$ is an analytic function in $D\setminus\{0\}$ and has a pole of order 1 at $z = 0$, and $w_n(0) = \infty$, $w_n(a) = 0$, $|w_n(1)| = 1$. From Corollary 4.4, it is not difficult to see that $\Phi_n(z)$, $\Psi_n(z)$, $\omega_n(z)$ and w_{nz} satisfy the estimates

$$C_\beta[z\Phi_n(z),\bar{D}] \le M_4 \ , \ C_\beta[\Psi_n,\bar{D}] \le M_5 \ , \tag{5.11}$$

$$L_{p_0}[\omega_n,\bar{D}] \le M_6 \ , \ L_{p_0}[(zw_n)_z,\bar{D}] \le M_7 \ , \tag{5.12}$$

where $M_j = M_j(q_0,p_0,D,\varepsilon_n)$, $j = 4,\dots,7$. If we use the integral representation $\tilde{T}\omega$ for the solution of Problem D in § 3 of Chapter 1 and the schlicht conformal mapping $\tilde{\Phi}_n(z)$ from D onto G , then the function $w_n(z)$ can be expressed as

$$w_n(z) = C[\tilde{\Phi}_n(z)+e^{i\theta_0}\tilde{T}\omega] \tag{5.13}$$

where C is real constant so that $|w_n(1)| = 1$. Applying the principle of contraction and the Schauder fixed point theorem, we can prove that there exists a solution of the type (5.13) for (5.8). This method of proof is simpler and more convenient than the method used to prove Theorem 4.6.

Next, we verify that $w_n(z)$ quasiconformally maps D onto G . From (5.10), we easily see that $w_n(z)$ is continuous in D and maps Γ_j onto the rectilinear slit L_j with oblique angles $\theta_j (j=0,1,\dots,N)$. Choosing arbitrarily a complex number

$$w_0 \notin L = \bigcup_{j=0}^{N} L_j \ ,$$

and by the argument principle, we have

$$\frac{1}{2\pi}\Delta_L \, arg[w_n(z)-w_0] = 0 = N_D - P_D \ .$$

Because the total number of poles is equal to 1, i.e. $P_D = 1$, we have $N_D = 1$. Thus, there exists a point $z_0 \in D$, so that $w_n(z_0) = w_0$. Besides, using the reductio ad absurdum, we can prove that $w_n(D) \cap L = \emptyset$. This shows that $w_n(z)$ quasiconformally maps D onto G as stated in Theorem 5.1.

Finally, because the solution $w_n(z) = \Phi_n(z) + \Psi_n(z)$ for (5.8) satisfies the estimates (5.11) and (5.12) and possesses the representation

$$w_n(z) = \Phi_n[\zeta_n(z)] \ , \ \zeta_n(0) = 0 \ , \ \zeta_n(1) = 1 \ , \tag{5.14}$$

where $\zeta_n(z)$ and its inverse function $z_n(\zeta)$ satisfy the estimate (2.20), we can select subsequences of $\{w_n(z)\}$ and $\{\zeta_n(z)\}$ converging uniformly to $w_0(z)$ and $\zeta_0(z)$ on any closed subset in $\bar{D} \setminus \{0\}$, respectively, and select a subsequece of $\{\Phi_n(\zeta)\}$ converging uniformly to $\Phi_0(\zeta)$ on any closed subset in $\zeta_0(D) \setminus \{0\}$. We have to extend $w_n(z)$ symmetrically into the outer part of D. Therefore, $\Phi_n(\zeta)$ is defined in $\zeta_0(\bar{D}) \setminus \{0\}$. Since $|\Phi_0(1)| = 1$, $\Phi_0[\zeta_0(a)] = 0$, it can be seen that $\Phi_0(\zeta)$ is a schlicht analytic function in $\zeta_0(D) \setminus \{0\}$ and $\Phi_0(\zeta)$ has a pole of order 1 at $z = 0$. By Theorem 2.6, it is clear that $w_0(z) = \Phi_0[\zeta_0(z)]$ is a homeomorphic solution of (5.3), which maps D onto G as was stated above. □

Theorem 5.3 Under the hypothesis of Theorem 5.2, there exists a homeomorphic solution $w(z)$ of (5.3) which quasiconformally maps D onto a domain G in $|w| < 1$ with N circular slits, whose centres are $w = 0$.

Proof. Let $\varphi(z)$ denote a schlicht analytic function which conformally maps D onto G as stated above, and $\varphi(0) = 0$, $\varphi(1) = 1$. Similarly to Theorem 5.2, we can find a solution $W_n(z)$ to the complex equation

$$W_{\bar{z}} = F_n(z, \varphi e^W, \varphi' e^W + \varphi e^W W_z)/\varphi e^W \tag{5.15}$$

satisfying the boundary condition

$$Re[W_n(t)] = h_j \ , \ t \in \Gamma_j \ , \ j = 0,1,\ldots,N \ , \ W_n(1) = 0 \ , \tag{5.16}$$

where $h_j (j=0,1,\ldots,N)$ are constants and $h_0 = 0$. Obviously,

$$w_n(z) = \varphi(z) e^{W_n(z)} \tag{5.17}$$

is a solution of (5.8) in D and satisfies the boundary contition

$$Re[\ell n w_n(t)] = Re[\ln\varphi(t)+W_n(t)] = h_j \ , \ t \in \Gamma_j \ , \qquad (5.18)$$
$$j = 0,1,\ldots,N \ , \ w_n(1) = 1 \ .$$

Consequently, $w_n(z)$ maps Γ_j onto the circular arc $L_j(j=0,1,\ldots,N)$ and $L_0 = w_n(\Gamma_0) = \{|w|=1\}$. For any complex number

$$w_0 \notin L = \bigcup_{j=0}^{N} L_j$$

and $|w_0| < 1$, we have $\Delta_\Gamma \, arg[w_n(z)-w_0] = 2\pi$. Thus, there exists a point $z_0 \in D$, so that $w_n(z_0) = w_0$. Besides, we can prove $w_n(D) \subset \{|w|<1\}$ and $w_n(D) \cap L = \emptyset$. This shows that $w_n(z)$ quasiconformally maps D onto G as stated above. Because $w_n(z)$ satisfies the estimate

$$C_\beta[w_n,\bar{D}] \le M_8 = M_8(q_0,p_0,D,\varepsilon_n) \ , \qquad (5.19)$$

we may choose a subsequence of $\{w_n(z)\}$, which uniformly converges to a solution $w_0(z)$ of (5.3) in D , and can verify that $w_0(z)$ is the required homeomorphic solution. □

Theorem 5.4 Suppose that the complex equation (5.3) satisfies Condition C. Then there exists a homeomorphic solution $w(z)$ which quasiconformally maps D onto the spiral slit domain G , and $w(a) = 0$, $w(0) = \infty$ and $w(1) = 1$, where $a(\neq 0)$ is a point in D .

A so-called spiral slit domain G means a domain whose boundary consists of N+1 logarithmic spiral slits with oblique angles $\theta_j(j=0,1,\ldots,N)$, respectively.

Proof. In analogy to the proof of Theorem 5.3, we first denote by $\varphi(z)$ a schlicht meromorphic function which maps D onto G as stated above, and $\varphi(0) = \infty$, $\varphi(a) = 0$ $(a \in D$ and $a \neq 0)$, $\varphi(1) = 1$ and substitute $\varphi(z)$ into the corresponding position of (5.15).

Next, we find out a solution $W_n(z)$ of (5.15) satisfying the boundary condition

$$Re\left[e^{-i(\theta_j+\frac{\pi}{2})}W_n(t)\right] = h_j \,,\ t\in\Gamma_j \,,\ j=0,1,\ldots,N \,,\ W_n(1)=0 \,, \tag{5.20}$$

where $h_j(j=0,\ldots,N)$ are all constants and $h_0 = 0$, and can verify that

$$w_n(z) = \varphi(z)e^{W_n(z)} \qquad \text{is}$$

a homeomorphic solution of (5.3), which maps D onto the above spiral slit domain G. Finally, we can derive the estimate

$$C_\beta[w_n\zeta_n,\bar{D}] \le M_9 = M_9(q_0,p_0,D,\varepsilon_n) \tag{5.21}$$

where $\beta = 1-2/p_0$, $\zeta_n(z)$ is a homeomorphic solution of the Beltrami equation as stated in (5.14). Let $w_0(z)$ be a limit function of a subsequence of $\{w_n(z)\}$ in $D\setminus\{0\}$. It is not difficult to prove that $w_0(z)$ is a homemorphic solution of (5.3) satisfying the requirements stated in this theorem. □

Existence theorems for nonlinear quasiconformal mappings from D onto a circular domain and onto a parallel strip will be considered in § 2 of Chapter 4.

2. Uniqueness theorems for nonlinear quasiconformal mappings for (5.3)

In order to prove the uniqueness of quasiconformal mappings for (5.3), we add one restriction to (5.3).

$F(z,w,U) = Q(z,w,U)U$ is required to satisfy the condition

$$|F(z,w_1,U) - F(z,w_2,U)| \le R(z,w_1,w_2,U)\ |w_1-w_2| \,, \tag{5.22}$$

where $w_1(z)$, $w_2(z)$ are any continuous functions in $\mathbb{C}\setminus\{0\}$, $U(z)$ is any measurable function in D, $R(z,w_1,w_2,U)$ is a measurable function and $R(z,w_1,w_2,U) \in L_{p_0}(D)$.

Theorem 5.5 Let (5.3) satisfy Condition C and (5.22). Then the quasiconformal mapping $w(z)$ from D onto the domain G as stated in Theorem 5.3 is unique, if $w(z)$ satisfies one of the following three conditions:

(1) $w(a_j) = a_j$, $j = 1,2,3$ where a_1 , a_2 , a_3 are three distinct points on $\Gamma_0 = \{|z|=1\}$.

(2) $w(a_1) = a_1$, $w(b_1) = b_1$, where $a_1 \in \Gamma_0$ and $b_1 \in D$.

(3) $w(b_j) = b_j$, $j = 1,2$, where $b_j (j=1,2)$ are two distinct points in D .

Proof. Let $w_1(z)$ and $w_2(z)$ be two homeomorphic solutions as stated in this theorem, and denote $L_j^{(k)} = w_k(\Gamma_j)$, $k = 1,2$, $j = 0,1,\ldots,N$. Assume that $w_1(z) \not\equiv w_2(z)$. We might as well assume that $\rho_j^{(1)} = \rho_j^{(2)}$, $j = 0,1,\ldots,m (m \le N)$ and $\rho_j^{(1)} \ne \rho_j^{(2)}$, $j = m+1,\ldots,N$, where $\rho_j^{(k)}$ is the radius of $L_j^{(k)}$, $j = 0,1,\ldots,N$, $k = 1,2$. Now, we extend $w_k(z)$ across $\Gamma_j (j=0,1,\ldots,m)$ into the outer part of $\bar{D}$, i.e. we set

$$W_k(z) = \begin{cases} w_k(z) , \quad z \in \bar{D} , \\ \rho_j^2 \overline{\left[w_k\left(\dfrac{\gamma_j^2}{\bar{z}-\bar{z}_j} + z_j\right) \right]}^{-1} = \rho_j^2 [\overline{w_k(\zeta)}]^{-1} , \quad \zeta = \dfrac{\gamma_j^2}{\bar{z}-\bar{z}_j} + z_j , \\ \qquad z \in D_j , \quad j=0,1,\ldots,m , \end{cases} \tag{5.23}$$

where $z_0 = w_0 = 0$, $\gamma_1 = \rho_1 = 1$, $D_0 = \{1 < |z| \le 1 + \eta\}$, $D_j = \{\gamma_j - \eta \le |z-z_j| < \gamma_j\}$, $j = 1,\ldots,m$, and η is a suitable positive constant.

When $1 < |z| \le 1 + \eta$ it is evident that $\zeta = \frac{1}{\bar{z}}$ satisfies $\frac{1}{1+\eta} \le |\zeta| < 1$. From

$$w_{k\bar{\zeta}}(\zeta) = \left[\frac{z}{\overline{W_k(z)}}\right]^2 \overline{W_{k\bar{z}}(z)} , \quad w_{k\zeta}(\zeta) = \overline{\left[\frac{z}{W_k(z)}\right]^2 W_{kz}(z)} ,$$

we obtain the complex equation

$$W_{k\bar{z}}(z) = \left[\frac{W_k(z)}{\bar{z}}\right]^2 \overline{F\left\{\frac{1}{\bar{z}}, \frac{1}{\overline{W_k(z)}}, \overline{\left[\frac{z}{W_k(z)}\right]}W_{kz}(z)\right\}} .$$

Let

$$f(z,W_k(z),W_{kz}) = \begin{cases} F(z,W_k(z),W_{kz}) , & z \in \bar{D} , \\ \left[\frac{W_k(z)}{\bar{z}}\right]^2 \overline{F\left\{\frac{1}{\bar{z}}, \frac{1}{\overline{W_k(z)}}, \overline{\left[\frac{z}{W_k(z)}\right]^2} W_{kz}(z)\right\}} , & z \in D_0 . \end{cases}$$

Then $W_k(z)$ is a homeomorphic solution of the complex equation

$$W_{k\bar{z}} = f[z,W_k(z),W_{kz}(z)] , \tag{5.24}$$

on $\tilde{D} = \bar{D} \cup D_0$. From Condition C and (5.22), we can see that $W(z) = W_1(z) - W_2(z)$ is a solution of the following equation

$$W_{\bar{z}} = Q(z)\ W_z + A(z)W \tag{5.25}$$

on $\tilde{D}$, where

$$Q(z) = \begin{cases} [f(z,W_1,W_{1z})-f(z,W_1,W_{2z})]/(W_{1z}-W_{2z}) , & \text{for } W_{1z} \neq W_{2z} , \\ 0 , \text{ for } W_{1z} = W_{2z} , \end{cases}$$

$$A(z) = \begin{cases} [f(z,W_1,W_{2z})-f(z,W_2,W_{2z})]/(W_1-W_2) , & \text{for } W_1(z) \neq W_2(z) , \\ 0 , \text{ for } W_1(z) = W_2(z) , \end{cases}$$

and $|Q(z)| \leq q_0 < 1$, $A(z) \in L_{p_0}(\tilde{D})$. Similarly, we can verify that $W(z) = W_1(z) - W_2(z)$ is a solution of the complex equation of the type (5.25) on D_j , $j = 1,\ldots,m$. Due to Theorem 2.2, $W(z)$ can be expressed as

$$W(z) = \Phi[\zeta(z)]e^{\varphi(z)} \tag{5.26}$$

on $\tilde{D}$. Provided η is sufficiently small, the zero points of $W(z)$

on $\tilde{D}$ are isolated, and $W(z) \not\equiv 0$ on D_j, $j = 0,1,\ldots,m$. By the argument principle, we have

$$\frac{1}{2\pi}\Delta_{\Gamma_\eta} \arg W(z) = N_D + N_\Gamma , \tag{5.27}$$

where

$$\Gamma_\eta = \{|z| = 1+\eta\} \cup \bigcup_{j=1}^{m} \{|z-z_j| = \gamma_j - \eta\} \cup \bigcup_{j=m+1}^{N} \{|z-z_j| = \gamma_j\}$$

and N_D, N_Γ are the total numbers of zero points of $W(z)$ on D, Γ, respectively. From $\Delta_{\Gamma_j} \arg W(z) = 0$, $j = m+1,\ldots,N$, and

$$\begin{aligned} W(z) &= W_1(z)-W_2(z) = \rho_j^2\left[\overline{w_1\left(\frac{\gamma_j^2}{\bar{z}-\bar{z}_j} + z_j\right)}\right]^{-1} - \rho_j^2\left[\overline{w_2\left(\frac{\gamma_j^2}{\bar{z}-\bar{z}_j} + z_j\right)}\right]^{-1} \\ &= -\rho_j^2\,\overline{W(\zeta)}\left[\overline{w_1(\zeta)w_2(\zeta)}\right]^{-1} \text{ for } |z-z_j| = \frac{\gamma_j^2}{|\zeta-z_j|}, \ j=0,1,\ldots,m , \end{aligned} \tag{5.28}$$

where $|z| = 1+\eta$ for $j = 0$, and $|z-z_j| = \gamma_j - \eta$, $j = 1,\ldots,m$, it follows

$$\begin{aligned} \frac{1}{2\pi}\Delta_{\Gamma_\eta} \arg W(z) &= \frac{1}{2\pi}[\Delta_{\tilde{\Gamma}_\eta} \arg w_1(\zeta)+\Delta_{\tilde{\Gamma}_\eta} \arg w_2(\zeta)-\Delta_{\tilde{\Gamma}_\eta} \arg W(\zeta)] \\ &= 2 - N_D , \end{aligned} \tag{5.29}$$

where

$$\tilde{\Gamma}_\eta = \{|z| = \frac{1}{1+\eta}\} \cup \bigcup_{j=1}^{m} \{|z-z_j| = \frac{\gamma_j^2}{\gamma_j-\eta}\} \cup \bigcup_{j+m+1}^{m} \{|z-z_j| = \gamma_j\} .$$

Thus

$$2N_D + N_\Gamma = 2 . \tag{5.30}$$

This formula is contradictory to each one of (1), (2) and (3). Hence, $W(z) \equiv 0$, i.e. $w_1(z) \equiv w_2(z)$ in D. □

<u>Theorem 5.6</u> Under the hypothesis of Theorem 5.5 the quasiconformal mapping $w(z)$ stated in Theorem 5.2 is unique, provided $w(z)$ possesses the representation

$$w(z) = \frac{C_{-1}}{\chi(z)} + C_0 + C_1\, \chi(z) + \dots, \tag{5.31}$$

in the neighbourhood of $z = 0$, where $C_{-1}, C_0, C_1, \dots$ are constants, $\chi(z)$ is a fixed homeomorphism of the type stated in (2.29).

<u>Proof.</u> Let $w_1(z)$ and $w_2(z)$ be two solutions as stated in this theorem and $w(z) = w_1(z) - w_2(z)$. Suppose that $w(z) \not\equiv 0$ in D. From Condition C and (5.22), we derive that $w(z)$ satisfies a complex equation of the type (5.25) and the boundary condition

$$\mathit{Re}[ie^{-i\theta_j} w(t)] = h_j \ , \ t \in \Gamma_j \ , \ j = 0,1,\dots,N \ . \tag{5.32}$$

There is no harm in assuming $h_j = 0\ (j=0,1,\dots,m)$ and $h_j \neq 0\ (j=m+1,\dots,N)$. Set

$$W(z) = \begin{cases} w(z) \ , \ z \in \bar{D} \ , \\ e^{2i\theta_j}\, \overline{w\left(\dfrac{\gamma_j^2}{\bar{z}-\bar{z}_j} + z_j\right)} \ , \ z \in D_j \ , \ j = 0,1,\dots,m \ , \end{cases} \tag{5.33}$$

where $D_j\,(j=0,1,\dots,m)$ are defined in (5.23). We can verify that $W(z)$ is a solution of the complex equation (5.25) with coefficients

$$Q(z) = \begin{cases} Q(z) \ , \ z \in \bar{D} \ , \\ \overline{Q\left(\dfrac{\gamma_j^2}{\bar{z}-\bar{z}_j} + z_j\right)}\, \left(\dfrac{z-z_j}{\bar{z}-\bar{z}_j}\right)^2 \ , \ z \in D_j \ , \ j = 0,1,\dots,m \ , \end{cases}$$

$$A(z) = \begin{cases} A(z) \ , \ z \in \bar{D} \ , \\ -\, A\left(\dfrac{\gamma_j^2}{\bar{z}-\bar{z}_j} + z_j\right)\, \left(\dfrac{\gamma_j}{\bar{z}-\bar{z}_j}\right)^2 \ , \ z \in D_j \ , \ j = 0,1,\dots,m \end{cases} \tag{5.34}$$

on $\tilde{D} = \bar{D} \cup \bigcup_{j=0}^{m} D_j$, and by the argument principle, we can obtain

$$2N_D + N_\Gamma = 0 . \tag{5.35}$$

However from $W(0) = 0$, it follows that $2N_D + N_\Gamma \geq 2 > 0$. This contradiction proves $W(z) \equiv 0$, i.e. $w_1(z) = w_2(z)$ in D . □

Using a similar method, we can prove the uniqueness theorem corresponding to Theorem 5.4.

Theorem 5.7 Under the hypothesis of Theorem 5.5 the quasiconformal mapping $w(z)$ stated in Theorem 5.4 is unique, provided $w(z)$ possesses the representation

$$w(z) = \frac{c_{-1}}{z} + c_0 + c_1 z + \ldots \tag{5.36}$$

in the neighbourhood of $z = 0$, where c_{-1} , c_0, c_1,... are constants.

§ 6. The compound boundary value problem for elliptic complex equations

1. Formulation of the compound boundary value problem

Let $\Gamma_0,\ldots,\Gamma_N$, $L_1,\ldots,L_M$ be the boundary contours of an (N+M+1)-connected domain D where $\Gamma_1,\ldots,\Gamma_N$, $L_1,\ldots,L_M$ are located inside Γ_0. In the interior of D there are some mutually exclusive contours $\gamma_1,\ldots,\gamma_n$, $\ell_1,\ldots,\ell_m$. Let

$$\Gamma = \bigcup_{j=0}^{N} \Gamma_j , L = \bigcup_{j=1}^{M} L_j , \gamma_j = \bigcup_{j=1}^{n} \gamma_j , \ell = \bigcup_{j=1}^{m} \ell_j \in C_\mu^1 (0 < \mu < 1)$$

and denote

$$D_\gamma^- = \bigcup_{j=1}^{n} D_{\gamma_j} , D_\ell^- = \bigcup_{j=1}^{m} D_{\ell_j} , D^- = (D_\gamma^- \cup D_\ell^-) \cap D ,$$

$$D^+ = D \setminus \overline{D^-} , D_\ell^+ = (D^+ \cup D^-) \setminus D_\ell^- ,$$

where D_{γ_j} , D_{ℓ_j} are the bounded domains surround by γ_j , ℓ_j , respectively. Without loss of generality, we assume that D is a circular domain, and $\Gamma_0 = \{|z| = 1\}$, $0 \in D^+$.

We suppose that the nonlinear complex equation (1.45) or (1.47) in $D^{\pm} \setminus \{0\}$ satisfies Condition C as stated in § 1.

Problem F. The compound boundary value problem for the complex equation (1.45) may be formulated as follows. Find a sectional solution $w^{\pm}(z)$ in $D^{\pm}$, continuous on $\overline{D^{\pm}}$ and satisfying the boundary conditions

$$w^+[\alpha(t)] = G(t)\, w^-(t) + g(t) \;,\; t \in \gamma \;, \tag{6.1}$$

$$w^+[\alpha(t)] = G(t)\, \overline{w^-(t)} + g(t) \;,\; t \in \ell \;, \tag{6.2}$$

$$w^+[\alpha(t)] = G(t)\, w^+(t) + g(t) \;,\; t \in L \;, \tag{6.3}$$

$$w^+[\alpha(t)] = G(t)\, \overline{w^+(t)} + g(t) \;,\; t \in \Gamma \;, \tag{6.4}$$

where $\alpha(t)$ maps Γ_j , L_j , ℓ_j , γ_j topologically onto itself which is a positive shift on $\Gamma \cup \gamma$ and a reserve shift on $L \cup \ell$, $\alpha(t)$ has fixed points $a_j \in \Gamma_j$, $j = 0,1,\ldots,N$, and $G(t)$, $g(t)$, $\alpha(t)$ satisfy

$$\left.\begin{aligned} &C_\nu[G,\partial D^{\pm}] \le d < \infty \;,\; C_\nu[g,\partial D^{\pm}] \le d \;,\; \tfrac{1}{2} < \nu < 1 \;, \\ &C^1_\mu[\alpha,\partial D^{\pm}] \le d \;,\; |\alpha'(t)| \ge d^{-1} > 0 \;,\; 0 < \mu < 1 \;, \end{aligned}\right\} \tag{6.5}$$

and $G(t) \neq 0$, $t \in \partial D^{\pm}$; $\alpha[\alpha(t)] = t$, $t \in L \cup \Gamma$; $G(t)G[\alpha(t)] \equiv 1$, $G(t)g[\alpha(t)] + g(t) \equiv 0$, $t \in L$; $G(t)\overline{G[\alpha(t)]} \equiv 1$, $G(t)\overline{g[\alpha(t)]} + g(t) \equiv 0, t \in \Gamma$.

Now let

$$K_{\Gamma_j} = \mathit{Ind}\, G(t)\Big|_{\Gamma_j} = \frac{1}{2\pi}\Delta_{\Gamma_j}\, \mathit{arg}\, G(t) \;,\; K_{L_j} = \mathit{Ind}\, G(t)\Big|_{L_j} \;,$$

$$K_{\gamma_j} = \mathit{Ind}\, G(t)\Big|_{\gamma_j} , \quad K_{\ell_j} = \mathit{Ind}\, G(t)\Big|_{\ell_j}$$

and

$$K_\Gamma = \sum_{j=0}^{N} K_{\Gamma_j} , \quad K_L = \sum_{j=1}^{M} K_{L_j} , \quad K_\gamma = \sum_{j=1}^{n} K_{\gamma_j} , \quad K_\ell = \sum_{j=1}^{m} K_{\ell_j} .$$

In addition, denote by f the total number of fixed points of $\alpha(t)$ with $G(t) = -1$ on L. Then

$$K = (K_\Gamma - K_L - f - 2K_\ell + 2K_\gamma)/2 \tag{6.6}$$

is called the index of Problem F.

In particular, if the contours $L_1,\ldots,L_M$, $\gamma_1,\ldots,\gamma_n$, $\ell_1,\ldots,\ell_m$ do not occur, and $\alpha(t) = t$ in (6.4), K_{Γ_j} $(j=0,1,\ldots,N)$ are even numbers, then Problem F is just Problem A. Actually, if we denote $\overline{\lambda(t)} = i\overline{[G(t)]}^{1/2}$, $\gamma(t) = i\overline{[G(t)]}^{1/2}g(t)/2$, $w(z) = w^+(z)$, the boundary condition (6.4) can be reduced to the boundary condition (4.1) of Problem A.

In the following, we shall discuss the solvability of Problem F for (1.45) or (1.47), and give a brief introduction to the method of the proof.

Theorem 6.1 A necessary and sufficient condition that

$$w(z) = \begin{cases} w^+(z) , & z \in D^+ \\ w^-(z) , & z \in D^- \end{cases}$$

is a solution of Problem F for (1.45) satisfying Condition C, is that

$$w^*(\zeta) = \begin{cases} w[z_1(\zeta)] , & \zeta \in D_\ell^{*+} = \zeta(D_\ell^+) \\ \overline{w[z_2(\bar\zeta)]} , & \zeta \in D_\ell^{*-} = \zeta(D_\ell^-) \end{cases}$$

is a solution of Problem G, which satisfies the nonlinear complex equation

$$w^*_{\bar\zeta} = F_1(\zeta, w^*, w^*_\zeta) \ , \ F_1 = \begin{cases} \overline{z_1'(\zeta) F[z_1(\zeta), w^*, w^*_\zeta / z_1'(\zeta)]} \ , \ \zeta \in D^{*+}_\ell \ , \\ \overline{z_2'(\bar\zeta) F[z_2(\bar\zeta), w^*, \overline{w^*}_\zeta / z_2'(\bar\zeta)]} , \zeta \in D^{*-}_\ell \end{cases} \tag{6.7}$$

and the boundary condition

$$w^{*+}(\zeta) = G^*(\zeta) w^{*-}(\zeta) + g^*(\zeta) \ , \ \zeta \in \gamma^* \cup \ell^* \cup L^* \ , \ \gamma^* = \zeta(\gamma) \ , \ \ell^* = \zeta(\ell) \ , \tag{6.8}$$

$$w^{*+}[\beta(\zeta)] = G^*(\zeta) \overline{w^{*+}(\zeta)} + g^*(\zeta) \ , \ \zeta \in \Gamma^* = \zeta(\Gamma) \ , \ L^* = \zeta(L) \ , \tag{6.9}$$

where

$$\zeta(z) = \begin{cases} \zeta_1(z) \ , \ z \in D^+_\ell \\ \overline{\zeta_2(z)} \ , \ z \in \overline{D^-_\ell} \end{cases}$$

is a homeomorphism in $D^\pm$ and $z(\zeta)$ is the inverse function of $\zeta(z)$, $\zeta(z) \in C^1_\mu(D^\pm_\ell)$, $z(\zeta) \in C^1_\mu(D^{*\pm}_\ell)$, $C_\nu[G^*, \partial D^{*\pm}] \le d_1$, $C_\nu[g^*, \partial D^{*\pm}] \le d_1$, $|G^*| \ge d_1^{-1} > 0$, $\beta(\zeta)$ is a positive shift on Γ^* similar to $\alpha(t)$ on Γ, and $\partial D^{*\pm} = \Gamma^* \cup \gamma^* \cup L^* \cup \ell^*$; $\Gamma^*, \gamma^*, L^*, \ell^* \in C^1_\mu$; $\Gamma^*, \gamma^*, \ell^*$ are a set of closed smooth contours and L^* is a set of open smooth arcs.

Proof. In [69] the conformal glue theorems with gluing conditions

$$\zeta^+[\alpha(t)] = \overline{\zeta^-(t)} \ , \ t \in \ell \ , \tag{6.10}$$

$$\zeta^+[\alpha(t)] = \zeta^-(t) \ , \ t \in \gamma \ , \tag{6.11}$$

$$\zeta^+[\alpha(t)] = \zeta^+(t) \ , \ t \in \Gamma \ , \tag{6.12}$$

respectively, were proved. We apply successively the above conformal gluing functions, and can complete the proof of this theorem. □

Theorem 6.2 Let $w^*(\zeta)$ be a solution of Problem G for (6.7) satisfying some Condition C. Then the function

$$W(\zeta) = [w^*(\zeta)-\Phi(\zeta)]/X(\zeta) \tag{6.13}$$

is a solution of Problem H, which satisfies the nonlinear complex equation

$$\begin{aligned} W_{\bar\zeta} &= f(\zeta,W,W_\zeta) \ , \\ f &= F\{\zeta,\Phi(\zeta)+X(\zeta)W(\zeta),\Phi'(\zeta)+X'(\zeta)W(\zeta)+X(\zeta)W_\zeta\}/X(\zeta) \end{aligned} \tag{6.14}$$

in $D^* = \zeta(D)$ and the boundary condition

$$\begin{aligned} &W^+[\beta(\zeta)] = G_1(\zeta)\overline{W^+(\zeta)} + g_1(\zeta) \ , \ G_1(\zeta) = G^*(\zeta)\overline{X^+(\zeta)}/X^+[\beta(\zeta)], \\ &g_1(\zeta) = \{g^*(\zeta)-\Phi^+[\beta(\zeta)] + G^*(\zeta)\overline{\Phi^+(\zeta)}\}/X^+[\beta(\zeta)] \ , \ \zeta\in\Gamma^* \ , \end{aligned} \tag{6.15}$$

where $\Phi(\zeta)$ is a sectionally analytic function in D^* satisfying the boundary condition (6.8), and $X(\zeta)$ is a canonical function in D^* satisfying the homogeneous boundary condition

$$X^+(\zeta) = G^*(\zeta)X^-(\zeta) \ , \ \zeta\in\gamma^*\cup\ell^*\cup L^* \ . \tag{6.16}$$

Conversely, let $W(\zeta)$ be a continuous solution of Problem H for (6.14). Then

$$w^*(\zeta) = \Phi(\zeta) + X(\zeta)W(\zeta) \tag{6.17}$$

is a solution of Problem G for (6.7).

Proof. The function $\Phi(\zeta)$, $X(\zeta)$ have been given by (1.82) in Chapter 1. By analogy with (3.34), we have

$$\begin{aligned} w^{*+}(\zeta) &= \Phi^+(\zeta) + X^+(\zeta)W^+(\zeta) = \\ &= G^*(\zeta)\ [\Phi^-(\zeta)+X^-(\zeta)W^-(\zeta)] + g^*(\zeta) \ , \ \zeta\in\gamma^*\cup\ell^*\cup L^* \ . \end{aligned} \tag{6.18}$$

By virtue of $X^-(\zeta) \neq 0$, it follows that $W^+(\zeta) = W^-(\zeta)$ for $\zeta\in\gamma^*\cup\ell^*\cup L^*$ except the set E of end points of L^* , and we can prove that

$$W(\zeta) = \begin{cases} W^+(\zeta) , & \zeta \in D^{*+} \cup \Gamma^* \\ W^-(\zeta) , & \zeta \in D^{*-} \end{cases}$$

is a continuous solution for (6.14) in $D^* \setminus E$.

The converse of the above result holds, too.

Let $K_j = 1/2 \, Ind \, G_1(\zeta) \Big|_{\Gamma_j^*}$, $j = 0,1,\ldots,N$. It is clear that $K = \sum_{j=0}^{N} K_j$. Without loss of generality, we may assume that $K_j (j=1,\ldots,N_0, N_0 \le N)$ are not integers and $K_j (j=N_0+1,\ldots,N)$ are integers. □

2. The modified compound problem and a priori estimates of its solutions

Since Problem F $(K \le 2N-2)$ may not have any solution for (1.45), we discuss the modified boundary value problem for (1.45), i.e. the boundary condition (6.4) is replaced by

$$w^+[\alpha(t)] = G(t)\overline{w^-(t)} + g(t) + H(t) , \quad t \in \Gamma , \tag{6.19}$$

where $H(t) = G(t) \, \overline{h(t)} - h(t)$ and

$$h(t) = \begin{cases} 0 , \ t \in \Gamma , \text{ for } K > N-1 , \\ \left. \begin{array}{l} ih_j , \ t \in \Gamma_j , \ j = 1,\ldots,N-K' \\ 0 , \ t \in \Gamma_j , \ j=0,N-K'+1,\ldots,N \end{array} \right\} \begin{array}{l} \text{for } 0 \le K \le N-1 , \\ K' = [K+1/2] , \end{array} \\ ih_j , \ t \in \Gamma_j , \ j=0,1,\ldots,N , \text{ for } K < 0 , \end{cases} \tag{6.20}$$

in which $h_j (j=0,1,\ldots,N)$ are unknown real constants to be determined appropriately. Moreover, if $K \ge 0$, we suppose that the solution $w(z)$ satisfies the conditions

$$
B_j w = \begin{cases}
\left.\begin{array}{l}
Re\overline{G(a_{j+N-N_0-K'})}^{\frac{1}{2}} w(a_{j+N-N_0-K'}) = b_j, j=1,\ldots,[K]+1, K' < N-N_0, \\
\left.\begin{array}{l} Re\overline{G(a_j)}^{\frac{1}{2}} w(a_j) = b_j, j = \begin{cases} 1,\ldots,N-N_0+I, K_* = [K]+1-N+N_0, \\ I = \begin{cases} 1, & K_* \text{ is odd}, \\ 0, & K_* \text{ is even}, \end{cases} \end{cases} \end{array}\right\} K' \geq N-N_0, \\
Re\, w(a_j) = b_j, j = N-N_0+I+1,\ldots,N-N_0+I+[K_*/2], \\
Im\, w(a_j) = b_j, j = N-N_0+I+[K_*/2]+1,\ldots,[K]+1,
\end{array}\right\} 0 \leq K \leq N-1, \\
\left.\begin{array}{l}
Re\overline{G(a_j)}^{\frac{1}{2}} w(a_j) = b_j, j=1,\ldots,N-N_0+I, I = \begin{cases} 1, & K^* \text{ is odd}, \\ 0, & K^* \text{ is even}, \end{cases} \\
\left.\begin{array}{l} Re\, w(a_j) = b_j, j = N-N_0+I+1,\ldots,N-N_0+I+[K^*/2], \\ Im\, w(a_j) = b_j, j = N-N_0+I+[K^*/2]+1,\ldots,2K-N+1, \end{array}\right\} K^* = 2K-2N+N_0+1,
\end{array}\right\} K > N-1,
\end{cases}
\tag{6.21}
$$

where $a_j \in \Gamma_{j+N_0}$, $j = 1,\ldots,N-N_0+I$, $a_j (j=N-N_0+I+1,\ldots,N-N_0+I+[K^*/2])$ are distinct points in D^+, $b_j (j=1,\ldots,2K-N+1)$ are all real constants and $|b_j| \leq d(j=1,\ldots,2K-N+1)$. If $K < 0$, we allow that the solution $w(z)$ possesses a pole of order $\leq K' - 1$ at $z = 0$. The above modified boundary value problem for (1.45) is called Problem F^*, and Problem F^* with $g(t) = 0$ and $A_3 = 0$ is called Problem F_0^*.

Theorem 6.3 Problem F^* of analytic functions has a unique solution.

Proof. (1) $K > N-1$. According to the result in [103]11), we know that the general solution of Problem F for analytic functions can be written in the form

$$
\Phi(z) = \Phi_0(z) + \sum_{m=1}^{2K-N+1} c_m \Phi_m(z), \tag{6.22}
$$

where $\Phi_0(z)$ is a special solution to Problem F and $\Phi_m(z)$ $(m=1,\ldots,2K-N+1)$ are linearly independent solutions of the corresponding homogeneous problem. We can choose real constants $c_m (m=1,\ldots,2K-N+1)$ so that the function $\Phi(z)$ in (6.22) satisfies

(6.21). Let $\Phi_1(z)$ and $\Phi_2(z)$ be two solutions as stated in this theorem, and denote $\Phi(z) = \Phi_1(z) - \Phi_2(z)$. If $\Phi(z) \not\equiv 0$ in $D^{\pm}$, it is obvious that $\Phi(z)$ satisfies the condition

$$B_j\Phi = 0 \ , \ j = 1,\ldots,2K - N+1 \qquad (6.23)$$

and $\Phi(z) = 0$ on the fixed points so that $\alpha(t) = t$ and $G(t) = -1$ on L. Using the argument principle, we may derive the absurd inequality

$$2K + 1 + f \le 2N_D + N_\Gamma = K + f \ . \qquad (6.24)$$

This contradiction proves that $\Phi(z) \equiv 0$, i.e. $\Phi_1(z) \equiv \Phi_2(z)$ in $D^{\pm}$.

(2) If $0 \le K \le N-1$, it is easy to see that the index of the boundary value problem F^* with the boundary conditions

$$\left.\begin{array}{l}\Psi^+[\alpha(t)]=G^*(t)\Psi^-(t)+g^*(t),t\in\gamma\ , \\ \Psi^+[\alpha(t)]=G^*(t)\Psi^+(t)+g^*(t),t\in L\ ,\end{array}\right\} \begin{array}{l}\cdot^*(t) = [\frac{\alpha(t)}{t}]^n G(t)\ , \\ g^*(t) = [\alpha(t)]^n g(t)\ ,\end{array} \qquad \begin{array}{r}(6.25)\\(6.26)\end{array}$$

$$\left.\begin{array}{l}\Psi^+[\alpha(t)]=G^*(t)\overline{\Psi^-(t)}+g^*(t),t\in\ell\ , \\ \Psi^+[\alpha(t)]=G^*(t)\overline{\Psi^+(t)}+g^*(t)+[\alpha(t)]^n h(t),t\in\Gamma\ ,\end{array}\right\} \begin{array}{l}G^*=[\frac{\alpha(t)}{\bar{t}}]^n G(t)\ , \\ g^*=[\alpha(t)]^n g(t)\ ,\end{array} \qquad \begin{array}{r}(6.27)\\(6.28)\end{array}$$

$$B_j^!\Psi=Re[\alpha(t)]^n G(t)^{\frac{1}{2}}\overline{\Psi(t)}\Big|_{t=a_j} = b_j, j=1,\ldots,N+1+2(K-[K+\tfrac{1}{2}]), n = N-K\ , \qquad (6.29)$$

is $N + K - [K+\frac{1}{2}]$. We know from the result in (1) that the above Problem F^* has a unique solution $\Psi_0(z)$. Furthermore, we can find $2n$ linearly independent solutions to the corresponding homogeneous problem, $\Psi_j(z)$ $(j=1,\ldots,2n)$, and $2n$ real constants $c_1,\ldots,c_{2n}$, so that

$$\Psi(z) = \Psi_0(z) + \sum_{m=1}^{2n} c_m \Psi_m(z) \qquad (6.30)$$

in D possesses a zero of n-th order at $z = 0$.

Therefore, the function

$$\Phi(z) = z^{-n}\Psi(z) \tag{6.31}$$

is a solution to Problem F^*. The uniqueness of the solution for Problem F^* is not difficult to prove.

(3) If $K = -1/2$ or -1, we may prove the existence and uniqueness of the solution for Problem F^* analogously to (2). As for the case $K < -1$, we introduce the function

$$\Psi(z) = z^{K'-1}\,\Phi(z)\ ,\ K' = [\,|K|+\tfrac{1}{2}]\ , \tag{6.32}$$

which is a solution of the corresponding Problem F^* with the index $-1/2$ or -1. Hence, $\Phi(z) = z^{1-K'}\Psi(z)$ is the unique solution of Problem F^* with the index K. This completes the proof. □

<u>Theorem 6.4</u> Let the nonlinear complex equation (1.45) satisfy Condition C. Then the solution $w(z)$ of Problem F^* for (1.45) satisfies the estimates

$$C_\beta[W,D^\pm] \le M_1\ ,\ L_{p_0}[\,|W_{\bar z}|+|W_z|,D^\pm] \le M_2\ , \tag{6.33}$$

where

$$W(z) = \begin{cases} w(z)\ , & \text{for } K \ge 0\ ,\\ w(z)[\zeta(z)]^{K'-1}\ , & \text{for } K < 0\ ,\end{cases}$$

and $\zeta(z)$ is a homeomorphic solution to the Beltrami equation with

$$\zeta(0) = 0\ ,\ \zeta(1) = 1\ ,\ \beta = 1-2/p_0\ ,\ 2 < p_0 < \min(\tfrac{1}{1-\nu})\ ,$$
$$M_j = M_j(q_0,p_0,k_0,\mu,\nu,d,\alpha(t),K,D^\pm)\ ,\ j = 1,2\ .$$

<u>Proof.</u> Let $w(z)$ be a solution of Problem F^* for (1.45). By the representation theorem for solutions to (1.45) in § 2, the above solution $w(z)$ to Problem F^* can be expressed as

$$w(z) = \Phi[\zeta(z)]e^{\varphi(z)} + \psi(z) , \tag{6.34}$$

where $\varphi(z)$, $\psi(z)$, $\zeta(z)$ and its inverse function $z(\zeta)$ are as stated in Theorem 2.4, satisfying the estimates (2.18) - (2.20); $\Phi(\zeta)$ is an analytic function in $H_*^{\pm} = \zeta(D^{\pm})$. Hence the solution $w(z)$ of Problem F* may be transformed into a solution to the corresponding boundary value problem for ananlytic functions. Using the reductio ad absurdum and Theorem 6.1 to Theorem 6.3, we can obtain the following estimates for $\Phi(\zeta)$.

$$C_{\nu\beta}[\Phi,\overline{D_*^{\pm}}] \leq M_3 = M_3(q_0,p_0,k_0,\mu,\nu,d,\alpha(t),K,D^{\pm}) \tag{6.35}$$

where $\beta = 1-2/p_0$. Analogously to the method of proof in Theorem 4.3, the estimates in (6.33) are derived. □

3. Solvability of the compound boundary value problem

We first assume that the coefficients of (1.45) are equal to zero in $D \setminus D_m$, where

$$D_m = \{z | z \in D^{\pm} , |z-t| \geq \frac{1}{m} , t \in \Gamma \cup \gamma \cup L \cup \ell \cup \{0\}\} ,$$

m is a positive interger, and we write this complex equation in the form

$$w_{\bar{z}} = F_m(z,w,w_z) . \tag{6.36}$$

From Theorem 6.4, it is not difficult to conclude the following results.

Lemma 6.5 Let $w(z)$ be a solution to Problem F^* for (6.36). Then $w(z)$ may be expressed as

$$w(z) = \Phi(z) + \Psi(z) = \Phi(z) - \frac{1}{\pi}\iint\limits_{D_m} \frac{\omega(\zeta)}{\zeta-z}\, d\sigma_\zeta , \tag{6.37}$$

where $\omega(z) = w_{\bar{z}} \in L_{p_0}(D_m)$, $\Phi(z)$ is analytic in D_m , and these functions satisfy

$$C_\beta[\Phi,D_m] \le M_4 \ , \ L_{p_0}[\Phi',D_m] \le M_5 \ , \tag{6.38}$$

$$C_\beta[\Psi,D_m] \le M_6 \ , \ L_{p_0}[|\Psi_{\bar z}|+|\Psi_z|,D_m] < M_7 \ , \tag{6.39}$$

where $\beta = 1-2/p_0$, $M_j = M_j(q_0,p_0,k_0,\mu,\nu,d,\alpha(t),K,D^{\pm},D_m)$, $j = 4,\ldots,7$.

Theorem 6.6 Let (1.45) satisfy Condition C in $D^{\pm}$. Then Problem F^* for (1.45) has a solution.

Proof. First of all, we prove the solvability of Problem F^* for (6.36). According to the representation (6.37), we only need to find a solution $\Psi(z) = T\omega$ of the complex equation

$$\Psi_{\bar z} = F_m[z,\Phi(z)+\Psi(z),\Phi'(z)+\Psi_z] \ , \tag{6.40}$$

where $w(z) = \Phi(z) + \Psi(z)$ satisfies the boundary condition of Problem F^*. Using a method similar to that in the proof of Theorem 4.6, we introduce the Banach space $B = L_{p_0}(D_m)$ and denote by B_M the set of measurable functions $\omega(z)$ satisfying the inequality

$$L_{p_0}[\omega,D_m] < M_7 \ , \tag{6.41}$$

where M_7 is the constant in (6.39). We arbitrarily choose $\omega \in B$ and suppose that

$$\Psi(z) = -\frac{1}{\pi}\iint_{D_m} \frac{\omega(\zeta)}{\zeta-z}\, d\sigma_\zeta \ .$$

From Theorem 6.3, we find an analytic function $\Phi(z)$ in $D^{\pm}$ so that $w(z) = \Phi(z) + \Psi(z)$ satisfies the boundary conditions (6.1) - (6.3), (6.19) and (6.21). We substitute $w(z)$, $\Phi'(z)$ into (6.40), and consider the integral equation with the parameter k

$$\Omega(z) = k\, F_m(z,\Phi(z)+T\omega,\Phi'(z)+\Pi\Omega) \ , \ 0 \le k \le 1 \ . \tag{6.42}$$

From Condition C and the contraction principle, this integral equation has a unique solution $\Omega(z) \in B$. Denote by $\Omega = S(\omega,k)$ $(0 \le k \le 1)$ this mapping from $\omega(z)$ onto $\Omega(z)$. We can verify that $\Omega = S(\omega,k)$ satisfies the conditions of the Leray-Schauder theorem. Hence, there exists $\omega(z) \in B_M$, so that $\omega = S(\omega,1)$. Thus, the function $w(z) = \Phi(z) + \Psi(z)$ in (6.37) is a solution of Problem F^* for (6.36).

By Theorem 6.4 and the principle of compactness for a sequence of bounded solutions of (6.36), we can derive the result on solvability of Problem F^* for (1.45). □

From the above theorem, we can obtain the solvability condition for Problem F for (1.45).

Theorem 6.7 Under the hypothesis of Theorem 6.6 the following holds.

(1) If $K > N-1$, Problem F for (1.45) is solvable.

(2) If $0 \le K \le N-1$, the total number of solvability conditions of Problem F is not greater than $N - [K + 1/2]$.

(3) If $K < 0$, Problem F for (1.45) has $N-2K-1$ solvability conditions (cf. [103]11)) .

III Boundary value problems for elliptic equations of second order

In this chapter, we first transform the linear and nonlinear uniformly elliptic equations of second order into a complex form, and then give some properties of solutions for the equations, i.e. some extremum principles, a representation theorem and the compactness principle of solutions for the above equations. On the basis of the above properties of solutions, a priori estimates of solutions, the method of continuity and the Leray-Schauder theorem, we consider the solvability of the regular and irregular oblique derivative boundary value problems for the linear and nonlinear uniformly elliptic equations of second order with some conditions.

§ 1. Reduction of elliptic equations of second order

1. Complex form of linear and quasilinear elliptic equations of second order

Let us introduce the general linear equation of second order

$$au_{xx} + 2bu_{xy} + cu_{yy} + du_x + eu_y + fu = g , \tag{1.1}$$

where the coefficients a, b, c, d, e, f, g are real functions of $(x,y) \in D$, D is a bounded domain in $\mathbb{R}^2$. The equation (1.1) is called elliptic, if the following inequality holds in D

$$K(\lambda) = a\lambda^2 + 2b\lambda + c > 0 , \tag{1.2}$$

in which λ is any real number. The above inequality is equivalent to

$$\Delta = ac - b^2 > 0 , \ a > 0 . \tag{1.3}$$

If the coefficients a, b, c are bounded in D and

$$\Delta = ac - b^2 \geq \Delta_0 > 0 , \quad a > 0 , \tag{1.4}$$

where Δ_0 is a real constant, then the equation (1.1) is called uniformly elliptic.

Noting the relations

$$\left.\begin{aligned}
&(\ )_{z\bar{z}} = \frac{1}{4}[(\ )_{xx} + (\ )_{yy}] , \ (\ )_{zz} = \frac{1}{4}[(\ )_{xx} - (\ )_{yy} - 2i(\ )_{xy}] , \\
&(\ )_{\bar{z}\bar{z}} = \frac{1}{4}[(\ )_{xx} - (\ )_{yy} + 2i(\ )_{xy}] , \ (\ )_{xy} = i(\ )_{zz} - i(\ )_{\bar{z}\bar{z}} , \\
&(\ )_{xx} = 2(\ )_{z\bar{z}} + (\ )_{zz} + (\ )_{\bar{z}\bar{z}} , \ (\ )_{yy} = 2(\ )_{z\bar{z}} - (\ )_{zz} - (\ )_{\bar{z}\bar{z}} ,
\end{aligned}\right\} \tag{1.5}$$

equation (1.1) can be transformed into the complex form

$$u_{z\bar{z}} = \mathcal{R}e[Q(z)u_{zz} + A_1(z)u_z] + A_2(z)u + A_3(z) , \tag{1.6}$$

where

$$Q(z) = (-a + c - 2bi)/(a + c) , \quad A_1(z) = -(d + ei)/(a + c) ,$$
$$A_2(z) = -f/2(a + c) , \quad A_3(z) = g/2(a + c) ,$$

and the ellipticity condition (1.3) can be reduced to

$$|Q(z)| < 1 , \text{ for } z \in D \tag{1.7}$$

and the uniform ellipticity condition (1.4) can be reduced to

$$|Q(z)| \leq q_0 < 1 , \text{ for } z \in D , \tag{1.8}$$

in which q_0 is a real constant.

If the coefficients of the uniformly elliptic equation (1.1) satisfy more conditions, then (1.1) can be transformed into the standardized form. In fact, we can find a complete homeomorphic solution $\zeta(z)$ of the Beltrami equation

$$\zeta_{\bar{z}} - q(z)\zeta_z = 0 \;,\; q(z) = Q(z)/(1+\sqrt{(1-|Q(z)|^2)}) \;, \tag{1.9}$$

which topologically maps the z-plane onto the ζ-plane and $\zeta(\infty) = \infty$ (cf. [97]2)), and if $a,b,c \in W^1_{p_0}(D)$, $p_0 > 2$, then $Q(z)$, $q(z) \in W^1_{p_0}(D)$, $\zeta(z) \in W^2_{p_0}(D)$ and its inverse function $z(\zeta) \in W^2_{p_0}(G)$, $G = \zeta[D]$. Thus, through the relation $U(\zeta) = u[z(\zeta)]$, the equation (1.1) can be transformed into the canonical form

$$U_{\zeta\bar{\zeta}} = Re[a_1(\zeta)U_\zeta] + a_2(\zeta)U + a_3(\zeta) \;, \tag{1.10}$$

where

$$a_1(\zeta) = (-2\zeta_{z\bar{z}} + Q\zeta_{zz} + \bar{Q}\zeta_{\bar{z}\bar{z}} + A_1\zeta_z + \bar{A}_1\zeta_{\bar{z}})/a \;,$$

$$a_2(\zeta) = A_2/a \;,\; a_3(\zeta) = A_3/a \;,\; a = (1+|q|^2 - Q\bar{q} - \bar{Q}q)|\zeta_z|^2 \;.$$

In § 3, we shall transform (1.10) with $a_2(\zeta) \geq 0$ into the following form

$$V_{\zeta\zeta} = Re[b_1(\zeta)V_\zeta] + b_2(\zeta) \;. \tag{1.11}$$

If the coefficients of (1.1) are functions of x , y , u , u_x , u_y , then the equation (1.1) is called quasilinear. In the domain D , the ellipticity and uniform ellipticity conditions are the same as those given for the linear equation (1.1). In this case, the complex form of (1.1) is as follows:

$$\begin{aligned} u_{z\bar{z}} = {} & Re[Q(z,u,u_z)u_{zz} + A_1(z,u,u_z)u_z] + \\ & + A_2(z,u,u_z)u + A_3(z,u,u_z) \end{aligned} \tag{1.12}$$

and its ellipticity condition and uniform ellipticity condition can be written as

$$|Q(z,u,u_z)| < 1 , \tag{1.13}$$

$$|Q(z,u,u_z)| \le q_0 < 1 \tag{1.14}$$

in D , respectively.

2. Complex form of nonlinear elliptic equations of second order

Let $\Phi(x,y,\zeta_1,\zeta_2,\ldots,\zeta_6)$ be a continuous real function in $(x,y) \in D$ and the real variables $\zeta_1,\zeta_2,\ldots,\zeta_6$, and possess continuous partial derivatives with respect to ζ_4 , ζ_5 and ζ_6 . The nonlinear equation of second order

$$\Phi(x,y,u,u_x,u_y,u_{xx},u_{xy},u_{yy}) = 0 \tag{1.15}$$

is called elliptic, if the following inequality holds:

$$I = 4\Phi_{u_{xx}}\Phi_{u_{yy}} - (\Phi_{u_{xy}})^2 > 0 \tag{1.16}$$

in D . If $\Phi_{u_{xx}}$, $\Phi_{u_{yy}}$ and $\Phi_{u_{xy}}$ are bounded in D and the following inequality holds uniformly in D ,

$$I = 4\Phi_{u_{xx}}\Phi_{u_{yy}} - (\Phi_{u_{xy}})^2 \ge I_0 > 0 , \tag{1.17}$$

where I_0 is a real constant, then (1.15) is called uniformly elliptic. From (1.5), it is clear that

$$\left.\begin{array}{l} \Phi_{u_{xy}} = \frac{1}{2i}(\Phi_{u_{zz}} - \Phi_{u_{\bar{z}\bar{z}}}) , \; \Phi_{u_{xx}} = \frac{1}{4}(\Phi_{u_{z\bar{z}}} + \Phi_{u_{zz}} + \Phi_{u_{\bar{z}\bar{z}}}) \\ \Phi_{u_{yy}} = \frac{1}{4}(\Phi_{u_{z\bar{z}}} - \Phi_{u_{zz}} - \Phi_{u_{\bar{z}\bar{z}}}) . \end{array}\right\} \tag{1.18}$$

So we have

$$I = \frac{1}{4}(\Phi_{u_{z\bar{z}}})^2 - |\Phi_{u_{zz}}|^2 . \tag{1.19}$$

Theorem 1.1 Let the nonlinear equation (1.15) be uniformly elliptic in D. Then (1.15) is solvable with respect to $u_{z\bar{z}}$ and the uniformly elliptic complex equation

$$u_{z\bar{z}} = F(z,u,u_z,u_{zz}) , \tag{1.20}$$

can be obtained where the uniform ellipticity condition is

$$|F_{u_{zz}}| + |F_{u_{\bar{z}\bar{z}}}| \le q_0 < 1 , \tag{1.21}$$

in which q_0 is a real constant.

Proof. Taking the uniform ellipticity condition (1.17) and (1.19) into account, we can obtain (1.20) from (1.15), where $F(z,u,u_z,u_{zz})$ possesses the continuous partial derivative $F_{u_{zz}}$. Let $F(z,u,u_z,u_{zz})$ be substituted into the position of $u_{z\bar{z}}$ in (1.15) then differentiating with respect to u_{zz} yields

$$\Phi_{u_{z\bar{z}}} F_{u_{zz}} + \Phi_{u_{zz}} = 0 \text{ , i.e. } \Phi_{u_{zz}} = -\Phi_{u_{z\bar{z}}} F_{u_{zz}} . \tag{1.22}$$

From (1.17) and (1.19), the uniformly ellipticity condition (1.21) can be derived.

If we denote the right-hand sides of (1.6) and (1.12) by $F(z,u,u_z,u_{zz})$, it is easy to see that the equations (1.6) and (1.12) are special cases of (1.20). By a method similar to that from (1.36) to (1.37) in Chapter 2, we can derive

$$|F(z,u,u_z,V_1) - F(z,u,u_z,V_2)| \le q_0|V_1-V_2| . \tag{1.23}$$

from (1.21). Obviously, the equation (1.20) can be written in the following form

$$
\left.\begin{aligned}
&u_{z\bar{z}} = F(z,u,u_z,u_{zz}) \ , \\
&F(z,u,u_z,u_{zz}) = Re[Q(z,u,u_z,u_{zz})\, u_{zz} + A_1(z,u,u_z)u_z] + \\
&\qquad + A_2(z,u,u_z)u + A_3(z,u,u_z) \ .
\end{aligned}\right\} \tag{1.24}
$$

Applying the mean value theorem for derivatives we can also transform (1.15) into a complex form. Actually, from (1.15) we have

$$
B_1\, u_{xx} + B_2\, u_{xy} + B_3\, u_{yy} + B_4 = 0 \ , \tag{1.25}
$$

where $B_j = B_j(x,y,u,u_x,u_y,U_1,U_2,U_3) = \Phi_{U_j}(x,y,u,u_x,u_y,tU_1,tU_2,tU_3)$, $0 \le t \le 1$, $j = 1,2,3$, $U_1 = u_{xx}$, $U_2 = u_{xy}$, $U_3 = u_{yy}$, $B_4 = \Phi(x,y,u,u_x,u_y,0,0,0)$. Making use of the uniform ellipticity condition (1.8), the equation (1.25) can be changed to the complex form (1.24), but the coefficients Q , A_j $(j=1,2,3)$ are functions of $z \in D$, u , u_z , u_{zz} , $u_{z\bar{z}}$. □

3. Conditions for uniformly elliptic equations of second order

Without loss of generality, we may assume that the domain D is an $(N+1)$-connected circular domain as stated in § 1, Chapter 2. In fact, through a conformal mapping $\zeta(z)$ from a general $(N+1)$-connected domain D with boundary $L \in C^2_\mu$ $(0 < \mu \le 1)$ onto the above circular domain G , the equation (1.24) can be reduced to the form

$$
\begin{aligned}
U_{\zeta\bar{\zeta}} &= |z'(\zeta)|^2\, F[z(\zeta),U,U_\zeta/z'(\zeta),U_{\zeta\zeta}/[z'(\zeta)]^2 - U_\zeta z''(\zeta)/[z'(\zeta)]^2] \\
&= f(\zeta,U,U_\zeta,U_{\zeta\zeta}) \ ,
\end{aligned} \tag{1.26}
$$

where $U(\zeta) = u[z(\zeta)]$, $z(\zeta)$ is the inverse function of $\zeta(z)$. It is not difficult to see that the equation (1.26) still satisfies the uniform ellipticity condition, i.e.

$$
|f(\zeta,U,U_\zeta,V_1) - f(\zeta,U,U_\zeta,V_2)| \le q_0 |V_1 - V_2| \tag{1.27}
$$

in G .

Now, we assume that the equation (1.24) satisfies the following condition.

Condition C

(1) $Q(z,u,u_z,V)$, $A_j(z,u,u_z)$ $(j=1,2,3)$ are measurable in $z \in D$ for all continuously differentiable functions $u(z)$ in D and all measurable functions $V(z) \in L_{p_0}(\bar{D})$ $(p_0 > 2)$, and saitsfy

$$L_p[A_j(z,u,u_z,\bar{D}] \le k_0 , \tag{1.28}$$

and $A_j = 0$, for $z \notin D$, $j = 1,2,3$, where $p\ (2 < p_0 < p < \infty)$, $k_0 (0 \le k_0 < \infty)$ are real constants.

(2) $Q(z,u,u_z,V)$, $A_j(z,u,u_z)$ $(j=1,2,3)$ are continuous in $u \in \mathbb{R}$ (the real numbers) and $u_z \in \mathbb{C}$ for almost every point $z \in D$ and $V \in \mathbb{C}$.

(3) The equation (1.24) satisfies the uniform ellipticity condition (1.23) for almost every point $z \in D$, $u \in \mathbb{R}$, u_z , V_1 , $V_2 \in \mathbb{C}$.

If the equation (1.24) satisfies Condition C and the following condition

$$A_2[z,u(z),u_z] \ge 0 , \tag{1.29}$$

for almost every point $z \in D$, then we say that (1.24) satisfies Condition C*.

A so-called solution $u(z)$ for (1.24) in D means a continuously differentiable solution satisfying the condition $u(z) \in W^2_{p_0}(D^*)$, $2 < p_0 < p$, for any closed set D^* in D .

§ 2. Extremum principles for elliptic equations of second order

If we substitute a solution $u(z)$ into the coefficients of the equation (1.24), then this equation becomes a linear equation (1.6). Hence, we shall discuss only the linear equation (1.6) with Condition C* in this section. We rewrite (1.6) in the following form

$$Lu \equiv u_{z\bar{z}} - Re[Q(z)\, u_{zz} + A_1(z)\, u_z] - A_2(z)\; u = A_3(z) \; . \tag{2.1}$$

Condition C^* with $A_3(z) \geq 0$ and $A_3 \leq 0$ are denoted by Condition C_+^* and Condition C_-^*, respectively. Moreover, if the coefficients $Q(z)$, $A_j(z)$ $(j=1,2,3)$ are continuous on the closed domain $\bar{D}$, Condition C_+^* and Condition C_-^* are denoted by Condition C_+ and Condition C_-, respectively.

1. Extremum principles for equation (2.1) with continuous coefficients

Lemma 2.1 Let the equation (2.1) satisfy Condition C_+ and let $u(z)$ be a classical solution to (2.1) on $\bar{S} = \{|z-z_1| \leq R(<\infty)\}$. A classical solution $u(z)$ means that $u(z)$ is continuously differentiable up to the second order in $\bar{S}$. If $u(z)$ attains its nonnegaive maximum at a point $z_0 \in L = \{|z-z_1| = R\}$ and $u(z_0) > u(z)$, for $z \in S = \{|z-z_1| < R\}$, then

$$\varliminf_{z \to z_0} \frac{u(z_0) - u(z)}{|z - z_0|} > 0 \; , \tag{2.2}$$

where z belongs to a line $\vec{\ell}$ through z_0 in D, the angle between the direction of $\vec{\ell}$ and the outer normal vector of L in z_0 is less than $\pi/2$.

Proof. We introduce a function

$$V(z) = e^{-\alpha r^2} - e^{-\alpha R^2} \; , \; r = |z-z_1| \; ,$$

where α is a positive constant. Obviously, $V(z) > 0$ for $z \in S$ and $V(z) = 0$ for $z \in L$. Through a computation, we obtain

$$LV = e^{-\alpha r^2} \{[-\alpha + \alpha^2 r^2 - \alpha^2 Re[Q(\overline{z-z_1})^2] + \\ + \alpha\, Re[A_1(\overline{z-z_1})]\} - A_2 V \; .$$

Noting $\alpha^2 r^2 - \alpha^2 Re[Q(\overline{z-z_1})^2] \geq \alpha^2 r^2(1-q_0)$ and choosing α sufficiently large, it is clear that $LV > 0$ on $\bar{K} = \{R/2 \leq r = |z-z_1| \leq R\}$. Set

$$w(z) = u(z) - u(z_0) + \varepsilon V(z) , \text{ for } z \in \bar{K} , \tag{2.3}$$

where ε is a positive number. If we select ε small enough, then

$$w(z) < 0 \text{ on } r = |z-z_1| = \frac{R}{2} ; \; w(z) \leq 0 , \text{ on } |z-z_1| = R .$$

Next, we prove that

$$w(z) \leq 0 , \text{ for } z \in \bar{K} . \tag{2.4}$$

Otherwise, there would exist a point z_2 in $K = \{R/2 < |z-z_1| < R\}$, so that

$$w(z_2) = \max_{z \in \bar{K}} w(z) > 0 .$$

By the necessary condition for the maximum of functions of two variables, we have

$$w_z = \frac{1}{2}(w_x - iw_y) = 0 , \; w_{z\bar{z}} = \frac{1}{4}(w_{xx}+w_{yy}) \leq 0 ,$$

$$w_{xy}^2 \leq w_{xx} w_{yy} , \text{ i.e. } |w_{z\bar{z}}| \geq |w_{zz}| \text{ at } z = z_2 .$$

So $-w_{z\bar{z}} = |w_{z\bar{z}}| \geq |w_{zz}| \geq -Re[Qw_{zz}]$. Noting $-A_2 w \leq 0$, $Lw \leq 0$ can be obtained at $z = z_2$. However, from (2.3) it follows that

$$Lw = L\,u(z) - L\,u(z_0) + \varepsilon LV(z) > 0 , \text{ for } z \in \bar{K} .$$

This contradiction proves that (2.4) is true.

From $V(z_0) = 0$ and $u(z_0) - u(z) \geq \varepsilon V(z)$ for $z \in K$, we conclude

$$\underline{lim}_{z \to z_0} \frac{u(z_0) - u(z)}{|z-z_0|} \geq -\varepsilon \frac{\partial V}{\partial n} = 2\,\varepsilon\,\alpha\,r\,e^{-\alpha r^2} > 0\ , \tag{2.5}$$

where the limit is taken on the line from z_1 to z_0 , i.e. $z-z_0 = |z-z_0|\,|z_0-z_1|^{-1}\,(z_0-z_1)$ and $\vec{n}$ is the outward normal vector of L at z_0 .

Let $\vec{\ell}$ be a vector at z_0 as stated in the lemma. It is clear that $cos(\vec{\ell},\vec{n}) > 0$, $cos(\vec{\ell},\vec{s}) > 0$, and for $z \in \vec{s}$ ($\vec{s}$ is a tangent vector of L at z_0)

$$\underline{lim}_{z \to z_0} \frac{u(z_0) - u(z)}{|z-z_0|} \geq 0\ .$$

Hence we obtain (2.2). □

Using the above lemma, it is not difficult to prove the following maximum principle.

Theorem 2.2 Let (2.1) satisfy Condition C_+ and $u(z)$ be a continuous classical solution of (2.1) on $\bar{D}$. Then $u(z)$ cannot attain a nonnegative maximum in D , unless $u(z)$ is a constant, i.e. if $u(z)$ is not a constant, then

$$u(z) < M = \max_{z \in \bar{D}} u(z) \tag{2.6}$$

in D .

Proof. Suppose that $u(z)$ attains the nonnegative maximum M at $z_2 \in D$ and $u(z)$ is not a constant. Set $E = \{z \mid u(z) = M\ , z \in \bar{D}\}$. Obviously, $E \neq \bar{D}$. Hence there exist a point $z_1 \in D$, so that $u(z_1) < u(z_2)$ and $\rho(z_1,\Gamma) > \rho(z_1,E) > 0$, where $\rho(z_1,\Gamma)$ and $\rho(z_1,E)$ denote the distances from z_1 to Γ and E , respectively. Let S be a disc $|z-z_1| < R$, so that $\bar{S} \subset D$ and $\bar{S} \cap E = z_0$. It is clear that

$$u_z(z_0) = \frac{1}{2}[u_x(z_0) - i\,u_y(z_0)] = 0 .$$

This formula contradicts (2.2). Thus the theorem is proved. □

Similarly, we can prove the following minimum principle.

Theorem 2.3 Let (2.1) satisfy Condition C_- and $u(z)$ be a continuous classical solution of (2.1) on $\bar{D}$. Then $u(z)$ cannot attain a nonpositive minimum in D, unless $u(z)$ is a constant, i.e. if $u(z)$ is not a constant, then

$$u(z) > m = \min_{z\in\bar{D}} u(z) \tag{2.7}$$

in D.

Proof. Set $v(z) = -u(z)$. It is easy to see that $v(z)$ is a continuously differentiable solution of the second order equation

$$Lv = v_{z\bar{z}} - Re[Q(z)v_{zz} + A_1(z)v_z] - A_2(z)v = -A_3(z) \tag{2.8}$$

on $\bar{D}$, which satisfies Condition C_+. Applying Theorem 2.2, we know that if $v(z)$ is not a constant, then

$$v(z) < -m = \max_{z\in\bar{D}} v(z) \tag{2.9}$$

in D. From (2.9) the inequality (2.7) follows. □

From the proofs of Lemma 2.1 and Theorem 2.2, we easily obtain the following corollary.

Corollary 2.4 (1) Suppose that the equation (2.1) satisfies Condition C_+ and $A_2(z) \equiv 0$. Then the continuous classical solution $u(z)$ on $\bar{D}$ cannot attain its maximum in D, unless $u(z)$ is a constant.

(2) Suppose that (2.1) satisfies Condition C_- and $A_2(z) \equiv 0$. Then

the continuous classical solution $u(z)$ on $\bar{D}$ cannot attain its minimum in D, unless $u(z)$ is a constant.

We mention that the condition $A_2(z) \geq 0$ in Theorem 2.2 is indispensable. For instance, the continuous classical solution $u = \sin x \sin y$ of $u_{z\bar{z}} + \frac{1}{2}u = 0$, $A_2(z) = -\frac{1}{2}$ on $\bar{D} = \{0 \leq x \leq \pi, 0 \leq y \leq \pi\}$ is positive in the domain D and is equal to zero on the boundary of D.

2. Extremum principles for equation (2.1) with measurable coefficients

Theorem 2.5 Let the equation (2.1) satisfy Condition C_+^* and $u(z)$ be a solution of (2.1) in D. Then $u(z)$ cannot attain a positive maximum in D, unless $u(z)$ is a constant.

The proof my be decomposed into several steps, formulated as lemmas.

Lemma 2.6 Let the homogeneous equation

$$u_{z\bar{z}} = Re[Q(z)u_{zz} + A_1(z)u_z] \tag{2.10}$$

satisfy Condition C^* and $u(z)$ be a solution of (2.10) in D. Then $u(z)$ cannot attain a maximum or minimum in D, unless $u(z)$ is a constant.

Proof. Set $w(z) = u_z$. It is easily seen that $w(z)$ is a solution of the uniformly elliptic complex equation

$$w_{\bar{z}} = Re[Q(z)w_z + A_1(z)w] \tag{2.11}$$

in D. By Theorem 2.2 in Chapter 2, $u(z)$ can be expressed as

$$w(z) = \Phi[\zeta(z)]\, e^{\varphi(z)}\ . \tag{2.12}$$

If $u(z)$ is not a constant in D, the zeros of $w(z)$ in D are isolated points. Let z_0 be an extremal point of $u(z)$ in D.

Then $u_z = 1/2(u_x - iu_y) = 0$ at z_0. Denote by L a continuously differentiable level curve of $u(z)$ in the neighbourhood of z_0, so that the interior of L includes the point z_0. Let ℓ denote the positive direction of the tangent of L, and α be the angle between ℓ and the x-axis. Then

$$\frac{\partial u}{\partial \ell} = 0 = u_x \cos\alpha + u_y \sin\alpha = (u_z + u_{\bar{z}}) \cos\alpha + i(u_z - u_{\bar{z}}) \sin\alpha \tag{2.13}$$

$$= \overline{\lambda(z)}\, u_z + \lambda(z)\, u_{\bar{z}} = 2Re[\overline{\lambda(z)}\, u_z] = 2Re[\lambda(z) w] ,$$

where $\overline{\lambda(z)} = \cos\alpha + i \sin\alpha = e^{i\alpha}$. Provided the level curve L is sufficiently close to the point z_0, such that $w(z) \neq 0$ on L, from (2.13) it follows that

$$\frac{1}{2\pi} \Delta_L \arg w(z) = -\frac{1}{2\pi} \Delta_L \arg \overline{\lambda(z)} = -1 .$$

However, from (2.12) and the argument principle, we have $(1/2\pi)\, \Delta_L \arg w(z) > 0$. This contradiction proves that $u(z)$ cannot attain a maximum or minimum in D, unless $u(z)$ is constant. □

<u>Lemma 2.7</u> Let the nonhomogeneous equation

$$u_{z\bar{z}} = Re[Q(z) u_{zz} + A_1(u) u_z] + A_3(z) \tag{2.14}$$

satisfy Condition C_+^* and $u(z)$ be a continuous solution on $\bar{D}$. Then there exists a point z_1 on Γ so that

$$u(z_1) = \max_{z \in \bar{D}} u(z) . \tag{2.15}$$

<u>Proof.</u> Let $\bar{S}$ be a closed disc in D. If we can prove that the maximum of $u(z)$ on $\bar{S}$ is the maximum of $u(z)$ on the boundary ℓ of S, then it is clear that the statement in this theorem is true. There is no harm in putting $S = \{|z| < 1\}$. We choose a sequence of equations

$$u_{z\bar{z}} = Re[Q_n(z) u_{zz} + A_1^{(n)}(z) u_z] + A_3^{(n)}(z) \quad (n=1,2,\ldots) \tag{2.16}$$

so that the coefficients $Q_n(z)$, $A_1^{(n)}(z)$, $A_3^{(n)}(z)$ are continuously differentiable on $\bar{S}$ and satisfy

$$|Q_n(z)| \le q_0 < 1 \ , \ L_p[A_j^{(n)},\bar{S}] \le k_0 \ , \ j = 1,3 \ , \ A_3^{(n)}(z) \ge 0 \ ,$$

and $L_p[Q_n-Q,\bar{S}] \to 0$, $L_p[A_j^{(n)}-A_j,\bar{S}] \to 0$, for $n \to \infty$ (cf.[90]). By Theorem 4.7 in Chapter 2, we can seek solutions $w_n(z) = u_{nz}$ of (2.16) satisfying the boundary condition

$$Re\ w_n(z) = Re\ u_z \ , \ Im\ w_n(1) = Im\ u_z(1) \ , \ n = 1,2,\dots \ . \tag{2.17}$$

As in the proof of Theorem 2.6 in Chapter 2, from $\{w_n(z)\}$ we can select a subsequence $\{w_{n_k}(z)\}$ converging uniformly to the solution $w(z) = u_z$ of (2.14) on $\bar{S}$ and

$$u_{n_k}(z) = u(1) + Re\int_1^z 2w_{n_k}(t)\,dt \tag{2.18}$$

converging uniformly to $u(z) = u(1) + Re\int_1^z 2w(t)\,dt$. Because (2.16) satisfies Condition C_+, from Corollary 2.4, it can be derived that the maximum of $u(z)$ on $\bar{S}$ is equal to the maximum of $u(z)$ on ℓ. This completes the proof. □

Lemma 2.8 Under the hypothesis in Lemma 2.7, a continuous solution $u(z)$ of (2.14) on $\bar{D}$ cannot attain a maximum in D, unless $u(z)$ is a constant.

Proof. Suppose that $u(z)$ attains the maximum M in D and is not a constant. We can find a point z_0 and $\bar{S} = \{|z-z_1| \le \varepsilon(\varepsilon>0)\}$ in D, so that $|z_0-z_1| < \varepsilon$, $u(z_0) = M$ and $m < M$, where m, M are the minimum and maximum of u on $\bar{S}$. Similarly to the method in the proof of Lemma 2.7, we can find the solution $u_0(z)$ of the homogeneous equation (2.10), satisfying the boundary condition

$$u_0(z) = u(z) \ , \ \text{for} \ |z-z_1| = \varepsilon \ .$$

In fact,

$$u_0(z) = u(z_1+\varepsilon) + \mathit{Re}\int_{z_1+\varepsilon}^{z} 2w(t)\,dt ,$$

where $w(z)$ is a solution of the Dirichlet problem of (2.11) on S satisfying the boundary condition

$$\mathit{Re}\, w(z) = \mathit{Re}\, u_z ,\ |z-z_1| = \varepsilon ,\ \mathit{Im}\, w(z_1+\varepsilon) = \mathit{Im}\, u_z(z_1+\varepsilon) .$$

By Lemma 2.6, $u_0(z) < M$ in S, and by Lemma 2.7 $u(z) \le u_0(z)$. This contradiction proves this lemma. □

Proof of Theorem 2.5 Suppose that $u(z)$ attains a positive maximum at $z_0 \in D$. Then $u(z) \ge 0$ in a sufficiently small neighbourhood of z_0. The equation (2.1) can be rewritten in the form

$$u_{z\bar{z}} = \mathit{Re}[Q(z)u_{zz} + A_1(z)u_z] + A(z) , \tag{2.19}$$

where $A(z) = A_2(z)\, u(z) + A_3(z) \ge 0$. Hence, by Lemma 2.8, $u(z)$ attains the maximum at z_0 so $u(z)$ is a constant on D. □

Similarly, we can prove the following theorem on the minimum.

Theorem 2.9 Let (2.1) satisfy Condition C^*_- and $u(z)$ be a solution of (2.1) in D. Then $u(z)$ cannot attain a negative minimum in D, unless $u(z)$ is a constant.

The following result is a special case of Theorem 2.5 and Theorem 2.9.

Corollary 2.10 Suppose that the equation

$$u_{z\bar{z}} = \mathit{Re}[Q(z)u_{zz} + A_1(z)u_z] + A_2(z)\, u \tag{2.20}$$

satisfies Condition C^*, and $u(z)$ is a continuous solution on $\bar{D}$.

If $u(z) \not\equiv$ constant, then $u(z)$ cannot attain a positive maximum or a negative minimum in D.

Corollary 2.11 Under the hypothesis in Theorem 2.5 and if $u(z)$ is a continuous solution of (2.1) on $\bar{S} = \{|z-z_1| \le R(<\infty)\}$, the result of Lemma 2.1 remains valid.

Proof. Using the method in the proof of Lemma 2.1 and the result of Theorem 2.5, this corollary can be proved. □

§ 3. The representation theorem and compactness principle of solutions for equations of second order

In this section, we shall discuss the representation theorem of solutions for the uniformly elliptic equation (1.24). From this we may see the relation between the solution of the second order elliptic equation and the solution of the first order elliptic complex equation. Besides, the compactness principle of solutions for the second order elliptic equation is a generalization of the compactness principle for harmonic functions. Finally, we shall give the Harnack inequality for solutions to the homogeneous equation (2.20).

1. Representation theorem of solutions for elliptic equations of second order

We first give some lemmas.

Lemma 3.1 Let equation (1.6) satisfy Condition C^* in the unit disc $\mathbb{D} = \{|z| < 1\}$. Then (1.6) possesses a unique solution of the type

$$\psi(z) = \Pi_0 \rho = \frac{2}{\pi} \iint_{\mathbb{D}} \ell n \left|\frac{\zeta - z}{1-\bar{\zeta}z}\right| \rho(\zeta)\, d\sigma_\zeta \tag{3.1}$$

where $\rho(z)$ is a real measurable function in $\mathbb{D}$, $\rho(z) \in L_{p_0}(\mathbb{D})$, $2 < p_0 < p$, and ψ satisfies the estimates

$$C^1_\beta [\psi, \bar{\mathbb{D}}] \le M_1 , \quad L_{p_0}[|\psi_{z\bar{z}}| + |\psi_{zz}|, \bar{\mathbb{D}}] \le M_2 , \tag{3.2}$$

in which $\beta = 1 - 2/p_0$, $M_j = M_j(q_0,p_0,k_0,\mathbb{D})$ $(j=1,2)$.

Proof. In § 4, Chapter 1, we gave some properties of $\Pi_0\rho$. Substituting $\psi = \Pi_0\rho$ into the equation (1.6), we have

$$\rho - Re[Q\Pi_2\rho] = Re[A_1\Pi_1\rho + A_2\Pi_0\rho] + A_3 \qquad (3.3)$$

where $\Pi_1\rho = (\Pi_0\rho)_z$, $\Pi_2\rho = (\Pi_0\rho)_{zz}$. If we can find a solution $\rho \in L_{p_0}(\overline{\mathbb{D}})$ of (3.3), then $\psi = \Pi_0\rho$ is just a solution of (1.6).

By Theorem 4.5 in Chapter 1, it is known that $I - Re[Q\Pi_2]$ possesses an inverse operator R in $L_{p_0}(\overline{\mathbb{D}})$, and

$$\begin{aligned} \rho &= R\{Re[A_1\Pi_1\rho]\} + R\{A_2\Pi_0\rho\} + R[A_3] \\ &= R_1\rho + RA_3 \end{aligned}$$

is equivalent to the integral equation (3.3). Because $\Pi_0\rho$, $\Pi_1\rho$ are complete continuous operators, R_1 is also complete continuous (cf. [97]2)). This shows that the Fredholm theorem can be applied to (3.3). Let $\rho \in L_{p_0}(\overline{\mathbb{D}})$ be a solution of the homogeneous equation

$$\rho - Re[Q\Pi_2\rho] = Re[A_1\Pi_1\rho] + A_2\Pi_0\rho . \qquad (3.4)$$

Then $u = \Pi_0\rho$ is a solution of

$$Lu = u_{z\bar{z}} - Re[Qu_{zz} + A_1u_z] - A_2u = 0 . \qquad (3.5)$$

Noting that $u(z) = \Pi_0\rho = 0$ on the boundary $\Gamma = \{|z| = 1\}$ of $\mathbb{D}$, from Corollary 2.10, it follows that $u(z) \equiv 0$ in $\mathbb{D}$ and $\rho(z) = u_{z\bar{z}} = 0$. By the Fredholm theorem, we see that the nonhomogeneous integral equation (3.3) possesses a unique solution $\rho(z) \in L_{p_0}(\overline{\mathbb{D}})$. Consequently, the nonhomogeneous differential equation (1.6) possesses a solution of the type (3.1). □

The estimate (3.2) for the general equation (1.6) will be proved after the proof of Lemma 3.2. Now, we discuss the special case where $A_2(z) = 0$, i.e. let $\psi_0(z)$ be a solution of the Dirichlet problem of the uniformly elliptic equation

$$u_{z\bar{z}} - Re[Q(z)u_{zz}] = Re[A_1(z)u_z] + A_3(z) \tag{3.6}$$

with the boundary condition

$$\psi_0(t) = 0 \ , \ t \in \Gamma \ . \tag{3.7}$$

Set $w(z) = \psi_{0z}$, then it is evident that $w(z)$ is a solution to

$$w_{\bar{z}} - Re[Q(z)w_z] = Re[A_1(z)w] + A_3(z) \tag{3.8}$$

with the boundary condition

$$Re[i\,t\,w(t)] = 0 \ , \ \text{for} \ \ t \in \Gamma \ . \tag{3.9}$$

By Theorem 4.3 in Chapter 2, $w(z)$ satisfies the estimate

$$C_\beta[w,\bar{\mathbb{D}}] \le M_3 \ , \ L_{p_0}[|w_{\bar{z}}| + |w_z| \ , \ \bar{\mathbb{D}}] \le M_4 \ , \tag{3.10}$$

where $\beta = 1 - 2/p_0$, $M_j = M_j(q_0,p_0,k_0,\bar{\mathbb{D}})$, $j = 3,4$. From

$$\psi_0(z) = Re\int_1^z 2w(t)dt$$

and (3.10), it is easy to see that $\psi_0(z)$ satisfies

$$C_\beta[\psi_0,\bar{\mathbb{D}}] \le M_5 \ , \ L_{p_0}[|\psi_{0\bar{z}}| + |\psi_{0z}|,\bar{\mathbb{D}}] \le M_6 \ , \tag{3.11}$$

where $M_j = M_j(q_0,p_0,k_0,\mathbb{D})$, $j = 5,6$.

Lemma 3.2 Let the homogeneous equation (3.5) satisfy Condition C^* in $\mathbb{D}$. Then (3.5) possesses a unique solution of the type

$$\Psi(z) = 1 + \Pi_0\rho \ , \ \rho(z) \in L_{p_0}(\overline{\mathbb{D}}) \ , \ 2 < p_0 < p \ , \tag{3.12}$$

and $\Psi(z)$ satisfies the estimates

$$C^1_\beta[\Psi,\overline{\mathbb{D}}] \le M_7 \ , \ L_{p_0}[|\Psi_{z\bar{z}}| + |\Psi_{zz}|,\overline{\mathbb{D}}] \le M_8 \ , \tag{3.13}$$

$$\Psi(z) \le 1 \ , \ \Psi(z) \ge M_9 > 0 \ , \tag{3.14}$$

where $\beta = 1 - 2/p_0$, $M_j = M_j(q_0,p_0,k_0,\mathbb{D})$, $j = 7,8,9$.

Proof. By Lemma 3.1, the nonhomogeneous equation

$$u_{z\bar{z}} = Re[Q(z)u_{zz} + A_1(z)u_z + A_2(z)u] + A_2(z) \tag{3.15}$$

possesses a unique solution $\psi(z) = \Pi_0\rho$, $\rho(z) \in L_{p_0}(\overline{\mathbb{D}})$. Hence, equation (3.5) possesses a unique solution $\Psi(z) = 1 + \Pi_0\rho$. From Corollary 2.10, it follows that $\Psi(z) = 1 + \Pi_0\rho$ satisfies

$$0 \le \Psi(z) \le 1 \ , \ \text{for} \ \ z \in \overline{\mathbb{D}} \ . \tag{3.16}$$

So $L_p[A_2\Psi,\overline{\mathbb{D}}] \le k_0$. According to (3.11), we can see that Ψ satisfies (3.13). It remains to show that Ψ satisfies the second inequality in (3.14). Applying Lemma 3.1, we can see that $v(z) = \Pi_0\rho$ is a solution of the equation

$$u_{z\bar{z}} - Re[Qu_{zz} + A_1u_z] = A_2$$

in $\mathbb{D}$, satisfying the estimates in (3.11). Setting $W(z) = e^{v(z)} - \Psi(z)$, we obtain

$$LW = Le^v - L\Phi = e^v\{v_{z\bar{z}} + |v_z|^2 - Re[Q(v_{zz}+v_z^2) - A_1v_z] - A_2\}$$

$$= e^v[|v_z|^2 - Re\ Qv_z^2] \ge 0$$

in $\mathbb{D}$ and $W(z) \le 0$ for $z \in \Gamma$. By Lemma 2.8, it is easy to see that

$$W(z) = e^{v(z)} - \Psi(z) \leq 0 \ , \text{ i.e. } \ \Psi(z) \geq e^{v(z)} \geq e^{-M_5} = M_9 > 0$$

in $\overline{\mathbb{D}}$. □

Now we return to prove (3.2). Let us denote the solution $\psi(z) = \Psi(z)\ \psi_0(z)$, where $\Psi(z)$ is a function stated in Lemma 3.2, and $\psi_0(z) = \Pi_0\rho$ is a solution of the equation

$$\psi_{0_{z\bar{z}}} - Re[Q\psi_{0_{zz}}] = Re\ \{[-2(\ell n\Psi)_{\bar{z}} + 2Q(\ell n\Psi)_z + A_1]\psi_{0_z}\} + A_3/\Psi$$

satisfying the estimates in (3.11). From (3.11) and (3.13), it follows that $\psi(z)$ satisfies the estimates in (3.2). □

Next, we give the representation theorem for solutions of (1.6) and (1.24).

Theorem 3.3 Suppose that the equation (1.24) satisfies Condition C^* and $u(z)$ is a solution of (1.24) in the (n+1)-connected circular domain D . Then $u(z)$ can be expressed as follows

$$u(z) = [Re \int_0^z 2w(t)\ dt + c_0]\ \Psi(z) + \psi(z) \ , \tag{3.17}$$

in which $\psi(z) = \Pi_0\rho$, $\Psi(z) = 1 + \Pi_0\sigma$, and $\psi(z)$, $\Psi(z)$ satisfy the estimates (3.2), (3.13) and (3.14), respectively, c_0 is a real constant, and

$$w(z) = \Phi[\zeta(z)]e^{\varphi(z)} \tag{3.18}$$

where

$$\varphi(z) = Tg - T_1 g = -\frac{1}{\pi}\iint_D \frac{(z-1)g(\zeta)}{(\zeta-z)(\zeta-1)}\, d\sigma_\zeta \ , \ g(z) \in L_{p_0}(\overline{D}) \ ,$$

$\zeta(z)$ is a homeomorphism which maps D onto a circular domain G similar to D , $\zeta(0) = 0$, $\zeta(1) = 1$, $\Phi(\zeta)$ is an analytic function in G , and $\varphi(z)$, $\zeta(z)$ and its inverse function $z(\zeta)$ satisfy the following estimates:

$$C_\beta[\varphi,\overline{D}] \le M_{10}\ ,\ L_{p_0}[|\varphi_{z\bar{z}}| + |\varphi_{zz}|,\overline{D}] \le M_{11}\ , \tag{3.19}$$

$$C_\beta[\zeta,\overline{D}] \le M_{12}\ ,\ L_{p_0}[|\zeta_{z\bar{z}}| + |\zeta_{zz}|,\overline{D}] \le M_{13}\ , \tag{3.20}$$

$$C_\beta[z,\overline{G}] \le M_{14}\ ,\ L_{p_0}[|z_{\zeta\bar{\zeta}}| + |z_{\zeta\zeta}|,\overline{G}] \le M_{15}\ , \tag{3.21}$$

where $\beta = 1 - 2/p_0$, $M_j = M_j(q_0,p_0,k_0,D)$, $j = 10,\ldots,15$.

Proof. We consider the following equations

$$\psi_{z\bar{z}} = Re[Q\psi_{zz} + A_1\psi_z + A_2\psi] + A_3\ , \tag{3.22}$$

$$\Psi_{z\bar{z}} = Re[Q\Psi_{zz} + A_1\Psi_z + A_2\Psi]\ , \tag{3.23}$$

$$w_{\bar{z}} = Re[Qw_z + A_0 w]\ , \tag{3.24}$$

where $Q = Q(z,u,u_z,u_{zz})$, $A_j = A_j[z,u,u_z]$, $j = 1,2,3$, $A_0 = -2(\ell n\Psi)_{\bar{z}} + 2Q(\ell n\Psi)_z + A_1$, in which $u(z)$ is a solution as stated in this theorem. By Lemma 3.1 and Lemma 3.2, we see that the equations (3.22), (3.23) have solutions $\psi(z) = \Pi_0\rho$, $\Psi(z) = 1 + \Pi_0\sigma$, respectively, and $\psi(z)$, $\Psi(z)$ satisfy (3.2), (3.13), and (3.14). Set

$$U(z) = [u(z) - \psi(z)]/\Psi(z)\ . \tag{3.25}$$

Through immediate computation, it is shown that $U(z)$ is a solution of the following equation

$$U_{z\bar{z}} = Re[QU_{zz} + A_0 U_z] \tag{3.26}$$

in D . So $w(z) = U_z$ is a solution of the complex equation of first order (3.24) in D . From Condition C^*, (3.13) and (3.14), we have

$$L_{p_0}[A_0,\overline{D}] = L_{p_0}[-2(\ell n\Psi)_{\bar{z}} + 2Q(\ell n\Psi)_z + A_1,\overline{D}] \tag{3.27}$$

$$\leq k_1 = k_1(q_0,p_0,k_0,D) \ .$$

In virtue of Theorem 2.2 in Chapter 2, $w(z)$ can be expressed as (3.18), where $\varphi(z)$, $\zeta(z)$ and its inverse function $z(\zeta)$ satisfy (3.19) to (3.21). From $w(z) = U_z$, we conclude

$$U(z) = Re\int_0^z 2w(t)\, d(t) + U(0)$$

and admit $c_0 = U(0) = [u(0) - \psi(0)]/\Psi(0)$. Thus, $u(z)$ possesses the representation (3.17). □

2. The compactness principle for sequences of solutions to second order equations

We consider a sequence of equations

$$u_{z\bar{z}} = F_n(z,u,u_z,u_{zz}) \ , \tag{3.28}$$

$$F_n(z,u,u_z,u_{zz}) = \sigma_n(z)\ F(z,u,u_z,u_{zz}) \ , \ n = 1,2,\ldots$$

where $\sigma_n(z)$, $F(z,u,u_z,u_{zz})$ are stated in (2.34) of Chapter 2 and (1.24), respectively. Moreover, set $Q_n = \sigma_n(z)\ Q(z,u,u_z,u_{zz})$, $A_j^{(n)} = \sigma_n(z)\ A_j(z,u,u_z)$ $(j=1,2,3)$.

Theorem 3.4 Let the second order equation (1.24) satisfy Condition C^*, and $u_n(z)$ $(n=1,2,\ldots)$ be solutions of (3.28). If $\{u_n(z)\}$, $\{u_{nz}(z)\}$ are uniformly bounded on any closed set D^* in D, then from $\{u_n(z)\}$ we can choose a subsequence $\{u_{n_k}(z)\}$ which converges uniformly to a solution $u(z)$ of (1.24) on D^*.

Proof. We introduce a nonnegative function $\eta_m(z)$ in the unit disc $\mathbb{D}$, which possesses continuous partial derivatives up to the second order and

$$\eta_m(z) = \begin{cases} 1 \ , & \text{for} \quad z \in D_m \ , \\ 0 \ , & \text{for} \quad z \notin D_{2m} \ , \end{cases} \tag{3.29}$$

where m is a positive integer, D_m is stated in (2.34) of Chapter 2. For a fixed positive integer m, we denote $V_n^{(m)}(z) = \eta_m(z)\, u_n(z)$ and set $V_n^{(m)}(z) = 0$, for $z \notin D_{2m}$. So the function $V_n^{(m)}(z)$ is defined in $\mathbb{D}$, and satisfies the equation

$$V^{(m)}_{nz\bar{z}} - Re[Q_n V^{(m)}_{nzz}] = B_n^{(m)} \tag{3.30}$$

for almost every point in $\mathbb{D}$, in which

$$B_n^{(m)} = \eta_m(z)\, F_n(z,u_n,u_{nz},0) + 2\,Re\,[\eta_{mz} u_{n\bar{z}}] + $$

$$+ \eta_{mz\bar{z}} u_n - Re[Q_n(2\eta_{mz} u_{mz} + \eta_{mzz} u_n)] \ .$$

It is easy to see that $B_n^{(m)}$ satisfies

$$L_{p_0}[B_n^{(m)}, \bar{\mathbb{D}}] \le k_2 = k_2(q_0,p_0,k_0,k_3,D,\eta_m) \tag{3.31}$$

where

$$k_3 = \max_{1 \le n < \infty} C^1[u_n, \overline{D_{2m}}] \ .$$

Because $V_n^{(m)}(z)$ is a solution of (3.30) in $\mathbb{D}$ satisfying the boundary condition $V_n^{(m)}(z) = 0$ for $|z| = 1$ it can be expressed as

$$V_n^{(m)}(z) = \Pi_0 \rho_n^{(m)} = \frac{2}{\pi} \iint_{\mathbb{D}} \ln \left|\frac{\zeta - z}{1 - \bar{\zeta} z}\right| \rho_n^{(m)}(\zeta)\, d\sigma_\zeta \ . \tag{3.32}$$

By Condition C* and Lemma 3.1, $V_n^{(m)}(z)$ satisfies the estimate

$$C^1_\beta[V_n^{(m)}, \bar{\mathbb{D}}] \le M_{16} = M_{16}(q_0,p_0,k_0,k_3,D,\eta_m) \ . \tag{3.33}$$

Thus we can select a subsequence $\{V_{n,m}(z)\}$ from $\{V_n^{(m)}(z)\}$, and

$\{V_{n,m}(z)\}$, $\{V_{n,mz}\}$ converge uniformly to $V_0(z)$, V_{0z} on $\mathbb{D}$, respectively. If for instance we take $m = 2$, then noting the definition of $\eta_m(z)$, $u_n^{(2)}(z) = u_n(z)$ for $z \in D_2$, $\{V_n^{(2)}(z)\}$ has a subsequence $\{u_{n,2}(z)\}$. Denote $u_0(z) = V_0(z)$ for $z \in D_2$. Next, we take $m = 3$. Of $\{u_{n,2}(z)\}$, there exists a subsequence $\{u_{n,3}(z)\}$, and $\{u_{n,3}(z)\}$, $\{u_{n,3z}\}$ converge uniformly to $u_0(z)$, u_{0z} on D_3 , respectively. Similarly, we can find a subsequence of $\{u_{n,m}(z)\}$ corresponding to $m = 4,5,\ldots$, and $\{u_{n,m}(z)\}$, $\{u_{n,mz}\}$ converge uniformly to $u_0(z)$, u_{0z} on D_m $(m=4,5,\ldots)$, respectively. Afterwards, from $\{u_{n,m}(z)\}$ $(m=2,3,\ldots)$, we select the diagonal sequence $\{u_{m,m}(z)\}$ $(m=2,3,\ldots)$, and $\{u_{m,m}(z)\}$, $\{u_{m,mz}\}$ converge uniformly to $u_0(z)$, u_{0z} on any closed set in D .

It remains to prove that $u_0(z)$ is a solution of (1.24) in D . We are free to choose a point $z_0 \in D$ and a disc $K = \{|z-z_0| \le r\}$ $(r > 0)$, so that $K \subset D$. Because $w_n(z) = u_{nz}$ is a solution of the first order complex equation

$$\left.\begin{aligned} &w_{n\bar{z}} - Re[Q_n w_{nz}] = B_n\ , \quad Q_n = Q_n(z,u_n,u_{nz},u_{nzz})\ , \\ &B_n = Re[A_1^{(n)} u_{nz}] + A_2^{(n)} u_n + A_3^{(n)}\ , \quad A_j^{(n)} = A_j^{(n)}(z,u_n,u_{nz}) \\ &\qquad\qquad\qquad\qquad\qquad\qquad\qquad (j=1,2,3) \end{aligned}\right\} \tag{3.34}$$

on K and

$$|B_n| \le k_4 = k_4(p_0,k_0,k_3,D,K)\ ,$$

$$B_n(z) \to B_0(z) = Re[A_1^{(0)} u_{0z}] + A_2^{(0)} u_0 + A_3^{(0)}\ , \quad \text{when} \quad n \to \infty$$

for almost every point on K . By Theorem 2.6 in Chapter 2, $w_0 = u_{0z}$ is a solution of the equation (3.34) with $n = 0$ on K . Hence $u_0(z)$ is a solution of the equation (1.24) on K . Noting that $z_0 \in D$ is arbitrary, we easily see that $u_0(z)$ is a solution of (1.24) in D . This completes the proof. □

We mention that the condition in Theorem 3.4 may be weakened because, using the method in the proof of Lemma 2.7, Theorem 3.3, and the result of Lemma 3.8, from the uniform boundedness of $\{u_n(z)\}$ the uniformly boundedness of $\{u_{nz}\}$ on any closed set in D can be derived. Thus we have

Corollary 3.5 Suppose that (1.24) satisfies Condition C^* and $\{u_n(z)\}$ is a sequence of solutions for (3.28). If $\{u_n(z)\}$ is uniformly bounded on any closed set D^* in D, then from $\{u_n(z)\}$ we can choose a subsequence $\{u_{n_k}(z)\}$ which converges uniformly to a solution $u_0(z)$ of (1.24) on D^*.

Using the same method as stated above we can prove

Corollary 3.6 Let (1.24) satisfy Condition C^* and $\{u_n(z)\}$ be a sequence of solutions for (1.24). If $\{u_n(z)\}$ is uniformly bounded on any closed set D^* in D, then there exists a subsequence $\{u_{n_k}(z)\}$, which converges uniformly to a solution $u(z)$ of (1.24) on D^*.

3. The Harnack inequality for positive solutions to (2.20)

To prove the Harnack inequality of positive solutions of (2.20), we need to discuss the linear case of (3.28) with $A_2^{(n)} = A_3^{(n)} = 0$ in $S = \{|z-z_0| < 1\}$. Set

$$u_{z\bar{z}} - Re[Q_n(z)u_{zz}] = Re[A_1^{(n)}(z)u_z] . \tag{3.35}$$

We shall give the Poisson integral formula for solutions of (3.35) in S. Let

$$R(z,t) = Re\left(\frac{t+z-2z_0}{t-z_0}\right), \quad z \in S \ , \ t \in \Gamma = \{|t-z_0| = 1\} . \tag{3.36}$$

It is easy to see that

$$B_n(z,t) = Re[Q_n(z)R_{zz}+A_1^{(n)}(z)R_z] \in L_p(\bar{S}) \ , \ p > 2 .$$

Lemma 3.7 Suppose that (3.35) satisfies Condition C^* in S. Then there exists the Poisson kernel

$$P(z,t) = R(z,t) + H(z,t) \ , \ z \in \bar{S} \ , \ t \in \Gamma , \tag{3.37}$$

where $H(z,t)$ is a solution of the type

$$H(z,t) = \frac{2}{\pi} \iint_S \ell n \left| \frac{\zeta - z}{1-(\zeta-\bar{z}_0)(z-z_0)} \right| \rho(\zeta,t)\, d\sigma_\zeta \tag{3.38}$$

of the nonhomogeneous equation

$$u_{z\bar{z}} - Re[Q_n(z)u_{zz}] = Re[A_1^{(n)}(z)u_z] + B_n(z,t) \ , \tag{3.39}$$

and the continuous solution $u(z)$ of (3.35) on $\bar{S}$ can be expressed as follows:

$$u(z) = \frac{1}{2\pi} \int_\Gamma P(z,t)\, u(t)\, d\theta \ , \ z \in \bar{S} \ . \tag{3.40}$$

Proof. From Lemma 3.1, we see that (3.35) has a unique solution of the type (3.38), and satisfies the estimates

$$C_\beta^1[H,\bar{S}] \le M_{17} \ , \ L_{p_0}[|H_{z\bar{z}}| + |H_{zz}|, S_m] \le M_{18} \ , \tag{3.41}$$

where $2 < p_0 < p$, $\beta = 1 - 2/p_0$, $M_j = M_j(q_0,p_0,k_0,S,S_m)$, $S_m = \{|z-z_0| \le 1 - \frac{1}{m}, m > 1\}$, $j = 17,18$. As in Theorem 3.3 in Chapter 1, we can prove that $H(z,t)$, $H_z(z,t)$ are continuous on $t \in \Gamma$, and the function $P(z,t)$ in (3.37) is a solution of (3.35) in S with the parameter $t \in \Gamma$ and $P(z,t) = 0$ on $\Gamma \setminus \{t\}$. Moreover, $P(z,t) > 0$ for $z \in S$. Hence we have (3.40). □

Lemma 3.8 Let (3.35) satisfy Condition C^* and $u(z)$ be a bounded solution of (3.35) on $\bar{S}$ with the condition $|u(z)| \le M_{19} < \infty$ in S . Then $u(z)$ satisfies the estimate

$$C^1[u,S_m] \le M_{20} = M_{20}(q_0,p_0,k_0,S,S_m,M_{19}) \ . \tag{3.42}$$

Proof. If we can prove the Poisson kernel $P(z,t)$ in (3.37) to satisfy

$$C[P_z,S_m] \le M_{21} = M_{21}(q_0,p_0,k_0,S,S_m,M_{19}) \ , \tag{3.43}$$

then from (3.40), (3.24) follows. For the sake of convenience, we assume $S = \{|z| < 1\}$ and $t = 1$. By Theorem 2.2 in Chapter 2, the solution $w(z) = P_z(z,1)$ of the complex equation of first order

$$w_{\bar{z}} - Re[Q_n(z)w_z] = Re[A_1^{(n)}(z)w] \tag{3.44}$$

can be represented by

$$w(z) = \Phi[\zeta(z)]e^{\varphi(z)} , \tag{3.45}$$

where $\zeta(z)$ topologically maps S onto $G=\{|\zeta|<1\}$, $\varphi(z)$, $\zeta(z)$ and its inverse function $z(\zeta)$ satisfy the estimates of the types (3.19) to (3.21). Noting that $Q_n(z) = A_1^{(n)}(z) = 0$, $z \notin S_n$, we know that $\varphi(z)$, $\zeta(z)$, $\Phi[\zeta(z)]$ are analytic in $S \setminus S_n$, and $\Phi[\zeta(z)]$ is continuous on $\bar{S} \setminus \{1\}$ possessing a pole of second order at $t = 1$. From $P(z,1) = 0$ on $\Gamma \setminus \{1\}$, it follows that

$$Re[izP_z(z,1)] = Im[zw(z)] = 0 \quad \text{for} \quad z \in \Gamma \setminus \{1\} , \tag{3.46}$$

and

$$Im\{z(\zeta)\Phi(\zeta)e^{\varphi[z(\zeta)]}\} = 0 \quad \text{for} \quad \zeta \in L \setminus \{1\} , \; L = \zeta(\Gamma) . \tag{3.47}$$

It is not difficult to find an analytic function $\Psi(\zeta)$ in S satisfying the boundary condition

$$Im\ \Psi(\zeta) = arg\ z(\zeta) - arg\ \zeta + Im\ \varphi[z(\zeta)] , \; Re\ \Psi(0) = 0 .$$

So $f(\zeta) = \Phi(\zeta)e^{\Psi(\zeta)}(1-\zeta)^2$ is analytic in G and its continuous on $\bar{G}$. Noting that $\zeta(1-\zeta)^{-2}$ is a real function on $L = \{|\zeta| = 1\}$ and $f(\zeta)$ satisfies the boundary condition

$$Im[f(\zeta)] = 0 , \; \zeta \in L = \zeta(\Gamma) , \tag{3.48}$$

we see that

$$f(\zeta) = \Phi(\zeta)e^{\Psi(\zeta)}(1-\zeta)^2 = a \quad (a \text{ is a real constant}) ,$$

and $w(z) = \Phi[\zeta(z)]e^{\varphi(z)} = a\ e^{\varphi(z)-\Psi[\zeta(z)]}[1-\zeta(z)]^{-2}$. Hence

$$|w(z)/a| \geq M_{22} > 0 , \; |w(z)| = |P_z(z,1)| \leq |a|M_{23}|1-z|^{-\frac{2}{\beta}} , \tag{3.49}$$

where $2 < p_0 < p$, $\beta = 1 - 2/p_0$, $M_j = M_j(q_0,p_0,k_0,S)$, $j = 22,23$. Now, let Γ_* be a differentiable curve on $\bar{S}$ from $z = 0$ to a point z on $|z| = 1/2$, so that the tangent of Γ_* is parallel to the direction of the vector $\overline{w(z)} = \overline{P_z(z,1)}$. Consequently

$$P(0,1) \geq P(0,1) - P(z,1) = Re \int_{\Gamma_*} 2w(t)\,dt = \tag{3.50}$$

$$= 2 \int_{\Gamma_*} |w(t)|\,|dt| \geq M_{24}\,|a| \ , \ |a| \leq M_{24}^{-1} P(0,1) \ ,$$

where $M_{24} = M_{24}(q_0, p_0, k_0, S, S_m) > 0$. If 1 is replaced by $t \in \Gamma$, (3.49) remains valid. From (3.40), (3.49) and (3.50) we conclude

$$|u_z| \leq \frac{1}{2\pi} \int_0^{2\pi} |P_z(z,t)|\,|u(t)|d\theta \tag{3.51}$$

$$\leq \frac{M_{23}}{2\pi M_{24}|1-z|^{2/\beta}} \int_0^{2\pi} P(0,t)|u(t)|d\theta$$

$$\leq \frac{M_{23}}{2\pi M_{24}|1-z|^{2/\beta}} \int_0^{2\pi} P(0,t)[2M_{19}+u(t)]d\theta$$

$$= \frac{[2M_{19}+u(0)]M_{23}}{M_{24}|1-z|^{2/\beta}} \ ,$$

and then (3.42) is derived. □

Finally, we prove the Harnack inequality.

Theorem 3.9 Let the second order equation (3.5) satisfy Condition C^* in $D = \{|z-z_*| < R\}$ $(0 < R < \infty)$ and $u(z)$ be a positive solution possessing Hölder continuous partial derivatives. Then

$$\frac{1}{q(r)} \leq \frac{u(z)}{u(z_*)} \leq q(r) \ , \ r = |z-z_*| \ , \ z \in D \ , \tag{3.52}$$

where $q(r)$ is a continuous positive function of r $(0 \leq r < 1)$, $q(0) = 1$ and $q(r) = q(r, q_0, p_0, k_0, D)$.

Proof. Without loss of generality, we may assume that D is the unit disc. Let $\Psi(z)$ be a positive solution of (3.5) as stated in (3.12).

Then $U(z) = u(z)/\Psi(z)$ is a positive solution of the equation

$$U_{z\bar{z}} = Re[QU_{zz}+A_0U_z] \ , \ A_0 = -2(\ell n\Psi)_{\bar{z}} + 2Q(\ell n\Psi)_z + A_1 \ . \tag{3.53}$$

By (3.51)

$$|U_z| \le M_{23} \ U(0)/[M_{24}|1-z|^{2/\beta}]$$

and from $U(z) - U(0) = 2Re\int_0^z U_\tau d\tau$, it follows that

$$\left|\frac{U(z)-U(0)}{U(0)}\right| \le \frac{2}{U(0)} \int_{\overline{0z}} |U_\tau| \ |d\tau| \le \frac{2M_{23}|z|}{M_{24}|1-z|^{2/\beta}} \ .$$

This shows that there exists a nonnegative continuous function $q_*(r)$ $(0 \le r < 1)$ with $q_*(0) = 1$ so that

$$U(z) \le q_*(r)U(0) \ , \ 0 \le r = |z| < 1 \ , \ z \in D \ . \tag{3.54}$$

As for $U(z) \ge u(0)/q_*(r)$, we may choose a linear fractional transformation $z = z(\zeta)$ which maps $\mathbb{D}$ onto itself and maps $\zeta_* \neq 0$ $(|\zeta_*| \le R_* < 1)$ onto $z = 0$, and consider the function $U_*(\zeta) = U[z(\zeta)]$. As before, it can be shown that $U_*(\zeta) \le q_*(r)U_*(0)$, and then $U(0) \le q_*(r)U(z)$. Hence $u(z) = U(z)\Psi(z)$ satisfies (3.52). □

§ 4. The regular oblique derivative Problem for elliptic equations of second order

In this section, we consider the regular oblique derivative problem for the uniformly elliptic equation (1.24) in an (N+1)-connected circular domain D . For this purpose we first prove the existence of solutions for the Dirichlet problem of (1.24).

1. The Dirichlet problem for second order equations

Now, we state the first boundary value problem for the equation (1.24) in an (N+1)-connected domain D .

Problem I. Find a continuous differentiable solution $u(z)$ of equation (1.24) on $\overline{D}$, satisfying the boundary condition

$$u(t) = r(t) \ , \ t \in \Gamma \ , \tag{4.1}$$

where $C_\alpha^1[r,\Gamma] \le \ell < \infty$, $0 < \alpha < 1$.

Problem I with $r(t) \equiv 0$ is called Problem I_0.

In order to prove the uniqueness of and estimates for solutions to Problem I for the nonlinear equation (1.24), we suppose that (1.24) satisfies

$$\begin{aligned} &F(z,u_1,u_{1z},V)-F(z,u_2,u_{2z},V) = \\ &= Re[a_1(u_1-u_2)_z + a_2(u_1-u_2)] \ , \end{aligned} \tag{4.2}$$

in which $a_j = a_j(z,u_1,u_2)$ $(j=1,2)$, $L_p[a_j,\overline{D}] \le k_1 < \infty$, $j = 1,2$, $0 \le a_2 \le k_1$, k_1 is a real constant. Obviously, the linear equation (1.6) with Condition C^* satisfies the condition (4.2).

Theorem 4.1 Let the equation (1.24) satisfy Condition C^* and (4.2). Then the solution to the Problem I for (1.24) is unique.

Proof. Let $u_1(z)$, $u_2(z)$ be two solutions to Problem I for (1.24) and $u(z) = u_1(z) - u_2(z)$. From Condition C^* and (4.2), $u(z)$ is a solution of the following uniformly elliptic equation

$$u_{z\bar{z}} = Re[q(z)u_{zz} + a_1u_z + a_2u] \ , \tag{4.3}$$

where

$$Re[q(z)u_{zz}] = F(z,u_1,u_{1z},u_{1zz})-F(z,u_1,u_{1z},u_{2zz}) \ .$$

Noting that $u(z)$ satisfies the boundary condition

$$u(t) = 0 \ , \ t \in \Gamma \ ,$$

and applying Corollary 2.10, it is easy to see that $u(z) = 0$, $z \in D$, i.e. $u_1(z) = u_2(z)$, $z \in D$. □

Theorem 4.2 Suppose that the equation (1.24) satisfies Condition C^* and $u(z)$ is a solution to Problem I for (1.24). Then $u(z)$ satisfies the estimates

$$C^1_{\alpha\beta^2}[u,\bar{D}] \le M_1 , \quad L_{p_0}[|u_{z\bar{z}}| + |u_{zz}|,\bar{D}_m] \le M_2 , \tag{4.4}$$

where $\beta = 1 - 2/p_0$, $2 < p_0 < p$, $M_1 = M_1(q_0,p_0,k_0,\alpha,\ell,D)$, $M_2 = M_2(q_0,p_0,k_0,\alpha,\ell,D_m)$, $D_m = \{z | z\in D , |z-t| \ge 1/m , t\in I\}$ (m is a positive integer).

Proof. We may assume $\alpha \le \beta$, and first prove that the solution $u(z)$ satisfies the estimate

$$C[u,\bar{D}] \le M_3 = M_3(q_0,p_0,k_0,\alpha,\ell,D) . \tag{4.5}$$

Substituting $u(z)$ into the coefficients of the equation (1.24) and denoting $Q(z) = Q(z,u,u_z,u_{zz})$, $A_j(z) = A_j(z,u,u_z)$, $j = 1,2,3$, we obtain a linear equation (1.6), which satisfies Condition C^*. By Theorem 3.3, the solution $u(z)$ of the equation (1.6) can be expressed as

$$u(z) = [Re \int_0^z 2w(s)\, ds + c_0]\Psi(z) + \psi(z) = U(z)\Psi(z) + \psi(z) , \tag{4.6}$$

in which $w(z)$, $\Psi(z)$, $\psi(z)$ and c_0 are as stated in (3.17), (3.18). From (4.1), it follows that $U(z)$ satisfies the boundary condition

$$U(t) = r_0(t) = [r(t) - \psi(t)]/\Psi(t) , \quad t \in \Gamma . \tag{4.7}$$

We differentiate (4.7) with respect to the argument θ_j of $z - z_j = \gamma_j e^{i\theta_j}$ on $\Gamma_j = \{|z-z_j| = \gamma_j\}$, and obtain

$$Re[\overline{\lambda(t)}w(t)] = \tau(t) = \frac{\partial r_0(t)}{\partial\theta_j} , \quad t \in \Gamma_j , \tag{4.8}$$

where $\overline{\lambda(t)} = i(t-z_j)$, $t \in \Gamma_j$, $j = 0,1,\ldots,N$, and $r_0(t)$, $\tau(t)$ satisfy the conditions

$$C^1_\alpha[r_0,\Gamma] \le \ell_1 , \quad C_\alpha[\tau,\Gamma] \le \ell_1 , \tag{4.9}$$

where $\ell_1 = \ell_1(q_0,p_0,k_0,\alpha,\ell,D)$. The function $U(z)$ is a solution of the equation

$$U_{z\bar{z}} = Re[QU_{zz} + AU_z] , \quad A = -2(\ell n\Psi)_{\bar{z}} + 2Q(\ell n\Psi)_z + A_1 \tag{4.10}$$

in D . From Corollary 2.10, it follows that $U(z)$ satisfies the estimate

$$C[U,\bar{D}] \le M_4 = M_4(q_0,p_0,k_0,\alpha,\ell,D) , \tag{4.11}$$

and then (4.5) holds.

Next we prove that $w(z) = u_z$ satisfies the estimate

$$C[w,\bar{D}] \le M_5 = M_5(q_0,p_0,k_0,\alpha,\ell,D) . \tag{4.12}$$

Otherwise, there exists a corresponding sequence of equations

$$w_{\bar{z}} = Re[Q_n w_z + A_n w] \quad (n=1,2,\ldots) \tag{4.13}$$

and their solutions $w_n(z) = U_{nz}$ $(n=1,2,\ldots)$, so that $C[w_n,\bar{D}] = H_n \to \infty$, for $n \to \infty$. We may assume $H_n \ge 1$ and consider $W_n(z) = w_n(z)/H_n$. Because $w_n(z)$ satisfies the boundary condition

$$Re[\overline{\lambda(t)}w_n(t)] = \tau_n(t) , \quad t \in \Gamma \tag{4.14}$$

similar to (4.8), and the index of $\lambda(t)$ is equal to $K = N-1$, by Theorem 4.2 in Chapter 2, we have

$$C_{\alpha\beta^2}[W_n,\bar{D}] \le M_6 = M_6(q_0,p_0,k_0,\alpha,\ell,D) . \tag{4.15}$$

Hence, we may choose a subsequence $\{W_{n_k}(z)\}$, which converges uniformly to $W_0(z)$ on $\overline{D}$. For the sake of convenience, we relabel the subsequence as $\{W_n(z)\}$. As $V_n(z) = U_n(z)/H_n$ is a solution of the complex equation

$$V_{n\bar{z}} = \overline{W_n(z)} \tag{4.16}$$

is D satisfying the boundary condition

$$Re[iV_n(t)] = 0 \ , \ t \in \Gamma \ , \tag{4.17}$$

we conclude

$$C_\alpha[V_n,\overline{D}] \le M_7 = M_7(q_0,p_0,k_0,\alpha,\ell,D) \ . \tag{4.18}$$

So $\{V_n(z)\}$, $\{V_{nz}\} = \{W_n(z)\}$ converge uniformly to 0 on $\overline{D}$. However, from $C[W_n,\overline{D}] = 1$, we can show that there is a point $t_* \in \overline{D}$ so that $|W_0(z_*)| = 1$. This contradiction shows that (4.12) is true.

By Theorem 4.2 in Chapter 2, we can obtain the first estimate in (4.4). It remains to show the second estimate in (4.4). Applying the Cauchy formula for analytic functions, we have

$$C[\Phi'(\zeta),G_m] \le M_8 = M_8(q_0,p_0,k_0,\alpha,\ell,D_m) \ , \tag{4.19}$$

where $G_m = \zeta[D_m]$, and from (4.6), the second estimate in (4.4) is derived. □

In order to use the Leray-Schauder theorem to prove the solvability for Problem I for (1.24), we give a lemma.

Lemma 4.3 Under the hypothesis in Theorem 4.2, the solution $u(z)$ of Problem I for (3.28) can be expressed as

$$u(z) = U(z) + \Psi(z) , \Psi(z) = H_0\rho = \frac{2}{\pi}\iint_{D_n} \ln|\zeta-z| \rho(\zeta)d\sigma_\zeta , \tag{4.20}$$

where $\rho(z) \in L_{p_0}(\overline{D}_n)$, $2 < p_0 < p < \infty$, $U(z)$ is a harmonic function, and $U(z)$, $\Psi(z)$, $\rho(z)$, u_z satisfy the estimates:

$$C^1_{\alpha\beta^2}[U,\overline{D}] \le M_9 , C^1_\beta[\Psi,\overline{D}] \le M_{10} , \tag{4.21}$$

$$L_{p_0}[\rho,\overline{D}_n] = L_{p_0}[u_{z\bar{z}},\overline{D}] < M_{11} , L_{p_0}[u_{zz},\overline{D}] \le M_{12} , \tag{4.22}$$

where $\beta = 1 - 2/p_0$, $2 < p_0 < p$, $M_j = M_j(q_0,p_0,k_0,\alpha,\ell,D_n)$, $j = 9,\ldots,12$.

Proof. Let the solution $u(z)$ be substituted into (3.28) and denote $\rho(z) = F_n(z,u,u_z,u_{zz})$. Noting $[u(z)-\Psi(z)]_{z\bar{z}} = u_{z\bar{z}} - \rho(z) = 0$, we see that $U(z) = u(z) - \Psi(z)$ is a harmonic function in D , i.e. (4.20) holds. Due to $\rho(z) = 0$ for $z \notin \overline{D}$ and (4.4), we have (4.22) and, therefore, (4.21). □

Theorem 4.4 Let (1.24) satisfy Condition C^*. Then Problem I for (1.24) is solvable.

Proof. We first prove that Problem I for the equation (3.28) has a solution. From the representation (4.20), we only find a solution $\Psi(z) = H_0\rho(\rho(z) \in L_{p_0}(D_n))$ of

$$\Psi_{z\bar{z}} = F_n[z,U+\Psi,U_z+\Psi_z,U_{zz}+\Psi_{zz}] , \tag{4.23}$$

so that $u(z) = U(z) + \Psi(z)$ satisfies the boundary condition (4.1). Or we find a solution $\rho(z) \in L_{p_0}(\overline{D}_n)$ of the integral equation

$$\rho = F_n(z,u,u_z,u_{zz} + \Pi\rho) , \Pi\rho = -\frac{1}{\pi}\iint_{D_n} \frac{\rho(\zeta)}{(\zeta-z)^2} d\sigma_\zeta \tag{4.24}$$

where $u(z) = U(z) + \Psi(z) = U(z) + H_0\rho$, $\rho(z) \in L_{p_0}(\overline{D}_n)$.

Let us introduce the Banach space $B = L_{p_0}(\overline{D})$, and let B_M be an open set in B , the elements of which are all real measurable functions satisfying the inequality

$$L_{p_0}[\rho, D_n] < M_{11} , \tag{4.25}$$

where M_{11} is the constant in (4.22). We are free to select $\rho \in B_M$, and then

$$\Psi(z) = H_0\rho = \frac{2}{\pi}\iint_{D_n} \ell n|\zeta - z|\ \rho(\zeta)d\sigma_\zeta \in C^1_\beta(\overline{D})$$

and $\Psi_{z\bar{z}} = \Pi\rho \in L_{p_0}(\overline{D})$. Moreover, we can find a unique harmonic function $U(z)$ in D satisfying the boundary condition

$$U(t) = r(t) - \Psi(t) \ , \ t \in \Gamma \ . \tag{4.26}$$

Substituting $u(z) = U(z) + \Psi(z)$, u_z , U_{zz} into the corresponding positions of (4.24) and using the principle of contracting mappings one can show that there exists a unique solution $\rho^*(z) \in B$ of the following integral equation with the parameter k :

$$\rho^*(z) = kF_n(z,u,u_z,u_{zz}+\Pi\rho^*) \ , \ 0 \le k \le 1 \ . \tag{4.27}$$

Denote by $\rho^* = S(\rho,k)$ $(0 \le k \le 1)$ this mapping from $\rho(z)$ onto $\rho^*(z)$. We can verify that it satisfies the three conditions of the Leray-Schauder theorem:

1) For every $k \in [0,1]$, $\rho^* = S(\rho,k)$ maps the Banach space B continuously into itself. Actually, we arbitrarily choose $\rho_m(z) \in \overline{B}_M$, $m = 0,1,2,\ldots$, so that $L_{p_0}[\rho_m - \rho_0, \overline{D}_n] \to 0$, for $m \to \infty$. Let $\rho^*_m = S(\rho_m,k)$, i.e.

$$\rho^*_m = kF_n(z,u_m,u_{mz},U_{mzz}+\Pi\rho^*_m) \ , \ 0 \le k \le 1 \ , \tag{4.28}$$

where $u_m(z) = U_m(z) + H_0\rho_m$. We can obtain

$$\rho_m^* - \rho_0^* = k[F_n(z,u_m,u_{mz},U_{mzz}+\Pi\rho_m^*) - \quad (4.29)$$
$$- F_n(z,u_m,u_{mz},U_{mzz}+\Pi\rho_0^*) + B_m(z)] \ ,$$

in which $B_m(z) = F_n(z,u_m,u_{mz},U_{mzz}+\Pi\rho_0^*) - F_n(z,u_0,u_{0z},u_{0zz},\Pi\rho_0^*)$. By Condition C^* and similarly to the proof of Theorem 4.6 in Chapter 2, we can show that $L_{p_0}[B_m,\overline{D}_n] \to 0$ for $m \to \infty$. From (4.29), we have

$$L_{p_0}[\rho_m^*-\rho_0^*,\overline{D}_n] \le \frac{1}{1-k\,q_0\,\Lambda_{p_0}} L_{p_0}[B_m,\overline{D}_n] \ . \quad (4.30)$$

So $L_{p_0}[\rho_m^*-\rho_0^*,\overline{D}_n] \to 0$ for $m \to \infty$. This shows that $\rho^* = S(\rho,k)$ continuously maps B into itself.

By analogy with the above, we can prove that $\rho^* = S(\rho,k)$ is completely continuous on B_M, and for $\rho(z) \in \overline{B}_M$, $S(\rho,k)$ is uniformly continuous with respect to $k \in [0,1]$.

2) If $k = 0$, it is obvious that $\rho^* = S(\rho,0) = 0 \in B_M$.

3) From (4.22), we easily see that $\rho^* = S(\rho,k)$ $(0 \le k \le 1)$ has no solution $\rho(z)$ on $\partial B_M = \overline{B}_M \setminus B_M$.

Hence, by the Leray-Schauder theorem, there exists a solution $\rho(z)$ $(\in B_M)$ to the integral equation (4.24). Thus $\Psi = H_0\rho$ is a solutions of (4.23), i.e. $u(z) = U(z) + \Psi(z)$ is a solution of (3.28).

Finally, because Problem I for the equations (3.28) $(m=1,2,\ldots)$ has solutions $u_m(z)$ $(m=1,2,\ldots)$, respectively, and the solutions $u_m(z)$ satisfy the estimates (4.4), we can choose a subsequence $\{u_{m_k}(z)\}$ which uniformly converges to $u_0(z)$ satisfying the boundary condition (4.1). By Theorem 3.4, $u_0(z)$ is a solution of (1.24) in D. Hence $u(z)$ is a solution of Problem I for (1.24). □

2. Formulation and uniqueness of the regular oblique derivative problem for (1.24)

Problem IV. The so-called regular oblique derivative problem for equation (1.24) is to find a continously differentiable solution to

(1.24) on $\overline{D}$ satisfying the boundary condition

$$\left.\begin{aligned} &\frac{\partial u}{\partial \nu} + 2\sigma(t)\, u(t) = 2[\tau(t)+h(t)]\ ,\ t \in \Gamma\ ,\ u(a_0) = b_0\ , \\ &\text{i.e. } Re[\overline{\lambda(t)}u_t] + \sigma(t)\, u(t) = \tau(t) + h(t)\ ,\ t \in \Gamma\ ,\ u(a_0) = b_0\ , \end{aligned}\right\} \quad (4.31)$$

where $cos(\nu,n) \geq 0$ on Γ and $\overline{\lambda(t)} = cos(\nu,x) + i\, sin(\nu,x)$, $\sigma(t)$, $\tau(t)$ and b_0 satisfy

$$C_\alpha[\lambda,\Gamma] \leq \ell\ ,\ \sigma(t) \geq 0\ ,\ C_\alpha[\sigma,\Gamma] \leq \ell\ , \qquad (4.32)$$
$$C_\alpha[\tau,\Gamma] \leq \ell\ ,\ |b_0| \leq \ell\ .$$

If $cos(\nu,n) = 0$ and $\sigma(t) = 0$ on Γ_j , we set

$$\int_{\Gamma_j} \tau(t)ds = 0\ ,\ u(a_j) = b_j\ ,\ 1 \leq j \leq N\ . \qquad (4.33)$$

In (4.32) and (4.33), $\alpha(1/2 < \alpha < 1)$ and $\ell(0 \leq \ell < \infty)$ are constant, $a_j \in \Gamma_j$, b_j is a real constant, $|b_j| \leq \ell$, $0 \leq j \leq N$, and

$$h(t) = \begin{cases} h_0\ , & t \in \Gamma_0\ , \\ 0\ , & t \in \Gamma - \Gamma_0\ , \end{cases}$$

h_0 is an unknown real constant to be determined appropriately. Just as stated in § 2 of Chapter 1, we may assume that $cos(\nu,n) = 0$, $\sigma(t) = 0$ on $\Gamma_* = \Gamma_1 \cup \ldots \cup \Gamma_{N_0}$ $(N_0 \leq N)$ and $cos(\nu,n)$ and $\sigma(t)$ do not both vanish identically on Γ_j $(N_0 < j \leq N$, $\Gamma_{**} = \bigcup_{j=N_0+1}^{N} \Gamma_j)$. The above boundary value problem includes the Dirichlet problem (Problem I), the Neumann problem (Problem II) and the third boundary value problem (Problem III) as special cases. In fact, if $cos(\nu,n) > 0$, Problem IV is Problem III.

In order to prove the uniqueness of solutions to Problem IV for the nonlinear equation (1.24) with Condition C^*, we suppose that (1.24) satisfies the condition (4.2).

Theorem 4.5 Let equation (1.24) satisfy Condition C^* and (4.2). Then Problem IV for (1.24) is uniquely solvable.

Proof. Let $u_1(z)$, $u_2(z)$ be two solutions of Problem IV for (1.24) and $u(z) = u_1(z) - u_2(z)$. From Condition C^* and (4.2), $u(z)$ is a solution of the following uniformly elliptic equation

$$u_{z\bar{z}} = Re[q(z)u_{zz} + a_1 u_z + a_2 u] , \tag{4.34}$$

where $q(z)u_{zz}$ is as stated in (4.3), and $|q(z)| \le q_0 < 1$, $L_p[a_j,\bar{D}] \le k_1$, $j = 1,2$, $a_2 \ge 0$. Because $u(z)$ satisfies the boundary condition

$$\frac{\partial u}{\partial \nu} + 2\sigma(t)\,u(t) = 2h(t) \ , \ t \in \Gamma \ , \ u(a_j) = 0 \ , \ j = 0,1,\ldots,N_0 \ , \tag{4.35}$$

if $u(z) \not\equiv 0$ in D , then by Corollary 2.10, we see that there exists a point t_* or $t_{**} \in \Gamma$ so that $M = u(t_*) = \max_{z\in\bar{D}} u(z) > 0$ or $m = u(t_{**}) = \min_{z\in\bar{D}} u(z) < 0$. Using the same method as that in the proof of Theorem 2.2, Chapter 1, we can derive a contradiction. Hence $u(z) \equiv 0$ for $z \in D$, i.e. $u_1(z) \equiv u_2(z)$ in D . □

3. A priori estimates and existence of solutions to Problem III for (1.24)

First of all, we shall prove the boundedness of solutions to Problem III for (1.24).

Lemma 4.6 Suppose that $u(z)$ is a solution for Problem III of the equation (1.24) satisfying Condition C^*. Then for every point $t \in \Gamma$, we have

$$\frac{\partial u}{\partial \nu} \text{ or } u(t) \ge -M_{13} \ , \ \frac{\partial u}{\partial \nu} \text{ or } u(t) \le M_{13} \ , \tag{4.36}$$

and

$$|h| \le M_{14} , \tag{4.37}$$

where $M_j = M_j(q_0,p_0,k_0,\alpha,\ell,D)$, $j = 13,14$.

Proof. Let the solution $u(z)$ be substituted into the coefficients of (1.24). Then (1.24) becomes a linear uniformly elliptic equation (1.6). We denote by $\psi(z)$, $\Psi(z)$ the solutions of (1.6), (2.20) with the boundary conditions $\psi(t) = 0$, $\Psi(t) = 1$, respectively. By Theorem 4.2, $\psi(z)$ and $\Psi(z)$ satisfy the estimate (4.4) and similar to the proof of Lemma 3.2, we can conclude

$$\Psi(z) \ge M_{15} = M_{15}(q_0,p_0,k_0,\alpha,\ell,D) > 0 . \tag{4.38}$$

It is clear that the function

$$U(z) = [u(z)-\psi(z)]/\Psi(z) - b_0 \tag{4.39}$$

is a solution of the homogeneous equation

$$LU = U_{z\bar{z}} - Re[QU_{zz}+AU_z] , \ A = -2(\ell n\Psi)_{\bar{z}} + 2Q(\ell n\Psi)_z + A_1 \tag{4.40}$$

satisfying the boundary condition

$$\frac{\partial U}{\partial \nu} + 2\sigma^*(t)\, U(t) = 2\tau^*(t) + h(t) , \ t \in \Gamma , \ U(a_0) = 0 , \tag{4.41}$$

where

$$2\sigma^*(t) = 2\sigma(t) + \frac{\partial \Psi}{\partial \nu} , \ 2\tau^*(t) = 2\tau(t) - \frac{\partial \psi}{\partial \nu}$$

and

$$C_\alpha[\sigma^*,\Gamma] \le \ell_2 , \ C_\alpha[\tau^*,\Gamma] \le \ell_2 = \ell_2(q_0,p_0,k_0,\alpha,\ell,D) .$$

By Corollary 2.10, there exist points b_1 , $b_2 \in \Gamma$, so that

$$M = \max_{z\in\bar{D}} U(z) = U(b_1) \ge 0 , \ m = \min_{z\in\bar{D}} U(z) = U(b_2) \le 0 . \tag{4.42}$$

Obviously, we have

$$\frac{\partial U}{\partial \nu} + 2\sigma^*(t)\ U(t) = 2\tau^*(t) + h \ge 0 \quad \text{at} \quad t = b_1\ , \text{ i.e. } \quad h \ge -2\tau^*(b_1)$$

and

$$\frac{\partial U}{\partial \nu} + 2\sigma^*(t)\ U(t) = 2\tau^*(t) + h \le 0 \quad \text{at} \quad t = b_2\ , \text{ i.e. } \quad h \le -2\tau^*(b_2)\ .$$

So (4.37) is true.

We denote $\Gamma^1 = \{t\,|\,U(t) \ge 0\ ,\ t \in \Gamma\}$ and $\Gamma^2 = \{t\,|\,U(t) < 0\ ,\ t \in \Gamma\}$, and can see that

$$\frac{\partial U}{\partial \nu} = -2\sigma^*(t)\,U(t) + 2\tau^*(t) + h \le 2[\tau^*(t) - \tau^*(b_2)] = M_{16} \quad \text{on} \quad \Gamma^1 \qquad (4.43)$$

and

$$\frac{\partial U}{\partial \nu} = -2\sigma^*(t)\,U(t) + 2\tau^*(t) + h \ge 2[\tau^*(t) - \tau^*(b_1)] = M_{17} \quad \text{on} \quad \Gamma^2\ . \qquad (4.44)$$

Hence (4.36) remains valid. □

Theorem 4.7 Under the hypothesis of Lemma 4.6, the solution $u(z)$ for Problem III of (1.24) satisfies the estimate

$$C[u,\overline{D}] \le M_{18} = M_{18}(q_0, p_0, k_0, \alpha, \ell, D)\ . \qquad (4.45)$$

Proof. As stated in the proof of Lemma 4.6, we discuss only the solution $U(z)$ of the equation (4.40) with the boundary condition (4.41).

Suppose that $U(z)$ is not a constant in D . Without loss of generality, we can assume $U(0) = 0$. Setting $M = \max_{z \in \overline{D}} U(z)$, it is easy to see that the function $M\Psi(z) - U(z)$ is a positive solution in D and from Lemma 3.2 and Theorem 3.9, we have

$$\frac{1}{q(r)} \le \frac{M\Psi(z) - U(z)}{M\Psi(0)}\ , \text{ i.e. } U(z) \le M[\Psi(z) - \frac{\Psi(0)}{q(r)}] \le M(1 - \frac{M_{15}}{2}) \qquad (4.46)$$

in $|z| \le r_0$, where M_{15} is a postitive constant as stated in (4.38) and r_0 is a sufficiently small posititve number. Moreover, we find a solution $\eta(z)$ of (4.40) in $D_0 = D \cap \{|z| > r_0\}$ with the boundary condition

$$\eta(t) = \begin{cases} 0 , & t \in \Gamma , \\ 1 , & t \in \Gamma_{N+2} = \{|t| = r_0\} , \end{cases} \tag{4.47}$$

and then $\eta(z) > 0$ in D_0 and

$$\gamma_k \frac{\partial\eta}{\partial s} = 2Re[i(t-z_k)\eta_t] = 0 , \gamma_k \frac{\partial\eta}{\partial n} = 2(-1)^m Re[(t-z_k)\eta_t] < 0 , t \in \Gamma ,$$

where $m = 1$, for $k = 1,\ldots,N$ and $m = 0$, for $k = N+1$. Thus $\eta_t \neq 0$, on $\Gamma + \Gamma_{N+2}$. It is clear that $w = \eta_z$ is a solution to the equation of first order

$$w_{\bar{z}} = Re[Qw_z + Aw] , A = -2(\ell n\Psi)_{\bar{z}} + 2Q(\ell n\Psi)_z + A_1 \tag{4.48}$$

in D_0 and satisfies the boundary condition $Re[i(t-z_k)\eta_t] = 0$, for $t \in \Gamma_k$, $k = 1,\ldots,N+2$ with index N. Hence, η_z has N zero points $c_1,\ldots,c_N$ in D_0. Using Theorem 3.9, we conclude

$$M_{19} = \max_{1\le k\le N} [\max_{|z-c_k|=r_k} U(z)] \le M(1-\frac{M_{15}}{2}) \tag{4.49}$$

where r_k $(k=1,\ldots,N)$ are sufficient small positive numbers. We appropriately choose a positive constant c, so that the function $\varphi(z) = e^{c\eta(z)}$ satisfies

$$\frac{\partial\varphi}{\partial\nu} = c\varphi[\frac{\partial\eta}{\partial n} \cos(\nu,n) + \frac{\partial\eta}{\partial s} \cos(\nu,s)] = c\varphi \cos(\nu,n) \frac{\partial\eta}{\partial n} < -1 \quad \text{for } t \in \Gamma .$$

Because

$$\varphi_z = c\varphi\eta_z , \varphi_{z\bar{z}} = c^2\varphi|\eta_z|^2 + c\varphi\eta_{z\bar{z}} , \varphi_{zz} = c^2\varphi\eta_z^2 + c\varphi\eta_{zz} , \tag{4.50}$$

and η_z has no zeros in $D^* = \bar{D} \cap \{|z| \ge r_0\} \cap\ldots\cap\{|z-c_N| \ge r_N\}$, we can choose the constant $c = c(q_0,p_0,k_0,D)$ sufficiently large, so that

$$L\varphi = c\varphi L\,\eta + c^2\varphi[|\eta_z|^2 - Re(Q\eta_z^2)] \geq 0 \quad \text{in} \quad D^* . \tag{4.51}$$

Now, introducing a function $V(z) = M_{16}\varphi(z) + U(z)$ it is easy to see that

$$LV = M_{16}L\varphi + LU \geq 0 , \tag{4.52}$$

and $\partial V/\partial\nu < -M_{16} + M_{16} = 0$ for all points t with the condition $\partial U/\partial\nu \leq M_{16}$ on Γ . So $V(z)$ cannot attain a maximum on these points. By Lemma 2.6, $V(z)$ cannot attain a maximum in $D^* \setminus \partial D^*$. Hence, there are two possible cases:

(1) $V(z)$ attains its maximum at a point t satisfying the condition $U(t) \leq M_{16}$ on Γ .

(2) $V(z)$ attains the maximum at a point

$$b_3 \in \bigcup_{k=0}^{N} \{|z-c_k| = r_k^*\} .$$

In this case, we have

$$V(b_1) = M_{16}\varphi(b_1) + M \leq M_{16}\varphi(b_3) + U(b_3) \leq$$

$$\leq M_{16}\varphi(b_3) + M(1-\frac{M_{15}}{2}) , \text{ i.e. } M \leq M_{16}[\varphi(b_3)-1] + M(1-\frac{M_{15}}{2})$$

where $M = \max_{z\in\overline{D}} U(z) = U(b_1)$, $b_1 \in \Gamma$. Thus

$$M = \max_{z\in\overline{D}} U(z) \leq \max\,[M_{16}, \frac{2M_{16}}{M_{15}}(\varphi(b_3)-1)] .$$

Similarly, we can obtain the estimate on a lower bound of $U(z)$ on $\overline{D}$.

This completes the proof. □

Theorem 4.8 Under the hypthesis of Lemma 4.6, the solution $u(z)$ for Problem III of (1.24) satisfies the estimates

$$C^1_\beta[u,\overline{D}] \le M_{17}\ ,\ L_{p_0}[|u_{z\bar{z}}| + |u_{zz}|,\overline{D}] < M_{18}\ , \tag{4.53}$$

where $\beta = 1 - 2/p_0$ $(<\alpha)$, $2 < p_0 < min(p,1/1-\alpha)$, $M_j = M_j(q_0,p_0,k_0,\alpha,\ell,D)$, $j = 17,18$.

Proof. Through the transformation

$$w(z) = u_z + B(z)\ u(z)\ ,\ \text{i.e. } u_{\bar{z}} = \overline{-B(z)}\ u(z) + \overline{w(z)}\ , \tag{4.54}$$

where $B(z)$ is a harmonic function with the boundary values $\lambda(t)\,\sigma(t)$, the boundary condition (4.31) can be reduced to the following boundary condition

$$Re[\overline{\lambda(t)}w(t)] = \tau(t) + h(t)\ ,\ t \in \Gamma\ . \tag{4.55}$$

Noting that

$$u_{z\bar{z}} = w_{\bar{z}} - Bu_{\bar{z}} - B_{\bar{z}}u = w_{\bar{z}} - B\bar{w} + (|B|^2 - B_{\bar{z}})u\ ,$$
$$u_{zz} = w_z - Bu_z - B_z u = w_z - Bw + (B^2 - B_z)u\ ,$$

the equation (1.24) can be transformed into the following complex equation

$$w_{\bar{z}} = Re[Qw_z + A_1 w - QBw] + B\bar{w} + A\ , \tag{4.56}$$
$$A = A_2 u + A_3 + Re[Q(B^2 - B_z)u - A_1 Bu] - (|B|^2 - B_{\bar{z}})u\ .$$

By Theorem 4.7, the above equation satisfies a similar Condition C. From Theorem 4.3 in Chapter 2, the solution $w(z)$ of (4.56) satisfies the estimates

$$C_\beta[w,\overline{D}] \le M_{19}\ ,\ L_{p_0}[|w_{\bar{z}}| + |w_z|,\overline{D}] \le M_{20}\ , \tag{4.57}$$

where $M_j = M_j(q_0, p_0, k_0, \alpha, \ell, D)$, $j = 19, 20$. Because $u(z)$ is a solution for the Dirichlet problem with the boundary condition

$$Re[iu(t)] = 0\ ,\ t \in \Gamma\ ,\ u(a_0) = b_0\ ,$$

for the first order complex equation (4.54), we also have the estimates

$$C_\beta[u, \overline{D}] \le M_{21}\ ,\ L_{p_0}[u_z, \overline{D}] \le M_{22}\ , \qquad (4.58)$$

in which $M_j = M_j(q_0, p_0, k_0, \alpha, \ell, D)$, $j = 21, 22$. From (4.57) and (4.58), it follows that the first estimate in (4.53) holds. The second estimate in (4.53) is easily derived by the first estimate. □

If we apply the method in the proof of the following Theorem 4.10, then the estimates in Theorem 4.8 can also be derived.

Finally, we prove the existence of solutions for Problem III of (1.24).

Theorem 4.9 Let the equation (1.24) satisfy Condition C^*. Then Problem III of (1.24) has a solution.

Proof. Using the result in Theorem 4.8, we can prove this theorem in accordance with Lemma 4.3 and Theorem 4.4, provided the constant M_{11} in (4.25) is replaced by M_{18} in (4.53), and the boundary condition (4.26) is replaced by

$$\begin{cases} \dfrac{\partial U}{\partial \nu} + 2\sigma(t)U(t) = 2\tau(t) + h(t) - \dfrac{\partial \Psi}{\partial \nu} - 2\sigma(t)\Psi(t)\ ,\ t \in \Gamma\ , \\ U(a_j) = b_j - \Psi(a_j)\ ,\ j = 0, 1, \ldots, N_0\ , \end{cases}$$

where $\Psi(z)$ is as stated in (4.20). □

4. A priori estimates and existence of solutions to Problem IV for (1.24)

We first give a priori estimates for the solutions.

Theorem 4.10 Suppose that the equation (1.24) satisifies Condition C^* and $u(z)$ is a solution of Problem IV for (1.24). Then $u(z)$ satisfies the following estimates:

$$C^1_\beta[u,\bar{D}] \le M_{23} \ , \ L_{p_0}[|u_{z\bar{z}}| + |u_{zz}|,\bar{D}] \le M_{24} \ , \tag{4.59}$$

where $\beta = 1 - 2/p_0 < \alpha$, $2 < p_0 < min(p,1/(1-\alpha))$, $M_j = M_j(q_0,p_0,k_0,\alpha,\ell,D)$, $j = 23,24$.

Proof. We first prove that the solution $u(z)$ of Problem IV for (1.24) satisfies the estimate

$$C^1[u,\bar{D}] \le M_{25} = M_{25}(q_0,p_0,k_0,\alpha,\ell,D) \ . \tag{4.60}$$

By Theorem 3.3, $u(z)$ can be expressed as

$$\left.\begin{aligned} u(z) &= [Re\int_{a_0}^{z} 2w(\tau)d\tau + b_0]\Psi(z) + \psi(z) \ , \\ w(z) &= \Phi[\zeta(z)]e^{\varphi(z)} \end{aligned}\right\} \tag{4.61}$$

where $\psi(z)$, $\Psi(z)$, $\varphi(z)$, $\zeta(z)$ and $\Phi(\zeta)$ are as stated in Theorem 3.3 with the boundary condition $\psi(t) = 0$, $\Psi(t) = 1$ on Γ . Noting that $\psi(z)$, $\Psi(z)$ satisfy the estimates (3.2), (3.13) and (3.14), respectively, and $\Psi(z)$ attains the maximum 1 on Γ and then $\partial\Psi/\partial\nu \ge 0$ on Γ , we see that $U(z) = [u(z)-\psi(z)]/\Psi(z)$ is a solution of the equation (4.40) satisfying the boundary condition

$$\left.\begin{aligned} &\frac{\partial U}{\partial \nu} + 2\sigma^*(t)\ U(t) = 2\tau^*(t) + h(t) \ , \\ &\text{i.e. } Re[\overline{\lambda(t)}U_t] + \sigma^*(t)U(t) = \tau^*(t) + h(t) \ , \ t \in \Gamma, \\ &U(a_j) = b_j \quad (j=0,1,\dots,N_0) \ , \end{aligned}\right\} \tag{4.62}$$

where $2\sigma^*(t) = 2\sigma(t) + \partial\Psi/\partial\nu$ and $2\tau^*(t) = 2\sigma(t) - \partial\psi/\partial\nu$ satisfy the conditions

$$C_\alpha[\sigma^*,\Gamma] \le \ell_3 \ , \ C_\alpha[\tau^*,\Gamma] \le \ell_3 \ , \tag{4.63}$$

where $\ell_3 = \ell_3(q_0,p_0,k_0,\alpha,\ell,D)$.

Suppose that the estimate (4.60) is not true. There exist sequences $\{\lambda_n(t)\}$, $\{\sigma_n(t)\}$, $\{\tau_n(t)\}$ and $\{b_j^{(n)}\}$ with the same conditions as $\lambda(t)$, $\sigma(t)$, $\tau(t)$ and b_j , so that $\{\lambda_n(t)\}$, $\{\sigma_n(t)\}$, $\{\tau_n(t)\}$ and $\{b_j^{(n)}\}$ uniformly converge to $\lambda_0(t)$, $\sigma_0(t)$, $\tau_0(t)$ and $b_j^{(0)}$ on Γ , respectively, and the corresponding equations

$$\left.\begin{aligned} U_{nz\bar z} &= Re[Q_nU_{nzz}+A_nU_{nz}] \ , \\ A_n &= -2(\ln\Psi_n)_{\bar z} + 2Q_n(\ln\Psi_n)_z \end{aligned}\right\} \tag{4.64}$$

have solutions $U_n(z)$ in D satisfying the boundary conditions

$$\left.\begin{aligned} &\frac{\partial U_n}{\partial\nu_n} + 2\sigma_n(t)U_n(t) = 2\tau_n(t) + h_n(t) \ , \ t \in \Gamma \\ &U_n(a_j) = b_j^{(n)} \ , \ j = 0,1,\ldots,N_0 \end{aligned}\right\} \tag{4.65}$$

and $C^1[U_n,\bar D] = H_n \to \infty$, for $n \to \infty$. Again we may assume $H_n \ge 1$, $n = 1,2,\ldots$ and denote $V_n(z) = U_n(z)/H_n$. It is clear that the function $W_n = V_{nz}$ is a solution of the complex equation

$$W_{n\bar z} = Re[Q_nW_{nz}+A_nW_n] \tag{4.66}$$

with the boundary condition

$$Re[\overline{\lambda_n(t)}W_n(t)] = -\sigma_n(t)V_n(t) + [\tau_n(t)+h_n(t)]/H_n \ , \ t \in \Gamma \ , \tag{4.67}$$

where the index of $\lambda_n(t) = \cos(\nu_n,x) - i\cos(\nu_n,y)$ is $N-1$, and $C[W_n,\bar D] \le 1$ shows that $W_n(z)$ is bounded on $\bar D$. According to

Theorem 4.3 in Chapter 2, we can obtain that $W_n(z) = V_{nz}$ satisfies the estimate

$$C_\beta[W_n,\overline{D}] \le M_{26}\ ,\ L_{p_0}[|W_{n\bar{z}}|+|W_{nz}|,\overline{D}] \le M_{27}\ , \tag{4.68}$$

and then $V_n(z) = 2Re\int_{a_0}^{z} W_n(\tau)d\tau + b_0/H_n$ satisfies

$$C_\beta^1[V_n,\overline{D}] \le M_{28}\ ,\ L_{p_0}[|V_{nz\bar{z}}|+|V_{nzz}|,\overline{D}] \le M_{29}\ , \tag{4.69}$$

where $M_j = M_j(q_0,p_0,k_0,\alpha,\ell,D)$, $j = 26,\ldots,29$. Hence, from $\{V_n(z)\}$, $\{V_{nz}\}$ we can select subsequences $\{V_{n_k}(z)\}$, $\{V_{n_k z}\}$, which uniformly converge to $V_0(z)$, V_{0z} , respectively, and $V_0(z)$ is a solution of the corresponding equation

$$V_{0\bar{z}} = Re[Q_0V_{0zz}+A_0V_{0z}]\ ,\ A_0 = -2(\ell n\Psi_0)_{\bar{z}} + 2Q_0(\ell n\Psi_0)_z \tag{4.70}$$

satisfying the corresponding boundary condition

$$\left.\begin{aligned} &\frac{\partial V_0}{\partial \nu} + 2\sigma_0(t)V_0(t) = h_0(t)\ ,\ t \in \Gamma\ ,\\ &V_0(a_j) = 0\ ,\ j = 0,1,\ldots,N_0\ . \end{aligned}\right\} \tag{4.71}$$

By Theorem 4.5, we see that $V_0(z) = 0$ in $\overline{D}$. However, from $C^1[V_n,\overline{D}] = 1$, we derive that $C^1[V_0,\overline{D}] = 1$. This contradiction proves that (4.60) is true. Afertwards, using the method of deriving (4.69) from $C^1[V_n,\overline{D}] \le 1$, we can obtain the estimates in (4.59). □

Similarly to the method in the proof of Theorem 4.9, we can prove

<u>Theorem 4.11</u> Under the hypothesis in Theorem 4.9, Problem IV for (1.24) is solvable.

§ 5. The irregular oblique derivative problem for elliptic equations of second order

1. Formulation and uniqueness for the irregular oblique derivative problem

Problem P^*. Find a solution to equation (1.24) in D, continuously differentiable on $\overline{D}$ and satisfying the boundary condition

$$\frac{\partial u}{\partial \mu} + 2[\sigma(t)+h(t)]\,u(t) = 2\tau(t)\ ,\ t \in \Gamma\ , \tag{5.1}$$

$$\text{i.e.}\quad Re[\lambda(t)u_t] + [\sigma(t)+h(t)]\,u(t) = \tau(t)\ ,\ t \in \Gamma\ ,$$

where $\vec{\mu}(t)$ is a given vector at the point $t \in \Gamma$, $\overline{\lambda(t)} = cos(\vec{\mu}(t),x) + i\,sin(\vec{\mu}(t),x)$, and $\lambda(t)$, $\sigma(t)$, $\tau(t)$ satisfy the conditions

$$\sigma(t)\,cos(\vec{\mu}(t),n) \ge 0\ ,\ C_\alpha[\lambda,\Gamma] \le \ell\ , \tag{5.2}$$
$$C_\alpha[\sigma,\Gamma] \le \ell\ ,\ C_\alpha[\tau,\Gamma] \le \ell\ ,$$

in which $\alpha(1/2 < \alpha < 1)$, $\ell(0 \le \ell < \infty)$ are real constants. In § 2 of Chapter 1, we gave a simple description of the above boundary condition, the main points are as follows: $cos(\vec{\mu},n) \ge 0$, $\sigma(t) \ge 0$, for $t \in \Gamma^+$ and $cos(\vec{\mu},n) \le 0$, $\sigma(t) \le 0$, for $t \in \Gamma^-$, $\Gamma^+ \cup \Gamma^- = \Gamma$, $\Gamma^+ \cap \Gamma^- = \emptyset$, and $\overline{\Gamma}^+ \cap \overline{\Gamma}^-$ contains a finite number of points $a_j(j=1,\dots,m)$ and $a_j'\ (j=1,\dots,m')$, $h(t) = 0$ for $m \ge 1$ and

$$h(t) = \begin{cases} h_0\ ,\ t \in \Gamma_0\ , \\ 0\ ,\ t \in \Gamma\setminus\Gamma_0\ , \end{cases} \quad \text{for}\quad m = 0\ ,$$

h_0 is an unknown real constant to be determined appropriately. If $cos(\vec{\mu},n) = 0$, $\sigma(t) = 0$, $t \in \Gamma_j$, $1 \le j \le N_0$, then we set

$$\int_{\Gamma_j} \tau(t)ds = 0\ ,\ u(a_j^*) = b_j^*\ ,\ a_j^*\ (\neq a_j') \in \Gamma\ , \tag{5.3}$$
$$|b_j^*| \le \ell\ ,\ 1 \le j \le N_0\ ,$$

and

$$u(a_j) = b_j \ , \ j = 1,\dots,J \ , \ J = \begin{cases} 1 \ , & \text{for} \quad m = 0 \ , \\ m \ , & \text{for} \quad m \geq 1 \ , \end{cases} \quad |b_j| \leq \ell \ . \tag{5.4}$$

The index of the above Problem P^* is

$$K = \frac{1}{2\pi} \Delta_\Gamma \, arg \, \lambda(t) = N-1 + (m-m')/2 \ . \tag{5.5}$$

If we allow that the above solution $u(z)$ is discontinuous at $a_j'(j=1,\dots,m')$, but is bounded on $D^* = \overline{D} \setminus \{a',\dots,a_{m'}'\}$, the problem is called Problem P_*, and Problem P_* with the conditions $A_3 = 0$ and $\sigma(t)\, u(t) = 0$ is called Problem Q^*.

Theorem 5.1 Let the equation (1.24) satisfy Condition C^* and (4.2). Then the solution for Problem P_* of (1.24) is unique.

Remark. Here, in addition, we assume that the coefficient $Q(z,u,u_z,u_{zz})$ satisfies the condition $Q \in W_p^1(D_{a_j'}^\varepsilon)$, where $D_{a_j'}^\varepsilon$ is is a neighbourhood of the point a_j' in $\overline{D}$, $p > 2$, $j = 1,\dots,m'$. If there is no point a_j satisfying the above condition, then we choose $a_1 = a_1^*$ for $N_0 \geq 1$ or a_1 is any point on Γ, but $a_1 \neq a_j'$ for $N_0 = 0$. In addition, if $cos(\vec{\mu},n) = 0$ on Γ_j $(N_0 < j \leq N)$, we first modify $\partial/\partial\mu$ into $\partial/\partial\nu$ on Γ_j, so that $cos(\vec{\nu},n) \geq \delta > 0$, and then cancel the condition.

Proof. We denote by $u_1(z)$ and $u_2(z)$ two solutions for Problem P_* of (1.24) and set $u(z) = u_1(z) - u_2(z)$. Similarly to the proof of Theorem 4.5, $u(z)$ is a solution of the equation (4.34), i.e.

$$u_{z\bar{z}} = Re[Q(z)u_{zz} + B_1 u_z] + B_2 u \tag{5.6}$$

with the condition

$$|Q(z)| \leq q_0 < 1 \ , \ L_p[B_j,\overline{D}] \leq k_1 \ , \ j = 1,2 \ , \ B_2 \geq 0 \ , \tag{5.7}$$

and satisfies the boundary condition

$$\left.\begin{aligned}&\frac{\partial u}{\partial \mu} + 2\sigma(t)\ u(t) = -2\sigma(t)\ h(t)\ ,\ t \in \Gamma\ ,\\ &u(a_j) = 0\ ,\ j = 1,\ldots,J\ .\end{aligned}\right\} \tag{5.8}$$

We shall prove $u(z) = 0$ in D. Otherwise, by Corollary 2.10, we see that there exist points t_*, $t_{**} \in \Gamma$, so that

$$M = u(t_*) = \sup_{z\in D^*} u(z) > 0 \quad \text{or} \quad m = u(t_{**}) = \inf_{z\in D^*} u(z) < 0\ , \tag{5.9}$$

where $D^* = \overline{D}\setminus\{a_1',\ldots,a_{m'}'\}$. If $M = u(t_*) > 0$, by the same reasoning as in the proof of Theorem 4.5, $t_* \notin \Gamma_*$. If $t_*(\neq a_k') \in \Gamma_j$ $(N_0 < j \le N)$, it is easy to see that $t_* \neq a_k (k=1,\ldots,J)$, and

$$\left.\frac{\partial u}{\partial \mu}\right|_{t=t_*} \ge 0 \quad \text{as} \quad \cos(\vec{\mu},n) \ge 0\ ,$$

and

$$\left.\frac{\partial u}{\partial \mu}\right|_{t=t_*} \le 0 \quad \text{as} \quad \cos(\vec{\mu},n) < 0\ .$$

So $\sigma(t_*) = 0$ and $\cos(\vec{\mu},n) = 0$ at $t = t_*$. Let $\tilde{\Gamma}$ be the longest circular arc of Γ including the point t_* so that $\cos(\vec{\mu},n) = 0$, $\sigma(t) = 0$ on $\tilde{\Gamma}$. Hence $\partial u/\partial\mu = 0$, $u(t) = M$ on $\tilde{\Gamma}$. It is clear that a_j is not an end point of $\tilde{\Gamma}$, and if a_j' is not an end point of $\tilde{\Gamma}$, then there exists a point $t_*' \in \Gamma\setminus\tilde{\Gamma}$, so that $\cos(\vec{\mu},n) > 0$, $\cos(\vec{\mu},s) > 0$, $\partial u/\partial n > 0$ and $\partial u/\partial s \ge 0$ at $t = t_*'$. Thus,

$$\left.\frac{\partial u}{\partial \mu}\right|_{t=t_*'} = \left.\left[\frac{\partial u}{\partial n}\cos(\vec{\mu},n) + \frac{\partial u}{\partial s}\cos(\vec{\mu},s)\right]\right|_{t=t_*'} > 0\ . \tag{5.10}$$

But, if $\cos(\vec{\mu},n) < 0$ at $t = t_*'$, in a similar way we can derive that $\left.\frac{\partial u}{\partial \mu}\right|_{t=t_*'} < 0$. These estimates are contradictory with (5.8). Thus, $\cos(\vec{\mu},n) = 0$ at $t = t_*'$, $\left.\frac{\partial u}{\partial \mu}\right|_{t=t_*'} \ge 0$ $\left(\left.\frac{\partial u}{\partial \mu}\right|_{t=t_*'} \le 0\right)$ and $\sigma(t_*') > 0$ (or $\sigma(t_*') < 0$). This is also impossible. If a_j' is an end point of $\tilde{\Gamma}$ or in the more general case where $u(a_j')$ or $u(a_j' \pm 0)$ is equal to M, and there exists an arc $\tilde{\Gamma}$ on Γ^+ or Γ^- with the end point a_j' or $a_j' \pm 0$ ($\tilde{\Gamma}$ is not a point) so that $cos(\vec{\mu},n) = 0$ on $\tilde{\Gamma}$, then using the same method as above a contradiction can be derived. If a_j' is an end point of $\tilde{\Gamma}$, and $\tilde{\Gamma}$ is only a point, denoting the coefficient $q(z) = Q(z) \in W_p^1(D_{a_j'}^{\varepsilon})$, $p > 2$, and finding a homeomorphic solution $\zeta(z)$ of the Beltrami equation

$$\zeta_{\bar{z}} - q^*(z)\zeta_z = 0 \ , \ q^*(z) = Q(z)/(1+\sqrt{1-|Q|^2})$$

the equation (5.6) can be transformed into the following equation

$$u_{\zeta\bar{\zeta}} = Re[A_1^* u_\zeta + A_2^* u] \ , \ \zeta \in \zeta[(D_{a_j'}^{\varepsilon})] \ , \tag{5.11}$$

where $A_j^* \in L_{p_0}[D_{a_j'}^{\varepsilon}]$, $j = 1,2$, and the boundary condition (5.8) can be reduced to

$$\left.\begin{aligned} &\frac{\partial u}{\partial \nu} + 2\sigma(t)\ [u(t)+h(t)] = 0 \ , \ t \in \Gamma \cap D_{a_j}^{\varepsilon} \ , \\ &\text{i.e. } \frac{\partial u}{\partial \nu^*} + 2\sigma[t(\zeta)]\ \{u[t(\zeta)] + h[t(\zeta)]\} = 0 \ , \\ &\qquad\qquad \zeta \in L_j^{\varepsilon} = \zeta[\Gamma \cap D_{a_j'}^{\varepsilon}] \ , \end{aligned}\right\} \tag{5.12}$$

where $\vec{\nu}^*(\zeta) = \vec{\nu}(t(\zeta))$. □

<u>Remark.</u> We have to extend $Q(z)$ continuously to the outside of $\bar{D}$ through Γ, so that the new coefficient satisfies the same condition as $Q(z)$ in $D_{a_j'}^{\varepsilon}$.

Denoting $a_j'' = \zeta(a_j')$, it is not difficult to see that a_j'' possesses a similar property to a_j'. Because $\sigma[z(a_j'')] = 0$, we can obtain

$$u[z(\zeta)] = c_j\ arg(\zeta - a_j'') + u(\zeta) \tag{5.13}$$

in $\zeta[D_{a_j'}^{\varepsilon}]$, $1 \le j \le m'$, where c_j is a real constant, and $u[z(\zeta)]$ is a Hölder continuous function in $\zeta[D_{a_j'}^{\varepsilon}]$. Hence, there exists a point t_*' near a_j', so that $cos(\vec{\nu}_*,\vec{n}_*) > 0$ (or < 0) . If $c_0 = 0$, then $u[z(\zeta)]$ is continuously differentiable in $\zeta[D^{\varepsilon}_{a'_j}]$. It is easy to see that

$$\frac{\partial u}{\partial \nu_*} = \frac{\partial u}{\partial n_*} \cos(\nu_*,n_*) + \frac{\partial u}{\partial s_*} \cos(\nu_*,s_*) > 0 \quad (\text{or} < 0)$$

at $t = t'_*$ and $\sigma(t'_*) \geq 0$ (or ≤ 0) , where n_*, s_* are the outer normal vector and the tangent vector of Γ at $t = t'_0$, respectively. This is a contradiction to

$$[\frac{\partial u}{\partial \nu_*} + 2\sigma(t)u(t)]\Big|_{t=t'_*} = 0 .$$

If $c_0 \neq 0$, then $\partial u/\partial n_* > 0$ (or < 0) at $t = t_*$, and $\partial u/\partial \nu_* > 0$ (or < 0) . This shows that $u(z)$ cannot attain a least upper bound on Γ_{**} . By the same reason, we can prove that $u(z)$ cannot attain a largest lower bound on $\Gamma \setminus \Gamma_0$. Therefore the least upper bound and the largest lower bound of $u(z)$ on $\overline{D}$ can be attained at least on Γ_0 . If $m \geq 1$, we have $h(t) = h_0 = 0$ for $t \in \Gamma_0$. If $m = 0$, we can prove $h_0 = 0$. In fact, due to

$$M = \sup_{z \in D^*} u(z) = \sup_{z \in \Gamma^*} u(z) > 0 , \text{ or } m = \inf_{z \in D^*} u(z) < 0 ,$$

and if $h_0 > 0$, then $M + h_0 = \sup_{z \in D^*} U(z) > 0$, where $U(z) = u(z) + h_0$. A contradiction can be derived to $\partial U/\partial \nu + 2\sigma(t)\, U(t) = 0$, for $t \in \Gamma_0$, so that $h_0 \leq 0$. Using a similar method, we can obtain $h_0 \geq 0$. Hence, $h_0 = 0$. Afterwards, according to the method used before, we conclude $u(z) = 0$, i.e. $u_1(z) = u_2(z)$ in D .

2. A priori estimate of solutions for the irregular oblique derivative problem

Theorem 5.2 Suppose that the equation (1.24) satisfies Condition C^* and $u(z)$ is a solution for Problem P_* of (1.24). Then $u(z)$ satisfies the estimates

$$C_\beta[V,\overline{D}] < M_1 , \; C[u,D_*] + C_\alpha[\sigma u,\Gamma] < M_2 , \tag{5.14}$$

where $\beta = 1 - 2/p_0 < \alpha$,

$$V(z) = u_z\ Y(z) = u_z \prod_{j=1}^{m_1} [\zeta(z)-\zeta(a_j')] \prod_{j=1}^{N} \prod_{j=m_i+1}^{m_{i+1}} [\frac{\zeta(z)-\zeta(a_j')}{\zeta(z)-\zeta_i}] ,$$

$$a'_{m_i+1},\ldots,a'_{m_{i+1}} \in \Gamma_i \ (i=0,1,\ldots,N , m_0 = 0 , m_{N+1} = m') ,$$

$\zeta(z)$ is a homeomorphic solution of the Beltrami equation mapping D onto a circular domain similar to D , so that $\zeta(0) = 0$, $\zeta(1) = 1$, and ζ_i is the centre of the circle $L_i = \zeta(\Gamma_i)$, $i = 1,\ldots,N$.

<u>Proof.</u> Using the same method as in the proof of Theorem 4.10, the solution $u(z)$ for Problem P_* of the equation (1.2) can be transformed into the solution $U(z) = [u(z)-\psi(z)]/\Psi(z)$ for Problem P_*' of the equation

$$U_{z\bar{z}} = Re[QU_{zz}+AU_z] , A = -2(\ln\Psi)_{\bar{z}} + 2Q(\ln\Psi)_z + A_1 \tag{5.15}$$

with the boundary condition

$$\left.\begin{aligned} &\frac{\partial U}{\partial \nu} + 2\sigma^*(t)U(t) = 2\tau^*(t) , \\ &\text{i.e. } Re[\overline{\lambda(t)U_t}] + \sigma^*(t)U(t) = \tau^*(t) , t \in \Gamma , \\ &U(a_j) = b_j , j = 1,\ldots,J , U(a_j^*) = b_j^* , j = 1,\ldots,N_0 , \end{aligned}\right\} \tag{5.16}$$

where $2\sigma^*(t) = 2\sigma(t) + \partial\Psi/\partial\nu$, $2\tau^*(t) = 2[\tau(t)-\sigma(t)h(t)] - \partial\psi/\partial\nu$, $t \in \Gamma$. We easily see that the equation (5.15) satisfies Condition C^*, $\sigma^*(t)$ and $\tau^*(t) + \sigma(t)h(t)$ satisfy conditions similar to (5.2), (5.3). From the properties of $\psi(z)$, $\Psi(z)$ and $u(z)$ in the neighbourhood of a_j' $(j=1,\ldots,m')$, it can be derived that U_z has a pole of order ≤ 1 at a_j' $(j=1,\ldots,m')$. Set

$$W(z) = U_z(z)\ Y(z) , \tag{5.17}$$

where $Y(z)$ is as stated in (5.14). Then $W(z)$ satisfies the boundary condition

$$Re[\overline{\Lambda(t)}W(t)] + \sigma^*(t)|Y(t)|U(t) = |Y(t)|\tau^*(t) \ , \ t \in \Gamma \ , \tag{5.18}$$

in which $\overline{\Lambda(t)} = \overline{\lambda(t)}|Y(t)|/Y(t)$, and the index of $\Lambda(t)$ equals $m/2 + N-1$; moreover, $\sigma^*(t)|Y(t)|U(t)$, $\tau^*(t)|Y(t)| \in C_\beta(\Gamma)$. Now, we prove

$$C[W,\overline{D}] \le M_3 = M_3(q_0,p_0,k_0,\alpha,\ell,D) \ . \tag{5.19}$$

Otherwise, there exists the solution $u_n(z)$ for the corresponding boundary value problem P_* and $W_n(z) = U_{nz} Y_n(z) = [(u_n(z)-\psi_n(z))/\Psi_n(z)]_z Y_n(z)$, so that $C[W_n,\overline{D}] = H_n \to \infty$ for $n \to \infty$. It is obvious that the functions $U_n^*(z) = U_n(z)/H_n$, $W_n^*(z) = W_n(z)/H_n$ satisfy the equations

$$U^*_{nz\bar{z}} = Re[Q_n U^*_{nzz} + A_n U^*_z] \tag{5.20}$$

and the boundary conditions

$$\left.\begin{aligned} &Re[\overline{\lambda_n(t)}U^*_{nt}] + \sigma^*_n(t)U^*_n(t) = \tau^*_n(t)/H_n \ , \ t \in \Gamma \ , \\ &\text{i.e. } Re[\overline{\Lambda_n(t)}W^*_n(t)] + \sigma^*_n|Y_n|U^*_n = \tau^*_n|Y_n|/H_n \ , \ t \in \Gamma \ , \end{aligned}\right\} \tag{5.21}$$

$$U^*_n(a_j) = b_j/H_n \ , \ j = 1,\dots,J \ , \ U^*_n(a^*_j) = b^*_j/H_n \ , \quad j = 1,\dots,N_0 \ . \tag{5.22}$$

Using the Theorem 2.3 in Chapter 4, $W^*_n(z)$ and $U^*_n(z)$ can be seen to satisfy the estimates

$$C_\beta[W^*_n,\overline{D}] \le M_4 \ , \ C[U^*_n,D_*] + C_\alpha[\sigma^*_n U^*_n,\Gamma] \le M_5 \tag{5.23}$$

where $M_j = M_j(q_0,p_0,k_0,\alpha,\ell,D)$, $j = 4,5$. Hence, we can choose subsequences $\{W^*_{n_k}\}$ and $\{U^*_{n_k}\}$, which uniformly converge to $W^*_0(z)$ on $\overline{D}$ and $U^*_0(z)$ on any closed set in D , respectively, and $W^*_0(z)$, $U^*_0(z)$ satisfy the corresponding equation

$$U^*_{0z\bar{z}} = Re[Q_0 U^*_{ozz} + A_0 U^*_0] \tag{5.24}$$

and the boundary conditions

$$\left.\begin{aligned} &Re[\overline{\lambda_0(t)U^*_{0t}}] + \sigma^*_0(t)U^*_{0t} = h_0(t)\ ,\ t \in \Gamma\ ,\\ &\text{i.e. } Re[\overline{\Lambda_0(t)W^*_0(t)}] + \sigma^*_0(t)|Y_0(t)|U^*_0(t) = h_0(t)|Y_0(t)|\ ,\\ &\qquad\qquad t \in \Gamma \end{aligned}\right\} \tag{5.25}$$

$$U^*_0(a_j) = 0\ ,\ j = 1,\ldots,J\ ,\ U^*_0(a^*_j) = 0\ ,\ j = 1,\ldots,N_0\ . \tag{5.26}$$

By Theorem 5.1, we have $U^*_0(z) = 0$, $W^*_0(z) = Y_0(z)U^*_{0z} = 0$ in D . However, from $C[W^*_n,\bar{D}] = 1$, we can show that there is a point $z^* \in D$, so that $|W^*_0(z^*)| = 1$. This contradiction proves the estimate (5.19). Afterwards, according to the proof of (5.23), we can obtain the estimates (5.14). This completes the proof of this theorem. □

3. Solvability of the irregular oblique derivative problem

We first prove the solvability of Problem Q^* and Problem P_* for harmonic functions, and then consider the solvability for Problem P_* and Problem P^* of the equation (1.24).

Lemma 5.3 Problem Q^* for harmonic functions is solvable.

Proof. Set

$$X(z) = \prod_{j=1}^{m_1} (z-a'_j) \prod_{i=1}^{N} \prod_{j=m_i+1}^{m_{i+1}} \left(\frac{z-a'_j}{z-z_i}\right)\ ,$$

where $a'_j \in \Gamma_i$ $(j=m_i+1,\ldots,m_{i+1}\ ,\ m_0 = 0\ ,\ m_{N+1} = m')$, $i = 0,1,\ldots,N$, and discuss the boundary value problem B^* for analytic functions with the boundary condition

$$Re[\overline{\Lambda(t)}w(t)] = r(t)\ ,\ t \in \Gamma\ , \tag{5.27}$$

in which

$$\overline{\Lambda(t)} = \overline{\lambda(t)}|X(t)|/X(t)\ ,\ r(t) = \begin{cases} |X(t)|\tau(t)\ , & \text{for } m \ge 1\ ,\\ |X(t)|[\tau(t)-\sigma(t)h(t)]\ , & \text{for } m = 0\ . \end{cases}$$

It is easy to see that $\Lambda(t)$ is discontinuous at $t = a_j'(j=1,\dots,m')$, and $Ind\,\Lambda(t) = m/2 + N-1$. By Theorem 1.5 in the next chapter, Problem B^* has a solution $w_0(z)$ which is analytic in D and satisfies (5.27) and $n = \begin{cases} m+N-1, & m \ge 1 \\ N, & m = 0 \end{cases}$ point conditions

$$Im[\overline{\Lambda(a_j^*)}w_0(a_j^*)] = b_j, \quad j = 1,\dots,n. \tag{5.28}$$

If $w_0(z)$ satisfies the condition

$$Re\int_{\Gamma_j} w_0(z)dz = 0, \quad j = N',\dots,N, \quad N' = N_0 + 1 \tag{5.29}$$

and

$$u_0(z) = b_1 + 2Re\int_{a_1}^{z} w_0(t)dt \tag{5.30}$$

satisfies (5.4), then the harmonic function $u_0(z)$ is a solution of Problem Q^*. Otherwise, we find n linearly independent solutions $w_k(z)$ $(k=1,\dots,n)$ of the homogeneous problem B_0^* for analytic functions with the point conditions

$$Im[\overline{\Lambda(a_j^*)}w_k(a_j^*)] = \delta_{kj} = \begin{cases} 1, & k = j, \\ 0, & k \neq j, \end{cases} \quad k = 1,\dots,n, \tag{5.31}$$

respectively. We can prove

$$I = \begin{vmatrix} u_1(a_1^*)\dots u_1(a_{N_0}^*) & I_{N'N'}\dots I_{NN'} & u_1(a_2)\dots u_1(a_m) \\ \vdots \quad\quad \vdots & \vdots \quad\quad \vdots & \vdots \quad\quad \vdots \\ u_n(a_1^*)\dots u_n(a_{N_0}^*) & I_{NN'}\dots I_{NN} & u_n(a_2)\dots u_n(a_m) \end{vmatrix} \neq 0, \tag{5.32}$$

where $I_{kj} = Re\int_{\Gamma_j} w_k(z)dz$, $j = N',\dots,N$,

$$u_k(a_j^*) = 2Re \int_{a_1}^{a_j^*} w_k(z)dz \ , \ k = 1,\dots,N_0 \ , \ u_k(a_j) = 2Re \int_{a_1}^{a_j} w_k(z)dz \ ,$$

$j = 2,\dots,m$, $k = 1,\dots,n$, in which we choose a fixed integral path in D with end points a_1 , a_j $(j=2,\dots,m)$. If $I = 0$, there exist real constants c_k $(k=1,\dots,n)$ which are not all equal to zero, so that

$$u(z) = \sum_{k=1}^{n} c_k u_k(z)$$

satisfies the condition

$$\left.\begin{aligned} &Re \int_{\Gamma_j} u_z dz = 0 \ , \ j = 1,\dots,N \ , \\ &u(a_j) = 0 \ , \ j = 1,\dots,m \ , \ u(a_j^*) = 0 \ , \ j = 1,\dots,N_0 \ , \end{aligned}\right\} \qquad (5.33)$$

i.e. $u(z)$ is a single-valued harmonic function in D . From Theorem 5.1, it follows that $u(z) = 0$, $w(z) = u_z = 0$. This is a contradiction to

$$w(a_j^*) = \sum_{k=1}^{n} c_k w_k(a_j^*) = c_j \ (j=1,\dots,n) \ .$$

So (5.32) is true. Thus, for the linear algebraic system

$$\left.\begin{aligned} &\sum_{k=1}^{n} c_k u_k(a_j^*) = -u_0(a_j^*) + b_j^* \ , \ j = 1,\dots,N_0 \ , \\ &\sum_{k=1}^{n} c_k I_{kj} = -I_{0j} = -Re \int_{\Gamma_j} w_0(z)dz \ , \ j = N',\dots,N \ , \\ &\sum_{k=1}^{n} c_k u_k(a_j) = -u_0(a_j) + b_j \ , \ j = 2,\dots,m \ , \end{aligned}\right\} \qquad (5.34)$$

we can seek a unique solution $c_1,\dots,c_n$, and

$$u(z) = b_1 + 2Re \int_{a_1}^{z} [w_0(z) + \sum_{k=1}^{n} c_k w_k(z)]dz \tag{5.35}$$

is a solution as stated in the theorem. □

Theorem 5.4 Problem P_* for harmonic functions has a soltuion .

Proof. Using the method of parameter extension, we prove the solvability of Problem P'_* for harmonic functions with the boundary condition

$$\frac{\partial u}{\partial \mu} + 2k\sigma(t)u(t) = s(t) \ , \ s(t) = 2[\tau_0(t) - \sigma(t)h(t)] \ , \ t \in \Gamma \ , \ 0 \le k \le 1 \ , \tag{5.36}$$

where $\tau_0(t) \in C_\alpha(\Gamma)$. If $k = 0$, from Lemma 5.3, Problem P'_* for harmonic functions is solvable. If $k = k_0$ $(0 \le k_0 < 1)$, Problem P'_* is solvable. Then according to the proof of Theorem 1.7 in Chapter 1, we can prove that there exists a positive constant $\delta > 0$ so that for every k satisfying the condition $|k-k_0| \le \delta$, $0 \le k \le 1$, Problem P'_* is solvable. Hence, from the solvability of Problem P'_* with $k = 0$, we can show the solvability of Problem P'_* with $k = \delta$, $2\delta, \ldots, [1/\delta]\delta, 1$ and in particular, that of Problem P'_* with $k = 1$, $\tau_0(t) = \tau(t)$ for harmonic functions. This completes the proof. □

In the following, we shall prove the solvability of Problem P_* and Problem P^* for equation (1.24).

Lemma 5.5 Let the nonlinear equation

$$\left.\begin{aligned} u_{z\bar{z}} &= F_0(z,u_z,u_{zz}) \ , \\ F_0(z,u_z,u_{zz}) &= Re[Q(z,u_{zz})u_{zz} + A_1(z)u_z] + A_3(z) \end{aligned}\right\} \tag{5.37}$$

satisfy Condition C^*. Then its Problem Q^* has a solution.

Proof. Let us introduce a bounded closed and convex set B_M in the Banach space $B = L_\infty(D^*) \times L_{p_0}(\overline{D}) \times L_{p_0}(\overline{D})$, in which the elements are vectors of three functions, $w = [Q(z), f(z), g(z)]$, satisfying the following conditions

$$L_\infty[Q,D] \le q_0 \ , \ L_{p_0}[f,\overline{D}] \le K_0 \ , \ L_{p_0}[q,\overline{D}] \le K_0 \ , \tag{5.38}$$

where q_0 is stated in (1.27), K_0 is an unknown constant to be determined appropriately. As in the proof of Theorem 2.4 of Chapter 4, using the method of parameter extention or the principle of contraction mapping, we can solve

$$\rho(z) - Q(z)\Pi\rho = Q(z) \ , \ \Pi\rho = -\frac{1}{\pi}\iint_D \frac{\rho(\zeta)}{(\zeta-z)^2}\, d\sigma_\zeta \ , \tag{5.39}$$

$$f^*(z) = F_0(z,w,\Pi f^*) - F_0(z,w,0) + Re\, A_1(z)\, Tf^* + A_3(z) \ , \tag{5.40}$$

$$\tilde{w}g^*(z) = F_0(z,w,\tilde{w}\Pi g^* + \Pi f^*) - F_0(z,w,\Pi f^*)\ f\ Re\, A_1(z)\, \tilde{w} \ , \tag{5.41}$$

$$\begin{aligned}\Phi'(z)\ \rho^*(z)e^{\varphi(z)} &= F_0(z,w,\Phi'(1+\Pi\rho^*)e^{\varphi} + \tilde{w}\Pi g^* + \Pi f^*) \\ &\quad - F_0(z,w,\tilde{w}\Pi g^* + \Pi f^*) \ , \ \Phi' = \{\Phi[\Psi(\chi)]\}'_\chi = \frac{d}{d\chi}\Phi[\Psi(\chi)] \ ,\end{aligned} \tag{5.42}$$

$$Q^* = \rho^*/(1+\Pi\rho^*) \ , \tag{5.43}$$

where $w = [Q(z), f(z), g(z)]$ is an element in B_M,

$$\psi(z) = Tf = -\frac{1}{\pi}\iint_D \frac{f(\zeta)}{\zeta-z}\, d\sigma_\zeta \ ,$$

$\varphi(z) = Tg$, $\chi(z) = z + T\rho(z)$, $\zeta(z) = \Psi[\chi(z)]$, $\zeta = \Psi(\chi)$ is a univalent analytic function which maps $\chi(D)$ onto a circular domain G similar to D, so that $\Psi[\chi(0)] = 0$, $\Psi[\chi(1)] = 1$, and $\Phi(\zeta)$ is an analytic function in G satisfying the boundary condition

$$\begin{aligned}Re\{\overline{\lambda[z(\zeta)]}\Phi(\zeta)e^{\varphi[z(\zeta)]}\} &= \tau[z(\zeta)] - \sigma[z(\zeta)]\, h[z(\zeta)] - \\ &\quad - Re\{\overline{\lambda[z(\zeta)]}\ \psi[z(\zeta)]\} \ , \ \zeta \in L = \zeta(\Gamma) \ ,\end{aligned} \tag{5.44}$$

where

$$h(\zeta) = \begin{cases} h_0 \ , & \zeta \in L_0 = \zeta(\Gamma_0) \ , \\ 0 \ , & \zeta \in L - L_0 \ , \ m = 0 \ , \end{cases} \qquad h(\zeta) = 0 \ , \ \zeta \in L \ , \ m \geq 1 \ .$$

By analogy with the proof of Lemma 5.3, we can obtain a single-valued function

$$u(z) = 2Re \int_{a_1}^{z} w(t)dt + b_1 \tag{5.45}$$

satisfying the point conditions

$$u(a_j) = b_j \ , \ j = 1,\ldots,J \ , \ u(a_j^*) = b_j^* \ , \ j = 1,\ldots,N_0 \ , \tag{5.46}$$

in which $w(z) = \Phi[\zeta(z)]e^{\varphi(z)} + \psi(z)$. In fact, if $u(z)$ is not single-valued in D or (5.46) does not hold, we can seek n linearly independent analytic functions $\Phi_k(z)$ $(k=1,\ldots,n)$ in G satisfying the boundary conditions

$$Re\{\overline{\lambda[z(\zeta)]}\Phi_k(\zeta)e^{\varphi[z(\zeta)]}\} = -\sigma[z(\zeta)]h[z(\zeta)] \ , \ \zeta \in L \ , \tag{5.47}$$

$$Im\{\overline{\lambda(a_j^*)}\Phi[\zeta(a_j^*)]e^{\varphi(a_j^*)}\} = \begin{cases} 0 \ , & j \neq k \ , \\ 1 \ , & j = k \ , \end{cases} \quad j = 1,\ldots,n \ . \tag{5.48}$$

Set

$$w_k(z) = \Phi_k[\zeta(z)]e^{\varphi(z)} \ , \ u_k(a_j) = 2Re \int_{a_1}^{a_j} w_k(z)dz \ , \ j = 2,\ldots,m \ ,$$

$$u_k(a_j^*) = 2Re \int_{a_1}^{a_j^*} w_k(z)dz \ , \ j = 1,\ldots,N_0 \ ,$$

and

$$I_{kj} = Re \int_{\Gamma_j} w_k(z)dz \ , \ j = N_0+1,\ldots,N \ .$$

Using reductio ad absurdum, $I \neq 0$ can be derived, but then

$$u(z) = 2Re \int_{a_1}^{z} \sum_{k=1}^{n} c_k w_k(t)dt$$

is a solution of the following equation

$$u_{z\bar{z}} = Qu_{zz} + (g - Q\Pi g)u_z \ , \ Q = \zeta_{\bar{z}}/\zeta_z \tag{5.49}$$

in D. Afterwards, we can find the solutions $f^*(z)$, $g^*(z)$, $Q^*(z)$ of (5.40), (5.41), (5.43), respectively, and determine the constant K_0 in (5.38). Moreover, we can prove that the mapping $\omega^* = S(\omega)$, from $\omega = [Q(z),f(z),g(z)]$ onto $\omega^* = [Q^*(z),f^*(z),g^*(z)]$ is a continuous mapping, whicl maps B_M into a compact set in B_M. Hence, there exists a vector $\omega = [Q(z),f(z),g(z)] \in B_M$ satisfying $\omega = S(\omega)$. We substitute the corresponding function $w(z) = \Phi[\zeta(z)]e^{\varphi(z)} + \psi(z)$ in (5.45), and the determined function $u(z)$ is a solution for Problem Q^* of the equation (5.37). □

Theorem 5.6 Let the equation (1.24) satisfy Problem C^*. Then Problem P_* is solvable.

Proof. We introduce a bounded and open set B_M in the Banach space $B = C^1(\bar{D}) \times L_\infty(D) \times L_{p_0}(\bar{D}) \times L_{p_0}(\bar{D})$, where the elements are all vectors of four functions, $\omega = [u(z),Q(z),f(z),g(z)]$, with norms $\|\omega\| = C[V,\bar{D}] + L_\infty[Q,D] + L_{p_0}[f,\bar{D}] + L_{p_0}[g,\bar{D}]$ satisfying the conditions

$$\left.\begin{aligned} &C[u,D^*] + C_\alpha[\sigma u,\Gamma] + C[V,\bar{D}] < M_1 + M_2 \ , \\ &L_\infty[Q,D] < (1+q_0)/2 \ , L_{p_0}[f,\bar{D}] < K_0 \ , L_{p_0}[g,\bar{D}] < K_0 \ , \end{aligned}\right\} \tag{5.50}$$

where $V(z) = u_z Y(z)$, $Y(z)$ is as stated in (5.14), M_1 and M_2 are constants stated in (5.14), K_0 is an unknown constant to be determined appropriately.

Next, we can solve the following system of integral equations with the parameter k $(0 \le k \le 1)$:

$$\rho(z) = Q(z)\Pi\rho = Q(z) , \tag{5.51}$$

$$f^*(z) = F(z,u,w,\Pi f^*) - F(z,u,w,0) + \\ + Re[A_1(z,u,w)Tf^*] + k\, A_2(z,u,w)u + A_3(z,u,w) , \tag{5.52}$$

$$\tilde{w}g^*(z) = F(z,u,w,\tilde{w}\Pi g^* + \Pi f^*) - F(z,u,w,\Pi f^*) \\ + Re\, A_1(z,u,w)\tilde{w} , \tag{5.53}$$

$$\Phi'\rho^*\, e^{\varphi(z)} = F(z,u,w,\Phi'(1+\Pi\rho^*)e^{\varphi} + \tilde{w}\Pi g^* + \Pi f^*) - \\ - F(z,u,w,\tilde{w}\Pi g^* + \Pi f^*) , \quad \Phi' = \{\Phi[\Psi(\chi)]\}'_{\chi} \tag{5.54}$$

$$Q^*(z) = \rho^*(z)/(1+\Pi\rho^*) , \tag{5.55}$$

where $\omega = [u(z),Q(z),f(z),g(z)]$ is any vector in B_M, $\psi(z)$, $\varphi(z)$, $\chi(z)$, $\zeta(z)$ are similar functions as stated in Lemma 5.5, $\Phi(\zeta)$ is an analytic function satisfying

$$\left.\begin{aligned} & Re\{\overline{\lambda[z(\zeta)]}\Phi(\zeta)e^{\varphi[z(\zeta)]}\} = -\sigma[z(\zeta)]\{ku[z(\zeta)] + h[z(\zeta)]\} + \\ & + \tau[z(\zeta)] - Re\{\overline{\lambda[z(\zeta)]}\psi[z(\zeta)]\} , \quad \Gamma \in L = \zeta(\Gamma) . \end{aligned}\right\} \tag{5.56}$$

Moreover, we may require that the function $u^*(z)$ determined by the integral

$$u^*(z) = 2Re\int_{a_1}^{z} w(t)dt + b_1 \tag{5.57}$$

is single valued in D, and satisfies

$$u^*(a_j) = b_j , \ j = 1,\dots,m , \ u^*(a_j^*) = b_j^* , \ j = 1,\dots,N_0 . \tag{5.58}$$

Denote by $\omega^* = [u^*(z),Q^*(z),f^*(z),g^*(z)] = S(\omega,k)$ $(0 \le k \le 1)$ the mapping from ω onto ω^*. We can verify that $\omega^* = S(\omega,k)$ satisfies the three conditions of the Leray-Schauder theorem. Hence, for $k = 1$, there exists a vector $\omega = [u(z),Q(z),f(z),g(z)]$, so that $\omega = S(\omega,1)$, and the corresponding function

$$u(z) = 2Re \int_{a_1}^{z} w(t)dt + b_1 \qquad (5.59)$$

is a solution for Problem P_* to the nonlinear equation. □

Theorem 5.7 Suppose that the equation (1.24) satisfies Condition C^*. Under m' conditions, Problem P^* possesses a continuous differentiable solution on $\overline{D}$.

Proof. Let $u(z)$ be a solution for the Problem P_* to the equation (1.24). If a_j' $(j=1,\ldots,m')$ are not poles of u_z, i.e. $c_j = 0$ $(j=1,\ldots,m')$ in (5.13), then $u(z)$ is a continuous function on $\overline{D}$. Thus, $u(z)$ is a solution for Problem P^* of (1.24) (cf. [103]26). □

§ 6. The Poincaré boundary value problem for elliptic equations of second order

In this section, we first consider the Poincaré problem for nonlinear uniformly elliptic equations of second order by using a priori estimates of solutions and the Schauder fixed point theorem, and then discuss the solvability for the Poincaré problem for the linear uniformly elliptic equation of second order by applying the Fredholm theorem of linear integral equations.

1. The Poincaré problem for nonlinear equations of second order

We rewrite the second order nonlinear equation (1.24) in the form

$$\left.\begin{aligned} &u_{z\bar{z}} = F(z,u,u_z,u_{zz}), \ F = Re[Qu_{zz} + A_1u_z + \varepsilon A_2u] + A_3, \\ &Q = Q(z,u,u_z,u_{zz}), \ A_j = A_j(z,u,u_z), \ j = 1,2,3, \end{aligned}\right\} \qquad (6.1)$$

where $\varepsilon(0 \le \varepsilon < 1)$ is a nonnegative constant. The so-called Poincaré boundary value Problem P is to find a continuously differentiable solution $u(z)$ on $\overline{D}$ satisfying the boundary condition

$$A(s)\frac{\partial u}{\partial s} + B(s)\frac{\partial u}{\partial n} + 2\varepsilon\, C(s)u = 2D(s) \ , \tag{6.2}$$

in which $A(s)$, $B(s)$, $C(s)$, and $D(s)$ are real functions of the arc length parameter s on the boundary Γ satisfying some regularity conditions, $\vec{n}$ is the outer normal vector on Γ (cf.[76]1). Noting the relations

$$\frac{\partial u}{\partial s} = \frac{\partial u}{\partial x}\cos\theta + \frac{\partial u}{\partial y}\sin\theta \ , \ \frac{\partial u}{\partial n} = \frac{\partial u}{\partial x}\sin\theta + \frac{\partial u}{\partial y}(-\cos\theta) \ , \tag{6.3}$$

where θ is the angle between the positive direction of the tangent of Γ and the x-axis, from (6.2) and (6.3) it follows that

$$a(t)u_x + b(t)u_y + 2\varepsilon C(t)\, u(t) = 2d(t) \ , \ t \in \Gamma \ , \tag{6.4}$$

where $a(t) = A(s)\cos\theta + B(s)\sin\theta$, $b(t) = A(s)\sin\theta - B(s)\cos\theta$, $c(t) = C(s)$, $d(t) = D(s)$, and $t = x + iy$ is the point on Γ corresponding to the arc length parameter s . From $u_x = u_t + u_{\bar{t}}$, $u_y = i(u_t - u_{\bar{t}})$, the boundary condition (6.4) can be reduced to

$$Re[\overline{\lambda(t)}u_t] + \varepsilon\,\sigma(t)\,u(t) = \tau(t) \ , \ t \in \Gamma \ , \tag{6.5}$$

in which $\overline{\lambda(t)} = a(t) + ib(t)$, $\sigma(t) = c(t)$, $\tau(t) = d(t)$. Next, we assume that the nonlinear equation (6.1) satisfies Condition C in the (N+1)-connected circular domain D and the coefficients $\lambda(t)$, $\sigma(t)$, $\tau(t)$ satisfy

$$|\lambda(t)| = 1 \ , \ C_\alpha[\lambda,\Gamma] \le \ell \ , \ C_\alpha[\sigma,\Gamma] \le \ell \ , \ C_\alpha[\tau,\Gamma] \le \ell \ , \tag{6.6}$$

where $\alpha(0 < \alpha < 1)$, $\ell(0 \le \ell < \infty)$, $\varepsilon(0 \le \varepsilon < 1)$ are nonnegative constants. The above boundary value problem P is the general irregular oblique derivative problem, and $K = 1/2\pi\, \Delta_\Gamma \arg\lambda(t)$ is called the index of Problem P.

In general, Problem P of (6.1) may not be solvable or the solution of Problem P may not be unique. To derive the result of solvability for Problem P of (6.1), we introduce the corresponding complex equation of first order

$$\left.\begin{array}{l} w_{\bar z} = f(z,u,u_z,w,w_z) \ , \ f = Re[Qw_z + A_1 w] + A \ , \\ Q = Q(z,u,u_z,w_z) \ , A = \varepsilon\, A_2 u + A_3 \ , A_j(z,u,u_z) \ , \ j = 1,2,3 \ , \end{array}\right\} \tag{6.7}$$

with the modified boundary conditions

$$\left.\begin{array}{l} Re[\overline{\lambda(t)}w(t)] = r(t) + h(t) \ , \ r(t) = \tau(t) - \varepsilon\sigma(t)u(t) \ , \ t \in \Gamma \ , \\ h(t) = \begin{cases} 0 \ , \ t \in \Gamma \ , \text{ for } K \ge N \ , \\ \left.\begin{array}{l} h_j \ , \ t \in \Gamma_j \ , \ j=1,\dots,N-K \ , \\ 0 \ , \ t \in \Gamma_j \ , \ j=N-K+1,\dots,N+1 \ , \end{array}\right\} \text{for } 0 \le K < N \ , \\ \left.\begin{array}{l} h_j \ , \ t \in \Gamma_j \ , \ j=1,\dots,N \ , \\ h_0 + Re \sum\limits_{m=1}^{-K-1} (\lambda_m^+ + i\lambda_m^-)t^m \ , \ t \in \Gamma_0 , \end{array}\right\} \text{for } K < 0 \ , \end{cases} \end{array}\right\} \tag{6.8}$$

and the relation

$$u(z) = b_0 + 2Re \int_{a_0}^{z} [w(t) - \sum_{j=1}^{N} \frac{id_j}{t-z_j}] \, dt \ , \tag{6.9}$$

where a_j is a point on Γ_j , $j = 1,\dots,N$ and a_j $(j=N+1,\dots,2K-N+1)$ are distinct points on Γ_0 $(a_{N+1}=a_0=1)$, and b_j $(j=0,1,\dots,2K-N+1)$ are real constants with the conditions $|b_j| \le \ell$, $j = 0,1,\dots,2K-N+1$, h_j $(j=1,\dots,N)$ and $\lambda_m^{\pm}$ $(m=1,\dots,|K|-1)$ are unknown real constants to be determined appropriately, and d_j $(j=1,\dots,N)$ are appropriate real constants so that the function determined by the integral in (6.9) is single-valued in D . The above modified boundary value problem is called Problem Q.

In order to give estimates of solutions for Problem Q of (6.7), we first prove a lemma.

Lemma 6.1 If $A[z,u,u_z]$ in (6.7) and $r(t)$ in (6.8) are replaced by $A_0[z,w(z)]$, $r_0(t)$, respectively, $A_0[z,u,u_z]$, $r_0(t)$ satisfy the conditions

$$L_p[A_0(z,u,u_z),\bar{D}] \le k_1 \ , \ C_\alpha[r_0,\Gamma] \le k_2 \ , \tag{6.10}$$

with positive constants k_1 and k_2, then any solution $[w(z),u(z)]$ for Problem Q of (6.7) must satisfy

$$C_{\alpha\beta^2}[w,\bar{D}] \le M_1(k_1+k_2) \ , \ C^1_{\alpha\beta^2}[u,\bar{D}] \le M_2(k_1+k_2) \ , \tag{6.11}$$

where $\beta = 1-2/p_0$, $2 < p_0 < p$, $M_j = M_j(q_0,p_0,k_0,\alpha,\ell,D)$, $j = 1,2$.

<u>Proof.</u> Put $W(z) = w(z)/(k_1+k_2)$, so that $W(z)$ satisfies the equation

$$W_{\bar{z}} = Re[QW_z+A_1W] + A_0/(k_1+k_2) \tag{6.12}$$

and the boundary condition

$$Re[\overline{\lambda(t)}W(t)] = r_0(t)/(k_1+k_2) + h(t) \ , \ t \in \Gamma \ . \tag{6.13}$$

Noting $L_p[A_0/(k_1+k_2),\bar{D}] \le 1$ and $C_\alpha[r_0/(k_1+k_2),\Gamma] \le 1$ and applying Theorem 4.2 of Chapter 2, the solution $w(z)$ for Problem A of the first order complex equation (6.7) satisfies the estimates

$$C_{\alpha\beta^2}[w,\bar{D}] \le M_3 \ , \tag{6.14}$$

where $M_3 = M_3(q_0,p_0,k_0,\alpha,\ell,D)$. So we have the first estimate in (6.11). From (6.9), it follows that the second estimate in (6.11) is also true. □

<u>Theorem 6.2</u> Let the equation (6.1) satisfy Condition C and the constant ε in (6.7) and (6.8) be sufficiently small. Then the solution $[w(z),u(z)]$ for Problem Q of (6.7) satisfies the estimates

$$\Delta_1 = C_{\alpha\beta^2}[w,\bar{D}] \le M_4 \ , \ \Delta_2 = C^1_{\alpha\beta^2}[u,\bar{D}] \le M_5 \ , \tag{6.15}$$

where $M_j = M_j(q_0,p_0,k_0,\alpha,\ell,D)$, $j = 4,5$.

Proof. Let the solution $[w(z),u(z)]$ in this theorem replace the corresponding functions in (6.7) - (6.9). From (6.7) - (6.9), we can derive

$$\Delta_2 \le M_6 \Delta_1 + \ell , \quad M_6 = M_6(p_0,D) , \tag{6.16}$$

$$L_p[A,\bar{D}] \le \varepsilon \Delta_2 L_p[A_2,\bar{D}] + L_p[A_3,\bar{D}] \le (\varepsilon\Delta_2+1)k_0 , \tag{6.17}$$

and

$$C_\alpha[r,\Gamma] \le C_\alpha[\tau,\Gamma] + \varepsilon \Delta_2 C_\alpha[\sigma,\Gamma] \le (\varepsilon\Delta_2+1)\ell . \tag{6.18}$$

Using Lemma 6.1 and admitting $k_1 = (\varepsilon\Delta_2+1)k_0$, $k_2 = (\varepsilon\Delta_2+1)\ell$, we have

$$\Delta_2 \le M_2(\varepsilon\Delta_2+1)(k_0+\ell) , \text{ i.e. } \Delta_2 \le M_2(k_0+\ell)/(1-\varepsilon M_2(k_0+\ell)) , \tag{6.19}$$

if the constant ε is sufficiently small, so that $1 - \varepsilon M_2(k_0+\ell) > 0$. Thus, the second estimate in (6.15) holds, and

$$\Delta_1 \le M_1(\varepsilon\Delta_2+1)(k_0+\ell) \le M_1(\varepsilon M_5+1)(k_0+\ell) = M_4 . \quad \square$$

Now, we use the Schauder fixed-point theorem to prove the existence of solutions for Problem Q of (6.7).

Theorem 6.3 Under the hypothesis in Theorem 6.2, Problem Q of the nonlinear equation (6.7) is solvable.

Proof. Let us introduce a closed and convex set B in the Banach space $C^1(\bar{D})$, in which the elements are the functions $u(z)$ satisfying the following condition

$$C^1[u,\bar{D}] \le M_5 , \tag{6.20}$$

where M_5 is a nonnegative constant as stated in (6.15). We arbitrarily select an element $u(z) \in B$ and substitute it into the corresponding positions of (6.7) and (6.8). By Theorem 4.1 and Theorem 4.7 in Chapter 2, we see that Problem Q of (6.7) has a unique solution $w(z)$. By Theorem 6.2, $w(z)$ satisfies the first estimate in (6.15). Substituting $w(z)$ into (6.9), and from Theorem 6.2, we obtain the function

$$u^*(z) = u_0 + 2Re\int_0^z [w(t) + \sum_{j=1}^{N} \frac{id_j}{t-z_j}]dt \tag{6.21}$$

satisfying the second estimate in (6.15). We denote by $w(z) = S_1[u(z)]$ and $u^*(z) = S[u(z)]$ the mappings from $u(z)$ onto $w(z)$ and from $u(z)$ onto $u^*(z)$, respectively.

From the above statement, we see that $u^* = S(u)$ maps B into a compact subset of itself. In the following, we shall show that $u^* = S(u)$ is a continuous mapping in B. We are free to choose a sequence of functions $u_n(z)$ $(n=0,1,2,\ldots)$ in B, so that $C^1[u_n(z)-u_0(z),\bar{D}] \to 0$, for $n \to \infty$. Denoting $w_n(z) = S_1[u_n(z)]$, $u_n^*(z) = S[u_n(z)]$, and noting that $w_n(z)$, $u_n^*(z)$ $(n=1,2,\ldots)$ satisfy the estimates in (6.15), we can prove that $C^1[w_n(z)-w_0(z),\bar{D}] \to 0$ and $C^1[u_n^*(z)-u_0^*(z),\bar{D}] \to 0$, for $n \to \infty$.

By Schauder fixed-point theorem, there exists a function $u(z)$, so that $u(z) = S[u(z)]$. Set $w(z) = S_1[u(z)]$. Then $[w,u]$ is just a solution for Problem Q of (6.7). □

If the solution $w(z)$ in the above theorem satisfies the N conditions

$$Re\int_{\Gamma_j} w(z)dz = 0\ ,\ j=1,\ldots,N\ , \tag{6.22}$$

then we amy admit $d_j = 0$ $(j=1,\ldots,N)$ and $u_z = w(z)$. Moreover, substituting $u(z)$ into the boundary condition (6.8), if $h(t) = 0$, then the solution $u(z)$ for Problem Q is also a solution for Problem P. Therefore, the following theorem holds.

Theorem 6.4 Suppose that the equation (6.1) satisfies Condition C and the constant ε is small enough. Then we have:

(1) If the index $K \geq N$, Problem P of the equation (6.1) has N solvability conditions.

(2) If $0 \leq K < N$, the total number of solvability conditions for Problem P is not greater than $2N-K$.

(3) If $K < 0$, under $2N-2K-1$ conditions, Problem P of (6.1) is solvable.

2. The Poincaré problem for linear equations of second order

Finally, we discuss the linear uniformly elliptic equation of second order

$$u_{z\bar{z}} - Re[Q(z)u_{zz}+A_1(z)u_z] = \varepsilon A_2(z)\ u(z) + A_3(z) \qquad (6.23)$$

and the boundary condition (6.5), where $\varepsilon\ (-\infty < \varepsilon < \infty)$, $\alpha\ (1/2 < \alpha < 1)$ are real constants and $Q(z)$, $A_j(z)\ (j=1,2,3)$ satisfy the conditions:

$$|Q(z)| \leq q_0 < 1\ ,\ L_p[A_j,\bar{D}] \leq k_0 < \infty\ ,\ j = 1,2,3\ . \qquad (6.24)$$

This boundary value problem is called Problem P.

We introduce the corresponding linear system of complex equations of first order

$$w_{\bar{z}} - Re[Q(z)w_z+A_1(z)w] = Re\ A_2(z)\ u(z) + A_3(z)\ , \qquad (6.25)$$

$$u_{\bar{z}} = \overline{w(z)} \qquad (6.26)$$

and the boundary conditions

$$Re[\overline{\lambda(t)}w(t)] = r(t) + h(t)\ ,\ r(t) = \tau(t) - \varepsilon\, Re\, \sigma(t)\ u(t)\ , \qquad (6.27)$$

$$Re[iu(t)] = H(t) = \begin{cases} 0\ ,\ t \in \Gamma_0\ , \\ H_j\ ,\ t \in \Gamma_j\ ,\ j = 1,\ldots,N\ , \end{cases} \qquad (6.28)$$

where $\sigma(t)$, $\tau(t)$, $h(t)$ are stated in (6.5), (6.8) and H_j $(j=1,\ldots,N)$ are all appropriate real constants. This boundary value problem is called Problem Q.

According to Theorem 4.9 in Chapter 2, the boundary value problem Problem Q_1 of the complex equation

$$w_{\bar{z}} - Re[Q(z)w_z+A_1(z)w] = A_3(z) \tag{6.29}$$

with the boundary condition

$$Re[\overline{\lambda(t)}w(t)] = r(t) + h(t) \ , \ t \in \Gamma \tag{6.30}$$

is solvable, and the general solution $\tilde{w}(z)$ can be expressed by

$$\tilde{w}(z) = w_0(z) + \sum_{m=1}^{J} c_m w_m(z) \ , \ J = \begin{cases} 2N-K+1 \ , \text{ for } K \geq N \ , \\ K+1 \ , \text{ for } 0 \leq K < N \ , \\ 0 \ , \text{ for } K < 0 \end{cases} \tag{6.31}$$

where $K = 1/2\pi \ \Delta_\Gamma \ arg \ \lambda(t)$ is the index of the boundary value problem, $w_0(z)$ is a special solution and $w_m(z)$ $(m=1,\ldots,J)$ are linearly independent solutions of the corresponding homogeneous problem. Next, let H_2u denote a solution of the complex equation

$$w_{\bar{z}} - Re[Q(z)w_z+A_1(z)w] = Re \ A_2(z)u \tag{6.32}$$

with the boundary condition

$$Re[\overline{\lambda(t)}w(t)] = -Re \ \sigma(t)u(t) + h(t) \ , \ t \in \Gamma \ , \tag{6.33}$$

$$Im[\overline{\lambda(a_j)}w(a_j)] = b_j \ , \ j = \begin{cases} 1,\ldots,2K-N+1 \ , \text{ for } K \geq N \ , \\ N-K+1,\ldots,N+1 \ , \text{ for } 0 \leq K < N \ . \end{cases} \tag{6.34}$$

It is easy to see that H_2 is a linear bounded operator from $u(z) \in C^1(\bar{D})$ onto $w(z) \in C_\alpha(\bar{D})$. Furthermore, we can seek a solution

$$u(z) = H_1w + c_0 \ , \ H_1w\big|_{z=0} = 0 \ , \tag{6.35}$$

of the complex equation

$$u_{\bar{z}} = \overline{w(z)} \tag{6.36}$$

with the boundary condition (6.28). In (6.35), c_0 is an arbitrary real constant. It is not difficult to verify that H_1 is a linear bounded and completely continuous operator from $C_\alpha(\bar{D})$ into $C^1(\bar{D})$. Applying the result of § 3, § 4, in Chapter 1, we can express $H_1 w$, $H_2 u$ in integral forms.

From (6.35) and $w(z) = \tilde{w}(z) + \varepsilon H_2 u$, the nonhomogeneous integral equation

$$u - \varepsilon H_1 H_2 u = H_1 w_0(z) + c_0 + \sum_{m=1}^{J} c_m H_1 w_m(z) , \tag{6.37}$$

can be obtained. $H_1 H_2$ is a completely continuous operator in $C^1(\bar{D})$. Hence, we may apply the Fredholm theorem to the integral equation (6.37). We denote by ε_j $(j=1,2,...)$ with $0 < |\varepsilon_1| \leq |\varepsilon_2| \leq ... \leq |\varepsilon_n| \leq |\varepsilon_{n+1}| \leq ...$ the eigenvalues of the homogeneous integral equation

$$u - \varepsilon H_1 H_2 u = 0 . \tag{6.38}$$

If in the case $K \geq N$, $\varepsilon \neq \varepsilon_j$ $(j=1,2,...)$, then the integral equation (6.37) has a solution $u(z)$, which includes $2K-N+2$ arbitrary real constants. Substituting $u(z)$ into the boundary condition (6.28) and putting $H_j = 0$, $j = 1,2,...,N$, we obtain a system of algebraic equations. Denoting by S the rank of the coefficients matrix of this system, $S \leq \min(N, 2K-N+2)$, we can determine S constants within the $2K-N+2$ arbitrary constants by S equalities from the algebraic system. If the other $N-S$ equalities hold, then from (6.25) and (6.26), it follows

$$\mathrm{Re}[iu_{z\bar{z}}] = 0 \text{ in } D , \text{ and } \mathrm{Re}[iu(t)] = 0 \text{ on } \Gamma . \tag{6.39}$$

So $\mathrm{Im}\, u(z) = 0$, for $z \in D$, i.e. $u(z)$ is a real function and a solution for Problem P of the second order equation (6.23). Thus, the general solution to Problem P includes $2K-N+2-S$ arbitrary constants. If ε is an eigenvalue of the integral equation (6.38) with the rank q, then according to the third Fredholm theorem (cf. [97]2), we can write down the solvability conditions of the nonhomogeneous equation and obtain a system of q algebraic equations to determine $2K-N+2$ arbitrary real constants $c_0,\ldots,c_{2K-N+1}$. Denoting by S_1 the rank of the corresponding coefficients matrix, $S_1 \le \min(q,2K-N+2)$; similarly, we can determine S_1 constants of $2K-N+2$ arbitrary constants and S_1 equalities of q algebraic equations. Afterwards we substitute the solution $u(z)$ including $2K-N+2+q-S_1$ arbitrary constants in (6.28) and put $H_j = 0$, $j = 1,\ldots,N$. Denoting by S_2 the rank of the corresponding coefficients matrix we have $S_2 \le \min(N,2K-N+2+q-S_1)$. For the same reason, we can determine S_2 equalities of N algebraic equations, and S_2 constants of $2K-N+2+q-S_1$ arbitrary constants. If the other $N-S_2$ equalities hold, then as above we see that $u(z)$ is a real function and a solution to Problem P for equation (6.23). Hence, Problem P for (6.23) has $N + q - S_1 - S_2$ solvability conditions, and, if these conditions are satisfied, the general solution includes $2K-N+2+q-S_1-S_2$ arbitrary real constants.

In the case $K < N$, using a similar method to that used before, we can obtain the result on solvability for the Problem P of the linear equation (6.23). Thus we have the following theorems.

Theorem 6.5 Suppose that the second order equation (6.23) satisfies the condition (6.24) and ε is not an eigenvalue of the corresponding homogeneous integral equation (6.38). Then we have:

(1) If $K \ge N$, Problem P of (6.23) has $N - S$ solvability conditions, $S \le \min(N,2K-N+2)$. Moreover, if these conditions are satisfied, then the general solution to Problem P for (6.23) includes $2K-N+2-S$ arbitrary constants.

(2) If $0 \le K < N$, Problem P of (6.23) has $2N-K-S$ solvability conditions, $S \le \min(2N-K,K+2)$. Moreover, if these conditions are

satisfied, then the general solution to Problem P for (6.23) includes $K+2-S$ arbitrary constants.

(3) If $K < 0$, Problem P has $2N-2K-1-S$ solvability conditions, $S = 0$ or 1 , and the general solution includes $1 - S$ arbitrary constants.

Theorem 6.6 Suppose that the second order equation (6.23) satisfies condition (6.24) and ε is an eigenvalue of the corresponding homogeneous integral equation (6.38) with rank q . Then we have:

(1) If $K \geq N$, Problem P for (6.23) has $N+q-S$ solvability conditions, $S \leq \min(N+q, 2K-N+2+q)$. If these conditions are satisfied, the general solution to Problem P includes $2K-N+2+q-S$ arbitrary constants.

(2) If $0 \leq K < N$, Problem P has $2N-K+q-S$ solvability conditions, $S \leq \min(2N-K+q, K+2+q)$. If these conditions are satisfied, the general solution to Problem P includes $K+2+q-S$ arbitrary constants.

(3) If $K < 0$, Problem P of (6.23) has $2N-2K-1+q-S$ solvability conditions, $S \leq \min(2N-2K-1+q, 1+q)$. If these conditions are satisfied, then the general solution includes $1+q-S$ arbitrary conditions (cf. [103]25)).

IV Boundary value problems with piecewise continuous coefficients for elliptic equations and systems

In the previous chapters, we discussed the boundary value problems with continuous coefficients in the boundary conditions. In this chapter, we shall first study some boundary value problems with piecewise continuous coefficients in the boundary conditions for analytic functions, and then consider some boundary value problems with piecewise continuous coefficients for the uniformly elliptic system of two equations of first order and the uniformly elliptic equation of second order.

§ 1. Boundary value problems with piecewise continuous coefficients for analytic functions

First of all, we shall discuss the Riemann-Hilbert boundary value problem in a simply connected domain; afterwards, we shall consider the above boundary value problem in a multiply connected domain and the Riemann boundary value problem with piecewise continuous coefficients for analytic functions.

1. The Riemann-Hilbert problem with piecewise continuous coefficients in a simply connected domain

Let Γ be a smooth closed curve, $\Gamma \in C_\mu (0 < \mu < 1)$ and $a(t)$, $b(t)$, $c(t)$ possess discontinuities of the first kind at the points $t_1,\ldots,t_m$, where m is a positive integer. We denote by D the bounded domain with boundary Γ , and $D^* = \overline{D}\setminus\{t_1,\ldots,t_m\}$, $\Gamma^* = \Gamma\setminus\{t_1,\ldots,t_m\}$.

We may allow $c(t_j) = \infty$ and $c(t) = O(|t-t_j|^{-\beta_j})$ in the neighbourhood of t_j in D , where $0 < \beta_j < 1$, $j = 1,\ldots,m$ (cf. the case of multiply connected domains).

Problem A_1^*. Find an analytic function $\Phi(z)$ in D, continuous on D^* and satisfying the boundary condition with piecewise continuous coefficients

$$Re[\overline{\lambda(t)}\Phi(t)] = au - bv = c \ , \ t \in \Gamma \ , \tag{1.1}$$

where $\overline{\lambda(t)} = a(t) + ib(t)$, $[a(t)]^2 + [b(t)]^2 \neq 0$. We assume that $\lambda(t)$, $c(t)$ satisfy the conditions

$$|\lambda(t)| = 1 \ , \ C_\alpha[\lambda,\Gamma^j] \le \ell \ , \ C_\alpha[c,\Gamma^j] \le \ell \ , \tag{1.2}$$

in which $\alpha(0<\alpha<1)$, $\ell(0\le\ell<\infty)$ are constants, $\Gamma^j(j=1,\ldots,m)$ are all arcs with end points in $\{t_1,\ldots,t_m\}$. The above problem is called Problem A_1^*, and Problem A_1^* with the condition $c(t) = 0$ is called Problem A_0^*.

Without loss of generality, we may assume that the simply connected domain D is the disc $|z| < 1$, and $\Gamma = \{|z|=1\}$.

Set

$$\gamma_j = \frac{1}{\pi i}\ell n\Big[\frac{\lambda(t_j-0)}{\lambda(t_j+0)}\Big] = \frac{\varphi_j}{\pi} - K_j \ , \ e^{i\varphi_j} = \frac{\lambda(t_j-0)}{\lambda(t_j+0)} \ , \ K_j = \Big[\frac{\varphi_j}{\pi}\Big] + J_j \ , \tag{1.3}$$

where $J_j = 0$ or 1, $j=1,\ldots,m$. It is easy to see that $0 \le \gamma_j < 1$, when $J_j = 0$, and $-1 \le \gamma_j < 0$, when $J_j = 1$. In the sequal $|\gamma_j| < 1$ will be assumed. Through the transformation

$$\Psi_0(z) = \Phi_0(z)/\Pi(z) \ , \ \Pi(z) = \prod_{j=1}^{m}(z-t_j)^{\gamma_j} \ , \tag{1.4}$$

the solution $\Phi_0(z)$ of Problem A_0^* can be reduced to the solution $\Psi_0(z)$ of the Problem A_0 with the boundary condition

$$Re[\overline{\lambda_1(t)}\Psi_0(t)] = 0 \ , \ \lambda_1(t) = \lambda(t)\overline{\Pi(t)} \ , \ t \in \Gamma \ . \tag{1.5}$$

The index of Problem A_0^* and Problem A_1^* is defined by

$$K = Ind\,\lambda_1(t) = \sum_{j=1}^{m} [\varphi_j/2\pi - \gamma_j/2] = \sum_{j=1}^{m} K_j/2 \; . \tag{1.6}$$

It is clear that the index K is not unique. Taking

$$\lambda_1(t_j-0)/\lambda_1(t_j+0) = \lambda(t_j-0)/[\lambda(t_j+0)e^{i\pi\gamma_j}] = \pm 1$$

into account, if $2K$ is even, provided we change the signs of $\lambda_1(t)$ on some arcs of $\Gamma^j(j=1,\dots,m)$, then the new function $\lambda_1^*(t)$ obtained is continuous on Γ and $K = Ind\,\lambda_1^*(t)$. If $2K$ is odd, we rewrite (1.5) in the form

$$Re[\overline{\lambda_1(t)}(t-t_0)\,\frac{\Psi_0(t)}{t-t_0}] = 0 \; , \; t \in \Gamma \; , \tag{1.7}$$

where t_0 is a point on Γ and $t_0 \neq t_j(j=1,\dots,m)$. Using the same method for $\lambda_1(t)\overline{(t-t_0)}/|t-t_0|$ as above, we obtain the new function $\lambda_1^*(t)$ continuous on Γ, and $K - 1/2 = Ind\,\lambda_1^*(t)$.

Applying the Schwarz formula, we can find an analytic function $\Psi_0^*(z)$ satisfying the boundary condition

$$Re[\overline{\lambda_1^*(t)}\Psi_0^*(t)] = 0 \; , \; t \in \Gamma \; . \tag{1.8}$$

In fact, let $S(z)$ be an analytic function in D, which satisfies the boundary condition

$$Re\; S(t) = arg\; \lambda_1^*(t)\overline{t}^{[K]} \; , \; t \in \Gamma \;\text{ and }\; Im\; S(0) = 0 \; .$$

Then

$$\Psi_0^*(z) = iz^{[K]}e^{iS(z)} \tag{1.9}$$

is a nontrivial solution to (1.8), and

$$\Psi_0(z) = \begin{cases} \Psi_0^*(z) \; , & \text{if } 2K \text{ is even} \; , \\ (z-t_0)\Psi_0^*(z) \; , & \text{if } 2K \text{ is odd} \end{cases} \tag{1.10}$$

is a solution of Problem A_0. Hence, Problem A_0^* has a standard solution

$$X(z) = \Pi(z)\Psi_0(z) = \begin{cases} iz^K \Pi(z) e^{iS(z)} , & \text{if } 2K \text{ is even} , \\ iz^{[K]}(z-t_0)\Pi(z) e^{iS(z)} , & \text{if } 2K \text{ is odd} . \end{cases} \qquad (1.11)$$

We can see that when $K \geq 0$, $X(z)$ has a zero point of order K at $z = 0$, if $K < 0$, $X(z)$ has a pole of order $|[K]|$ at $z = 0$, if $2K$ is odd, $X(z)$ has a zero point of order 1 at $z = t_0$.

Because $X(z)$ satisfies the homogeneous boundary condition

$$Re[\overline{\lambda(t)}X(t)] = 0 , \quad t \in \Gamma , \qquad (1.12)$$

the function $i\overline{\lambda(t)}X(t)$ is real on Γ. Let us divide the nonhomogeneous boundary condition (1.1) by $i\overline{\lambda(t)}X(t)$. We obtain the Dirichlet boundary condition for the function $\Phi(z)/iX(z)$,

$$Re[\Phi(t)/iX(t)] = c(t)/[i\overline{\lambda(t)}X(t)] = \lambda(t)c(t)/iX(t) , \quad t \in \Gamma . \quad (1.13)$$

Using the Schwarz formula in $D = \{|z| < 1\}$, the general solution $\Phi(z)$ of Problem A_1^* $(K \geq 0)$ can be expressed as

$$\left.\begin{aligned} &\Phi(z) = \frac{X(z)}{2\pi i}\left[\int_\Gamma \frac{(t+z)\lambda(t)c(t)}{(t-z)tX(t)}\,dt + Q(z)\right] , \\ &Q(z) = i\sum_{j=0}^{[K]}(c_j z^j + \overline{c}_j z^{-j}) + \begin{cases} 0 , & 2K \text{ is even} , \\ c_*(t_0+z)/(t_0-z) , & 2K \text{ is odd} , \end{cases} \end{aligned}\right\} \qquad (1.14)$$

where c_*, c_0 are arbitrary real constants, $c_j (j=1,\ldots,[K])$ are arbitrary complex constants. This shows that $\Phi(z)$ includes $2K + 1$ arbitrary real constants.

When $2|K|$ is odd, we note $\dfrac{z-t_0}{(t-z)(t-t_0)} = \dfrac{1}{t-z} - \dfrac{1}{t-t_0}$, thus the integral in (1.14) is understood as a difference of two integrals of Cauchy type.

If $K < 0$, we have to take

$$Q(z) = \begin{cases} 0 , & 2|K| \text{ is even} , \\ c_*(t_0+z)/(t_0-z) , & 2|K| \text{ is odd} \end{cases}$$

and require that the function in the square bracket of (1.14) has a zero point of order $|[K]|$ at $z = 0$, i.e. the following $-2K-1$ conditions hold:

$$\int_\Gamma \frac{\lambda(t)c(t)}{X(t)t^n} dt = \begin{cases} 0 , & n = 1,\ldots,-K , \ 2|K| \text{ is even} , \\ -c_* t_0^{-n+1} , & n = 2,\ldots,[-K]+1 , \ 2|K| \text{ is odd} , \end{cases} \tag{1.15}$$

where

$$c_* = -\int_\Gamma \frac{\lambda(t)c(t)}{X(t)t} dt$$

is a real constant. Thus, the function $\Phi(t)$ is analytic in D. The above results can be summarized in the following theorem.

Theorem 1.1 (1) If the index $K \geq 0$, the general solution of Problem A_1^* can be expressed by (1.14), which includes $2K+1$ arbitrary real constants.

(2) If $K < 0$, Problem A_1^* for analytic functions has $-2K-1$ solvability conditions as stated in (1.15). If these conditions are satisfied, the solution of the Problem A_1^* possesses the representation

$$\Phi(z) = \frac{X(z)z^{|[K]|}}{\pi i} \left[\int_\Gamma \frac{c(t)\lambda(t)dt}{t^{|[K]|}(t-z)X(t)} + \frac{c_*}{t_0^{|[K]|-1}(t_0-z)} \right] , \tag{1.16}$$

where

$$c_* = -\int_\Gamma \frac{\lambda(t)c(t)}{X(t)t} dt ,$$

if $2K$ is odd, and $c_* = 0$, if $2K$ is even. In order to discuss the boundedness and the integrability of the solution in the neighbourhood of the points of discontinuity $t_1,\ldots,t_m$, we need some theorems from §§ 22-25 of [76]1).

Let $L = ab$ be a smooth curve with end points a and b, and let the function $f(t)$ possess the following form

$$f(t) = f^*(t)/(t-c)^{\gamma}, \quad \gamma = \alpha + i\beta, \quad 0 \le \alpha < 1, \tag{1.17}$$

where $f^*(t) \in C_{\nu}(L)$, $0 < \nu < 1$, α, β, ν are real constants, $(t-c)^{\gamma} = e^{\gamma \ln(t-c)}$ is a continuous branch on $L \setminus \{c\}$. The integral

$$F(z) = \frac{1}{2\pi i} \int_L \frac{f(t)}{t-z}\, dt \tag{1.18}$$

is called an integral of Cauchy type.

<u>Lemma 1.2</u> (1) If $\gamma = 0$, then

$$F(z) = \mp \frac{f(c)}{2\pi i} \ln(z-c) + F_0(z), \tag{1.19}$$

in a neighbourhood $D_{\varepsilon} \setminus L$ of c and $z \notin L$, where the signs $-$, $+$ are determined by $c = a$ and b respectively, $F_0(z)$ is analytic in $D_{\varepsilon} \setminus L$, and $F_0(z) \to A = \text{constant}$, if $z \to c$.

(2) If $\gamma \neq 0$, then

$$F(z) = \pm \frac{e^{\pm i\gamma\pi}}{2i \sin \gamma\pi} \frac{f^*(c)}{(z-c)^{\gamma}} + F_0(z) \quad \text{in } D_{\varepsilon} \setminus L, \tag{1.20}$$

where the signs $+$, $-$ are determined by $c = a$ and b, respectively, the branch of $(z-c)^{\nu}$ is taken so that $(z-c)^{\nu} = (t-c)^{\nu}$ on the left side of L, if $\alpha > 0$, $F_0(z)$ is analytic and $(z-c)^{\alpha-\delta} F_0(z)$ is bounded in $D_{\varepsilon} \setminus L$ (δ is a positive number), and if $\alpha = 0$, $F_0(z) \to A = \text{constant}$, if $z(\in D_{\varepsilon} \setminus L) \to c$.

(3) If $\gamma = 0$, $t_0(\neq a,b) \in L$, then

$$F(t_0) = \mp \frac{f(t_0)}{2\pi i} \ln(t_0-c) + F_*(t_0), \tag{1.21}$$

where $F_*(t_0)$ is Hölder-continuous on $D_{\varepsilon} \cap L$, the signs $+$, $-$ are chosen to be the same as in (1).

(4) If $\gamma \neq 0$, $t_0(\neq a,b) \in L$, then

$$F(t_0) = \pm \frac{\operatorname{ctg}\gamma\pi}{2i} \frac{f^*(c)}{(t_0-c)^\gamma} + F_*(t_0) , \tag{1.22}$$

if $\alpha > 0$, $(t_0-c)^{\alpha-\delta}F_*(t_0)$ (δ is a positive number) is Hölder-continuous on $D_\varepsilon \cap L$, if $\alpha = 0$, $F_*(t)$ is Hölder-continuous on $D_\varepsilon \cap L$, and the signs + , - are chosen to be the same as in (2).

Applying the above lemma, we can see that if $J_j = 0$ and $\gamma_j > 0$, then the solution $\Phi(z)$ of Problem A_1^* is bounded in the neighbourhood D_j of t_j on D^* , if $J_j = 0$, $\gamma_j = 0$ or $J_j < 0$, then the solution $\Phi(z)$ of Problem A_1^* may be unbounded in D_j , and

$$\Phi(z) = \begin{cases} O(|z-c_j|^{\gamma_j}) , & \text{for } \gamma_j < 0 , \\ O(\ell n|z-c_j|) , & \text{for } \gamma_j = 0 , \end{cases} \tag{1.23}$$

but the integral $\int_0^z \Phi(\zeta)d\zeta$ is bounded in D_j , $j=1,\ldots,m$.

In particular, if $m = 2n$, $t_1,\ldots,t_{2n}$ are $2n$ points on Γ , arranged in accordance with the positive direction, we denote by Γ_j the arc from t_j to $t_{j+1}(j=1,\ldots,2n,t_{2n+1}=t_1)$ on Γ , and suppose that

$$\lambda(t) = \begin{cases} 1 , & t \in \Gamma_{2k-1} , \\ -i , & t \in \Gamma_{2k} , \end{cases} \quad k = 1,\ldots,n . \tag{1.24}$$

In this case, the boundary condition (1.1) becomes

$$\left. \begin{aligned} u(t) &= \mathit{Re}\,\Phi(t) = c(t) , \quad t \in \Gamma_{2k-1} , \\ v(t) &= \mathit{Im}\,\Phi(t) = c(t) , \quad t \in \Gamma_{2k} , \end{aligned} \right\} \quad k = 1,\ldots,n . \tag{1.25}$$

This is the so-called mixed boundary value problem, Problem M, for analytic functions. From $\lambda(t_{2k-1}+0) = \lambda(t_{2k-1}-0) = 1$, $\lambda(t_{2k-1}-0) = \lambda(t_{2k}+0) = i$, $k = 1,\ldots,n$,

$$\gamma_j = \frac{1}{\pi i} \ln\left[\frac{\lambda(t_j-0)}{\lambda(t_j+0)}\right] = \begin{cases} \frac{1}{2} - K_j \ , \ j=1,3,\ldots,2n-1 \ , \\ -\frac{1}{2} - K_j \ , \ j=2,4,\ldots,2n \ , \end{cases} \tag{1.26}$$

we may distinguish three cases:

(1) If $K_{2k-1} = 0, K_{2k} = -1$, $k = 1,\ldots,n$, then $\gamma_j = 1/2 > 0$, $j = 1,\ldots,2n$, $\Pi_1(z) = \sum_{j=1}^{2n} (z-t_j)^{1/2}$ and the index $K = \sum_{j=1}^{2n} K_j/2 = -n/2$.

(2) If $K_j = 0$, $j = 1,\ldots,2n$, then $\gamma_{2k-1} = 1/2 > 0$, $\gamma_{2k} = -1/2 < 0$, $k = 1,\ldots,n$, $\Pi_2(z) = \prod_{k=1}^{n} \left(\frac{z-t_{2k-1}}{z-t_{2k}}\right)^{1/2}$ and the index $K = 0$.

(3) If $K_{2k-1} = 1$, $K_{2k} = 0$, $k = 1,\ldots,n$, then $\gamma_j = -1/2$, $j = 1,\ldots,2n$, $\Pi_3(z) = \prod_{j=1}^{2n} (z-t_j)^{-1/2}$ and $K = n/2$. Thus, by the previous discussion, we can obtain the following results as a corollary.

Corollary 1.3 The solvability of Problem M for analytic functions in D is as follows:

(1) Under $-2K-1 = n-1$ solvability conditions as stated in (1.15), Problem M has a bounded solution in D, and the solution $\Phi(z)$ can be expressed as (1.14).

(2) There exist solutions of Problem M in D, which are bounded in the neighbourhoods D_{2k-1} of $t_{2k-1} (k=1,\ldots,n)$ and the integrals of which are bounded in the neighbourhoods $D_{2k} (k=1,\ldots,n)$ in D, and the general solution $\Phi(z)$ can be expressed by (1.14), which includes an arbitrary real constant.

(3) There exist solutions of Problem M, the integrals of which are bounded in $D_j (j=1,\ldots,2n)$, and the general solution $\Phi(z)$ can be expressed by (1.14), which includes $2K+1 = n+1$ arbitrary constants. In the above cases (1), (2), (3), the function $\Pi(z)$ in

(1.11) is replaced by $\Pi_1(z)$, $\Pi_2(z)$, $\Pi_3(z)$ respectively.

Through a conformal mapping from the disc onto the upper-half plane G , we can derive the results on the mixed boundary value problem for analytic functions and harmonic functions in G .

2. The Riemann-Hilbert problem with piecewise continuous coefficients in a multiply connected domain

Let D be a multiply connected circular domain stated as in § 1 of Chapter 1, whose boundary is $\Gamma = \bigcup_{j=0}^{N} \Gamma_j$, $\Gamma_j = \{|z-z_j| = \gamma_j\}$, $j = 1,\ldots,N$, $\Gamma_0 = \Gamma_{N+1} = \{|z| = 1\}$. We consider the Riemann-Hilbert problem with the boundary condition (1.1), where $t_1,\ldots,t_m (\in\Gamma)$ are all discontinuities of the first kind of $\lambda(t)$ and $c(t) = \prod_{j=1}^{m} |t-t_j|^{-\beta_j} c_0(t)$, $0 \le \beta_j < 1$, $j = 1,\ldots,m$, *) $\lambda(t)$, $c_0(t)$ satisfy the conditions:

$$|\lambda(t)| = 1 , \quad C_\alpha[\lambda,\Gamma^j] \le \ell , \quad C_\alpha[c_0,\Gamma^j] \le \ell , \tag{1.27}$$

in which $\Gamma^j (j=1,\ldots,m')$ are all arcs on $\Gamma^* = \Gamma \setminus \{t_1,\ldots,t_m\}$ with end points in $\{t_1,\ldots,t_m\}$, $\alpha(0 < \alpha < 1)$, $\ell(0 \le \ell < \infty)$ are constants. The above boundary value problem with piecewise continuous coefficients is called <u>Problem A*</u>, and the index of Problem A* is $K = \sum_{j=1}^{m} K_j/2$, where

$$K_j = [\frac{\varphi_j}{\pi}] + J_j , \ J_j = 0 \text{ or } 1 , \ e^{i\varphi_j} = \frac{\lambda(t_j-0)}{\lambda(t_j+0)} , \ \gamma_j = \frac{\varphi_j}{\pi} - K_j , \ j = 1,\ldots,m. \tag{1.28}$$

In order to discuss the solvability of Problem A*, we introduce a properly modified boundary value problem, <u>Problem B*</u>, with the boundary condition

$$Re[\overline{\lambda(t)}\Phi(t)] = c(t) + |Y(t)|\, h(t) , \quad t \in \Gamma , \tag{1.29}$$

---

*) If $\lambda(t)$ is continuous on $\Gamma_j (0 \le j \le N)$, then we add an arbitrarily chosen point $t^* \in \Gamma_j$ to $\{t_j,\ldots,t_m\}$, and relabel this set as $\{t_1,\ldots,t_m\}$.

where

$$h(t) = \begin{cases} 0 \ , \ t \in \Gamma \ , \ \text{for} \quad K > N-1 \ , \\ \left.\begin{array}{l} h_j \ , \ t \in \Gamma_j \ , j = 1,\dots,N-K', K' = [|K|+1/2], \\ 0 \ , \ t \in \Gamma_j \ , j = N-K'+1,\dots,N+1 , \end{array}\right\} \text{for} \quad 0 \le K \le N-1 \ , \\ h_j \ , \ t \in \Gamma_j \ , j = 1,\dots,N+1 \ , \text{for} \quad K < 0 \ , \end{cases} \qquad (1.30)$$

$h_j (j=1,\dots,N+1)$ are all unknown constants to be determined appropriately and $h_{N+1} = h_0 = 0$, if $2K$ is an odd integer, and

$$Y(z) = \prod_{j=1}^{m_1} (z-t_j)^{\gamma_j} \prod_{j=m_1+1}^{m_2} \left(\frac{z-t_j}{z-z_1}\right)^{\gamma_j} \cdots \prod_{j=m_N+1}^{m} \left(\frac{z-t_j}{z-z_N}\right)^{\gamma_j} \ , \qquad (1.31)$$

in which $t_{m_k+1},\dots,t_{m_{k+1}}$ are discontinuous points of $\lambda(t)$ on Γ_k , $k = 1,\dots,N$, $m_0 = 1$, $m_{N+1} = m$. There is no harm in assuming $K_j = \mathit{Ind}\ \lambda(t)|_{\Gamma_j}$ $(j=1,\dots,N_0, N_0 \le N)$ to be integers, and $K_j = \mathit{Ind}\ \lambda(t)|_{\Gamma_j}$ $(j=N_0+1,\dots,N)$ are not integers. If $K < 0$, we permit the solution $\Phi(z)$ to have a pole of order $\le K' - 1 = [|K|+1/2]-1$ at $z = 0 \in D$. If $K \ge 0$, we may require that the solution $\Phi(z)$ satisfies

$$\left.\begin{array}{l} Im[\overline{\lambda(a_j)}\Phi(a_j)] = |Y(a_j)| b_j \ , \\ j \in \{j\} = \begin{cases} 1,\dots,2K-N+1 \ , \ \text{for} \quad K > N-1 \ , \\ \left.\begin{array}{l} N_0+1,\dots,N_0+[K]+1 \ , \ 0 \le K' \le N-N_0 , \\ N-K'+1,\dots,N-K'+[K]+1 \ , \ N-N_0 < K' \le N-1 , \end{array}\right\} \text{for} \quad 0 \le K \le N-1 \ , \end{cases} \end{array}\right\} \qquad (1.32)$$

where $a_j (\ne t_k) \in \Gamma_j$, $j = 1,\dots,N_0$, $a_j (\ne t_k) \in \Gamma_0$, $j = N_0+1,\dots,2K-N+1$, and $b_j (j=1,\dots,N_0)$ are all real constants with conditions $|b_j| \le \ell (j=1,\dots,2K-N+1)$. In addition, we may require that the solution

$$\Phi(z) = O(|z-t_j|^{-\tau_j}) \ , \ \tau_j = \begin{cases} \beta_j + \delta , \text{for } \gamma_j \ge 0 , \text{or } \gamma_j < 0 \text{ and } \beta_j > |\gamma_j| \ , \\ |\gamma_j| + \delta \ , \text{ for } \gamma_j < 0 \quad \text{and} \quad \beta_j \le |\gamma_j| , \end{cases} \qquad (1.33)$$

in the neighbourhood of t_j on $D^* = \overline{D} \setminus \{t_1, \ldots, t_m\}$, in which δ is an arbitrarily small positive number.

<u>Theorem 1.4</u> If the constants $\gamma_j (j=1,\ldots,m)$ are determined, then the solution of Problem B^* for analytic functions is unique.

<u>Proof.</u> Let $\Phi_1(z)$, $\Phi_2(z)$ be two solutions of Problem B^* and denote $\Phi(z) = \Phi_1(z) - \Phi_2(z)$. Then the function $\Psi(z) = \Phi(z)/Y(z)$ satisfies the boundary condition

$$Re[\overline{\Lambda(t)}\Psi(t)] = h(t) \ , \ \overline{\Lambda(t)} = \overline{\lambda(t)}Y(t)/|Y(t)| \ , \ t \in \Gamma \qquad (1.34)$$

and

$$Im[\overline{\Lambda(a_j)}\Psi(a_j)] = 0 \ , \ j \in \{j\} \ . \qquad (1.35)$$

Suppose that $\Psi(z) \not\equiv 0$; it follows by a similar method to that used in the proof of Lemma 1.1 of Chapter 1

$$2N_D + N_\Gamma - 2P_D = 2K \ , \qquad (1.36)$$

where N_D and N_Γ are the total number of zero points of $\Phi(z)$ in D and on Γ, respectively; P_D is the total number of poles in D. Moreover, the total number of zero points on Γ_j is odd or even depending whether $2K_j$ is odd or even. Thus, we derive the following absurd inequalitties:

$$\left.\begin{array}{l} 2K+1 = 2N_0+N-N_0+2K-N+1-N_0 \le 2N_D+N_\Gamma = 2K \ , \text{ for } K \ge N-1/2 \ , \\ 2K+1 \le K'+[K]+1 = N-(N-K')+[K]+1 \le 2N_D+N_\Gamma = 2K \ , \text{ for } 0 \le K' \le N-N_0 \ , \\ 2K+1 \le K'+[K]+1 = 2[N_0-(N-K')]+N-N_0+N-K'+[K]+1-N_0 \le 2N_D+N_\Gamma \le 2K \ , \\ \qquad\qquad \text{for } N-N_0 \le K' \le N-1 \ , \ 0 \le K \le N-1 \ , \\ 2K+1 \le -2[|K|+1/2]+2 \le 2N_D+N_\Gamma-2P_D = 2K \ , \text{ for } K < 0 \ . \end{array}\right\} \qquad (1.37)$$

This contradiction proves $\Psi(z) = 0$, $\Phi(z) = 0$, i.e. $\Phi_1(z) = \Phi_2(z)$ in D. □

Theorem 1.5 Problem B* for analytic functions is solvable.

Proof. Following a similar method to that in § 1 of Chapter 1, we first find a solution $S(z)$ of the modified Dirichlet problem satisfying the boundary condition

$$Re\ S(t) = S_1(t) - \theta(t) \ , \ t \in \Gamma \ , \ Im\ S(1) = 0 \ , \tag{1.38}$$

where $\theta(t) = \begin{cases} 0 \ , \ t \in \Gamma_0 \ , \\ \theta_j \ , \ t \in \Gamma_j \ , \ j = 1,\dots,N \ , \end{cases}$ $\theta_j(j=1,\dots,N)$ are real constants and

$$S_1(t) = \begin{cases} arg\ \lambda(t) - K_* \ arg\ t + arg\ [Z(t)\Pi(t)/Y(t)] \ , \ t \in \Gamma_0 \ , \\ arg\ \lambda(t) + arg[Z(t)\Pi(t)/Y(t)] \ , \ t \in \Gamma_j \ , \ j = 1,\dots,N \ , \end{cases} \tag{1.39}$$

in which

$$K_* = [K-(N-N_0)/2] \ , \ \Pi(z) = \prod_{j=1}^{N} (z-z_j)^{[K_j]} \ ,$$

$$Z(t) = \prod_{j=1}^{N} \prod_{i=1}^{\ell_j} \left(\frac{t-d_{ji}}{t-z_j}\right)^{(-1)^i} \prod_{i=1}^{\ell_0} (t-d_{0i})^{(-1)^i} \ ,$$

where $d_{ji}(i=1,\dots,\ell_j)$ are all discontinuous points of Γ_j so that $\Lambda(t_k-0) = -\Lambda(t_k+0)$, $k = 0,1,\dots,N$. From

$$\Psi(z) = e^{-iS(z)}\Pi(z)\Phi(z)/Y(z) \ , \ z \in D \ , \tag{1.40}$$

the boundary condition (1.29) can be transformed into the boundary condition

$$Re[\overline{\Lambda_*(t)}\Psi(t)] = R(t) + H(t) \ , \ t \in \Gamma \tag{1.41}$$

where $R(t) = c(t)e^{Im S(t)}|\Pi(t)/Y(t)|$, $H(t) = h(t)e^{Im S(t)}|\Pi(t)|$ and

$$\overline{\Lambda_*(t)} = \begin{cases} \bar{t}^{K^*} e^{i arg\ Z(t)} \ , \ t \in \Gamma_0 \ , \\ e^{-i\theta_j + i arg Z(t)} \ , \ t \in \Gamma_j \ , \ j = 1,\dots,N \ . \end{cases}$$

It is easy to see that

$$Ind\, \Lambda_*(t)|_{\Gamma_j} = 0\ ,\ j=1,\dots,N_0\ ,\ Ind\, \Lambda_*(t)|_{\Gamma_j} = 1/2\ ,\ j=N_0+1,\dots,N\ ,$$

and

$$Ind\, \Lambda_*(t)|_{\Gamma_0} = K_* + J = [K-(N-N_0)/2]+J\ ,\ J = \begin{cases} 0\ , \text{ when } \ell_0 \text{ is even,} \\ 1/2\ , \text{ when } \ell_0 \text{ is odd,} \end{cases} \quad t \in \Gamma_0\ ;$$

the condition (1.32) can be reduced to the form

$$Im[\overline{\Lambda_*(a_j)}\Psi(a_j)] = B_j = b_j\ |\Pi(a_j)|e^{ImS(a_j)}\ ,\ j \in \{j\}\ . \tag{1.42}$$

The boundary value problem (1.41) and (1.42) for analytic functions is called <u>Problem C^*.</u>

(1) If $K \geq N-1/2$, it is clear that $H(t) = 0$ in (1.41). We can find an analytic function $\Psi_j(z)$ in D_j: $|z-z_j| > \gamma_j$ satisfying the boundary condition

$$Re[\overline{\Lambda_*(t)}\Psi_j(t)] = R(t)\ ,\ t \in \Gamma_j\ ,\ 1 \leq j \leq N\ . \tag{1.43}$$

In fact, using the transformation $\zeta = \gamma_j/(z-z_j)$, the above boundary value problem can be reduced to the corresponding boundary value problem in $|\zeta| < 1$, its solution can be obtained by Theorem 1.1. So we have the function $\Psi_j(z)$ as stated before.

If $\Lambda_*(t)$ is continuous on $\Gamma \setminus \Gamma_1$ and is discontinuous on Γ_1 , we denote by $\Psi_1(z)$ the analytic functions in $|z-z_j| > \gamma_1$ satisfying the boundary condition

$$Re[\overline{\Lambda_*(t)}\Psi_1(t)] = R(t)\ ,\ t \in \Gamma_1\ ,$$

then $\Psi_1^*(z) = \Psi(z) - \Psi_1(z)$ satisfies the boundary condition

$$Re[\overline{\Lambda_*(t)}\Psi_1^*(t)] = \begin{cases} 0\ ,\ t \in \Gamma_1\ , \\ R(t) - Re[\overline{\Lambda_*(t)}\Psi_1(t)]\ ,\ t \in \Gamma \setminus \Gamma_1\ , \end{cases} \tag{1.44}$$

$$Im[\overline{\Lambda_*(a_j)}\Psi_1^*(a_j)] = B_j - Im[\overline{\Lambda_*(a_j)}\Psi_1(t)] \ , \ j = 1,\ldots,2K-N+1 \ . \tag{1.45}$$

Applying the same method as in the proof of Theorem 1.1 and Theorem 1.8 in Chapter 1, the boundary value problem (1.44), (1.45) for analytic functions has a solution $\Psi_1^*(z)$, and

$$\Psi(z) = \Psi_1(z) + \Psi_1^*(z) \tag{1.46}$$

is a solution of Problem C^* for analytic functions.

If $\Lambda_*(t)$ is continuous on $\Gamma\setminus\{\Gamma_1 \cup \Gamma_2\}$ and is discontinuous on $\Gamma_1 \cup \Gamma_2$, using the above method, we may transform the boundary condition (1.41) into the homogeneous boundary condition on Γ_2 and then change $\Lambda_*(t)$ to a new continuous function on $\Gamma \setminus \Gamma_1$, which is not equal to zero. This is the case discussed above. Therefore, we can find a solution of Problem C^*.

Using induction, we can prove the existence of solutions of Problem C^*. From (1.40), we see that the analytic function $\Phi(z) = e^{-iS(z)}Y(z)\Psi(z)/\Pi(z)$ is just a solution of Problem B^*.

(2) If $-1 \le K \le N-1$ and $Ind\ \Lambda(t) = K = N-1$, then $Ind\ \Lambda_1(t) = Ind\ \Lambda(t)/\bar{t} = N \ge N-1$. From the result in (1), we can obtain a solution $\Psi(z)$ of the boundary value problem with the boundary conditions

$$Re[\overline{\Lambda_1(t)}\Psi(t)] = R(t) \ , \ t \in \Gamma \ , \tag{1.47}$$

$$Im[\overline{\Lambda_1(a_j)}\Psi(a_j)] = |Y(a_j)|b_j \ , \ j = 1,\ldots,N+1 \ . \tag{1.48}$$

If $\Psi(0) = 0$, then $\Phi(z) = \Psi(z)/z$ is a solution of Problem B^* with the index $K = N-1$. Otherwise, $\Psi(0) \neq 0$, we find two linearly independent analytic functions $\Psi_1(z)$ and $\Psi_2(z)$, which satisfy

$$Re[\overline{\Lambda_1(t)}\Psi_1(t)] = \begin{cases} |Y(t)| , & t \in \Gamma_1 , \\ 0 , & t \in \Gamma \setminus \Gamma_1 , \end{cases} \quad Im[\overline{\Lambda_1(a_j)}\Psi_1(a_j)] = 0 , \; j = 1,\dots,N+1 , \tag{1.49}$$

$$Re[\overline{\Lambda_1(t)}\Psi_2(t)] = 0 , \; t \in \Gamma , \quad Im[\overline{\Lambda_1(a_j)}\Psi_2(a_j)] = \begin{cases} |Y(a_j)| , & j = 1 , \\ 0 , & j = 2,\dots,N+1. \end{cases} \tag{1.50}$$

It is not difficult to prove

$$\begin{vmatrix} Re\ \Psi_1(0) & Re\ \Psi_2(0) \\ Im\ \Psi_1(0) & Im\ \Psi_2(0) \end{vmatrix} \neq 0 .$$

Hence, there exist two real constants c_1, c_2 so that

$$\left.\begin{aligned} c_1\ Re\ \Psi_1(0) + c_2\ Re\ \Psi_2(0) &= Re\ \Psi(0) , \\ c_1\ Im\ \Psi_1(0) + c_2\ Im\ \Psi_2(0) &= Im\ \Psi(0) , \end{aligned}\right\} \tag{1.51}$$

and $\Phi(z) = [\Psi(z)-c_1\Psi_1(z)-c_2\Psi_2(z)]/z$ is just a solution of Problem B^* with $K = N-1$. Similarly to the method used above, we can prove that Problem B^* with $K = N-3/2 , N-2,\dots,1/2 , 0 , -1/2 , -1$ is solvable.

(3) If $K < -1$, we rewrite the boundary condition (1.29) of Problem B^* with the index $K < -1$ in the form

$$Re[\overline{\Lambda_2(t)}\Psi(t)] = c(t) + |Y(t)|\ h(t) , \; t \in \Gamma , \tag{1.52}$$

where the index of $\Lambda_2(t) = \lambda(t)\bar{t}^{1-K'}$ $(K'=[|K|+1/2])$ equals -1 or $-1/2$. By the result in (2), the boundary value problem with the boundary condition (1.52) has a solution. So $\Phi(z) = z^{1-K'}\Psi(z)$ is a solution of Problem B^* with the index K. This completes the proof. □

Let $\Phi(z)$ be a solution of Problem B^* and substitute it into the boundary condition (1.29). If $h(t) = 0$, for $K \geq 0$, then the solution of Problem B^* is also a solution of Problem A^*. If $K < 0$ and the above function $h(t) = 0$ and the solution $\Phi(z)$ of Problem B^* is analytic in D, i.e. the main part $\sum_{j=1}^{K'-1} c_{-j} z^{-j}$ of the Laurent series of $\Phi(z)$ in the neighbourhood of $z = 0$ equals zero, then the solution of Problem B^* is also the solution of Problem A^*. Thus we obtain the following theorem.

Theorem 1.6 The solvability of Problem A^* for analytic functions is as follows:

(1) If $K \geq N-1/2$, Problem A^* is solvable and the general solution $\Phi(z)$ includes $2K-N+1$ arbitrary real constants.

(2) If $0 \leq K \leq N-1$, the total number of solvability conditions of Problem A^* $\leq N-K' = N-[K+1/2]$. If these conditions are satisfied, then the general solution of Problem A^* includes $[K]+1$ arbitrary real constants.

(3) If $K < 0$, Problem A^* for analytic functions has $N-2K-1$ solvability conditions (cf. [103]24).

3. Integral representation and estimates of solutions of Problem B^* in a multiply connected domain

In order to obtain the integral expression of solutuons of Problem B^* for analytic functions, we first discuss Problem B_* with the boundary conditions

$$Re[\overline{\Lambda(t)}\Psi(t)] = c(t)/|Y(t)| + h(t) \ , \ t \in \Gamma \ , \tag{1.53}$$

$$Im[\overline{\Lambda(a_j)}\Psi(a_j)] = b_j \ , \ j \in \{j\} \ , \tag{1.54}$$

and give the integral expressions of analytic functions $\Psi_j(z)$ on the unbounded domains: $\gamma_j < |z-z_j| < \infty$, $j = 1,\ldots,N$ with boundary conditions

$$Re[\overline{\Lambda(t)}\Psi_j(t)] = c(t)/|Y(t)| \ , \ t \in \Gamma_j \ , \ j = 1,\ldots,N \ , \tag{1.55}$$

where $\Lambda(t)$, $Y(t)$ are stated in (1.31) and (1.34). Set $K_j = Ind\ \lambda(t)|_{\Gamma_j}$ and when $K_j < 0$, the analytic function $\Psi_j(z)$

has a pole of order $\leq |[K_j]|$ at $z = \infty$, $j = 1,\ldots,N$. Through the transformation $\zeta = \gamma_j/(z-z_j)$, and by (1.14), we see that there exists the function $\Psi_j(z)$ and $\Psi_j(z)$ can be expressed as

$$\Psi_j(z) = \frac{1}{z\pi} \int_{\Gamma_j} P_j(z,t) \frac{c(t)}{|Y(t)|} d\theta , \tag{1.56}$$

where $t = z_j + \gamma_j e^{i\theta}$. If $K_j \geq 0$, we set all arbitrary constants equal to zero in (1.14) and, if $K_j < 0$, we let $\Psi_j(z)$ have a pole of order $\leq |[K_j]|$ at $z = z_j$, $j = 1,\ldots,N$. For $j = 0$, through a conformal mapping and by (1.14), we similarly represent the analytic function $\Psi_0(z)$ satisfying the boundary condition

$$Re[\overline{\Lambda(t)}\Psi_0(t)] = c(t)/|Y(z)| , \ t \in \Gamma_0 , \tag{1.57}$$

by

$$\Psi_0(z) = \frac{1}{2\pi} \int_{\Gamma_0} P_0(z,t)c(t)/|Y(t)|d\theta , \tag{1.58}$$

where $t = e^{i\theta}$ and $\Psi_0(z)$ has a pole of order $\leq |[K_0]|$ at $z = z_1$, when $K_0 < 0$. Afterwards, taking the fact that $P_j(z,t)$ is analytic on $\Gamma_k (k \neq j)$ into account and from Theorem 1.4, we can find a solution $P_*(z,t)$ of Problem B^*, which is analytic in D or $D \setminus \{0\}$ and satisfies the boundary condition

$$Re[\overline{\Lambda(z)}P_*(z,t)] = -Re[\overline{\Lambda(z)} \sum_{\substack{k=0 \\ k\neq j}}^{N} P_j(z,t)] + h(z,t) , \ z \in \Gamma_j , \ j = 0,\ldots,N . \tag{1.59}$$

According to the method in § 3, Chapter 1, we derive that a solution $\Psi(z)$ of Problem B^* has the integral expression

$$\Psi(z) = \frac{1}{2\pi} \int_{\Gamma} P(z,t)c(t)/|Y(t)|d\theta + \Psi_0(z) , \tag{1.60}$$

where $\Psi_0(z)$ is a solution of the corresponding homogeneous problem.

Hence, we have the following theorem.

Theorem 1.7 The solution $\Phi(z)$ of Problem B^* for analytic functions can be expressed as

$$\Phi(z) = \frac{1}{2\pi}\int_{\Gamma} T(z,t)c(t)d\theta + \Phi_0(z) , \tag{1.61}$$

where $T(z,t) = Y(z) P(z,t)/|Y(z)|$ is the Schwarz kernel, $\Phi_0(z) = Y(z) \Psi_0(z)$ is a solution of the homogeneous problem corresponding to Problem B^*; $Y(z)$ is as stated in (1.31).

Finally, we shall give some estimates for solutions of Problem B^* for analytic functions.

Theorem 1.8 The solution $\Phi(z)$ of Problem B^* satisfies the estimates

$$C[\Phi^*,D_n] + C[\Phi^{*\prime},D_n] \le M_1 , \tag{1.62}$$

$$C_{\beta_0}[Z\Phi^*,\overline{D}] \le M_2 , \tag{1.63}$$

where

$$\Phi^*(z) = \begin{cases}\Phi(z) , & K \ge 0 , \\ \Phi(z)z^{K'-1} , & K<0 ,\end{cases} \quad D_n=\{z|\,|z-t_j|\ge 1/n, j=0,1,\ldots,m, t_0=0, n>0 , z\in\overline{D}\} ,$$

$M_1 = M_1(\lambda,c,D,D_n)$, $M_2 = M_2(\lambda,c,D,\delta,\beta_0)$, $0 < \beta_0 < \delta$,

$$\tau_j = \begin{cases}\beta_j + \varepsilon , & \gamma_j \ge 0 \text{ or } \gamma_j < 0 \text{ and } \beta_j \ge |\gamma_j| , \\ |\gamma_j| + \delta , & \gamma_j < 0 \text{ and } \beta_j < |\gamma_j| ,\end{cases}$$

δ is a small positive number, $Z(z) = \prod_{j=1}^{m} (z-t_j)^{\tau_j}$.

Proof. By Lemma 1.2 and Theorem 1.7, it is not difficult to see that $Z(z)\,\Phi^*(z) \in C_{\beta_0}(\overline{D})$. Suppose that the estimate

$$C[Z\Phi^*,\overline{D}] \le M_3 = M_3(\lambda,c,D,\delta,\beta_0) \tag{1.64}$$

is not true. Then there exists sequences of functions $\{\lambda_n(t)\}$, $\{c_n(t)\}$ and $\{b_j^{(n)}\}$, which satisfy the same conditions as $\lambda(t)$, $c(t)$ and b_j in (1.27), (1.32), so that when $n \to \infty$, $\{\lambda_n(t)\}$, $\{c_0^{(n)}(t)\}$ and $\{b_j^{(n)}\}$ converge uniformly to $\lambda_0(t)$, $c_0^{(0)}(t)$ and $b_j^{(0)}$ on Γ^j, $j = 1,\ldots,m'$, and there exist solutions $\{\Phi_n(z)\}$ of Problem B^* for analytic functions satisfying the boundary conditions

$$Re[\overline{\lambda_n(t)}\Phi_n(t)]=c_n(t)+|Y(t)|h_n(t), c_n=\prod_{j=1}^{m}(t-t_j)^{-\beta_j}c_0^{(n)}(t), t\in\Gamma, \quad (1.65)$$

$$Im[\overline{\lambda_n(a_j)}\Phi_n(a_j)] = |Y(a_j)|b_j, \ j \in \{j\}, \quad (1.66)$$

so that $C[Z\Phi_n^*] = H_n \to \infty$, $H_n \geq 1$, $\Phi_n^*(z) = \begin{cases}\Phi_n(z), & K \geq 0, \\ z^{K'-1}\Phi_n(z), & K < 0.\end{cases}$

Set $\varphi_n^*(z) = \Phi_n^*(z)/H_n$. Then $C[Z\varphi_n^*,\overline{D}] = 1$ and $\varphi_n^*(t)$ satisfies the boundary conditions

$$Re[\overline{\lambda_n(t)}\varphi_n^*(t)] = c_n(t)/H_n + |Y(t)|h_n(t)/H_n, \ t \in \Gamma, \quad (1.67)$$

$$Im[\overline{\lambda_n(a_j)}\varphi_n^*(a_j)] = |Y(a_j)|b_j/H_n, \ j \in \{j\}. \quad (1.68)$$

By Lemma 1.2, Theorem 1.1 and Theorem 1.7, and using the method in the proof of Theorem 1.4 in Chapter 1, we can obtain the estimate

$$C_{\beta_0}[Z\varphi_n^*,\overline{D}] \leq M_4 = M_4(\lambda,c,D,\delta,\beta_0) \quad (1.69)$$

(cf. [7]2)). Hence, we may choose a subsequence of $\{\varphi_{n_k}(z)\} = \{Z(z)\varphi_{n_k}^*(z)\}$ from $\{Z(z)\varphi_n^*(z)\}$ which converges uniformly to $\varphi_0(z) = Z(z)\varphi_0^*(z)$ on any closed set D^* in $\overline{D}\setminus\{t_1,\ldots,t_m,0\}$, and $\varphi_0^*(z)$ satisfies the boundary condition

$$Re[\overline{\lambda_0(t)}\varphi_0^*(t)] = |Y(t)|h_0(t), \ t \in \Gamma, \quad (1.70)$$

$$Im[\overline{\lambda_0(a_j)}\varphi_0^*(a_j)] = 0, \ j \in \{j\}. \quad (1.71)$$

Due to Theorem 1.4, $\varphi_0^*(z) = 0$, $z \in D$. However, from $C[Z\varphi_n^*, \overline{D}] = 1$ it follows that $C[Z\varphi_0^*, \overline{D}] = 1$. This contradiction proves that (1.64) is true. Consequently, we have (1.63), and (1.62) is easily derived from (1.63). □

In addition, we can discuss the solvability of the Riemann boundary value problem with piecewise continuous coefficients for analytic functions and uniformly elliptic complex equations of first order, which will be considered in § 3 of this chapter.

§ 2. The Riemann-Hilbert problem with piecewise continuous coefficients for complex equations of first order

In this section, we discuss the nonlinear complex equation

$$w_{\bar{z}}=F(z,w,w_z), F(z,w,w_z)=Q_1(z,w,w_z)\bar{w}_{\bar{z}}+Q_2(z,w,w_z)w_z+A_1(z,w)w+ \\ +A_2(z,w)\bar{w}+A_3(z,w), \text{ or } w_{\bar{z}} = Q(z,w,w_z)w_z+A_1(z,w)w+A_2(z,w)\bar{w}+A_3(z,w) \tag{2.1}$$

with Condition C as stated in § 1 of Chapter 2. Before studying the Riemann-Hilbert problem with piecewise continuous coefficients for (2.1), we first give an existence theorem for solutions to (2.1) in a multiply connected domain D , which includes an existence theorem for nonlinear quasiconformal mappings from D onto an (N+1)-connected circular domain.

1. Existence theorem for solutions to the nonlinear elliptic complex equation of first order

First of all, we introduce two bounded and closed sets in the Banach space $L_{p_0}(\overline{D})\times L_{p_0}(\overline{D})\times L_{p_0}(\overline{D})$, $2 < p_0 < p$.

Definition. (1) Let B be the Banach space $L_{p_0}(\overline{D})\times L_{p_0}(\overline{D})\times L_{p_0}(\overline{D})$ $(2 < p_0 < p)$, and B_1 be a bounded subset in B , the elements of B_1 are all vectors of the measurable functions $q = [Q(z), f(z), g(z)]$, which satisfy the conditions

$$|Q(z)| \le q_0 < 1 , L_{p_0}[f,\overline{D}] \le k_1 , L_{p_0}[g,\overline{D}] \le k_1 , \tag{2.2}$$

where q_0 , k_1 are nonnegative constants stated in (1.44) and (2.17) of Chapter 2, respectively. It is clear that B_1 is a closed and convex set in B .

(2) Let B_2 be a bounded subset in B , the elements of which are all vectors of measurable functions $\omega = [f(z), g(z), h(z)]$ satisfying the following conditions

$$L_{p_0}[f,\overline{D}] \le k_1 , L_{p_0}[g,\overline{D}] \le k_1 , |h(z)| \le q_0|1+\Pi h| ,$$

where q_0 , k_1 are nonnegative constants as stated in (2.2) and $\Pi h = \frac{1}{\pi}\iint_D \frac{h(\zeta)}{(\zeta-z)^2} d\sigma_\zeta$. It can be verified that B_2 is a closed set and is not convex.

Now, we state and prove an existence theorem for solutions to (2.1) with Condition C.

Theorem 2.1 Suppose that the nonlinear complex equation (2.1) satisfies Condition C in the (N+1)-connected circular domain D , and $\Phi(\zeta)$ is an analytic function in $0 < |\zeta| < 1$. Then the equation (2.1) has a solution $w(z)$ of the following type.

$$w(z) = \Phi[\zeta(z)]e^{\phi(z)} + \psi(z) , \tag{2.3}$$

where

$$\psi(z) = Tf - T_1 f , \quad Tf = \frac{1}{\pi}\iint_D \frac{f(\zeta)}{\zeta - z} d\sigma_\zeta , \quad T_1 f = Tf\Big|_{z=1} ,$$

$$\phi(z) = Tg - T_1 g , \chi(z) = z + Th - T_0 h , T_0 h = Th\Big|_{z=0} ,$$

$\zeta = \Psi(\chi)$ is a conformal mapping which maps the domain $\chi(D)$ onto an (N+1)-connected circular domain in $|\zeta| < 1$ similar to D , $\zeta(z) = \Psi[\chi(z)]$, $\zeta(0) = 0$, $f(z)$, $g(z)$, $h(z)$, $\psi(z)$, $\phi(z)$, $\zeta(z)$ and its inverse function $z(\zeta)$ satisfy the estimates (2.17)-(2.20) in Chapter 2.

Proof. To apply the Schauder fixed-point theorem, we introduce five integral equations as follows:

$$h(z) = Q(z)\Pi h + Q(z) \quad , \quad \Pi h = -\frac{1}{\pi}\iint_D \frac{h(\zeta)}{(\zeta-z)^2}\,d\sigma_\zeta \ , \tag{2.4}$$

$$\begin{aligned} f^*(z) &= F(z,w,\Pi f^*) - F(z,w,0) + A_1(z,w)(Tf^*-T_1f^*) \\ &\quad + A_2(z,w)\overline{(Tf^*-T_1f^*)} + A_3(z,w) \ , \end{aligned} \tag{2.5}$$

$$\begin{aligned} Wg^*(z) &= F(z,w,W\Pi g^*+\Pi f^*) - F(z,w,\Pi f^*) \\ &\quad + A_1(z,w)W + A_2(z,w,)\overline{W} \ , \end{aligned} \tag{2.6}$$

$$\begin{aligned} S'(\chi)h^*e^{\varphi} &= F[z,w,S'(\chi)(1+\Pi h^*)e^{\varphi} + W\Pi g^* + \Pi f^*] \\ &\quad - F(z,w,W\Pi g^*+\Pi f^*) \ , \end{aligned} \tag{2.7}$$

$$Q^*(z) = h^*(z)/(1+\Pi h^*) \ , \tag{2.8}$$

where $q = [Q(z),f(z),g(z)] \in B$. By the principle of contracting mappings, the equation (2.4) has a unique solution $h(z) \in L_{p_0}(\overline{D})$. Putting $\chi(z) = z + Th - T_0h$, $\psi(z) = Tf - T_1f$, $\phi(z) = Tg - T_1g$, and introducing $\zeta(z) = \Psi[\chi(z)]$, $S(\chi) = \Phi[\Psi(\chi)]$, $w(z) = \Phi[\zeta(z)]e^{\phi(z)} + \psi(z) = W(z) + \psi(z)$ as stated in this theorem, we substitute $w(z)$, $W(z)$, $S'(\chi)$, $\phi(z)$ into the corresponding positions in (2.5)-(2.8). Using the method in the proof of Theorem 3.3 in Chapter 2, we can seek solutions $f^*(z)$, $g^*(z)$, $h^*(z)$, $Q^*(z) \in L_{p_0}(D)$ of the integral equations (2.5)-(2.8), successively. Let $\omega^* = [f^*(z),g^*(z),h^*(z)]$ and $q^* = [Q^*(z),f^*(z),g^*(z)]$ and denote by $h(z) = S_1(q)$, $w(z) = S_2(q)$, $\omega^* = S_3(q)$, and $q^* = S(q)$ the mappings from q to $h(z)$, $w(z)$, ω^* and q^*, respectively. Now, we shall verify the following statements.

(1) $\omega^* = S_3(q)$ is a continuous mapping from B_1 into B_2. We arbitrarily choose $q_n = [Q_n(z),f_n(z),g_n(z)] \in B_1 (n=0,1,2,\ldots)$, so that when $n \to \infty$,

$$\|q_n-q_0\| = L_{p_0}[Q_n-Q_0,\overline{D}] + L_{p_0}[f_n-f_0,\overline{D}] + L_{p_0}[g_n-g_0,\overline{D}] \to 0 \ . \tag{2.9}$$

Set $h_n(z) = S_1(q_n)$, $w_n(z) = S_2(q_n)$, $\omega_n^* = [f_n^*(z), g_n^*(z), h_n^*(z)] = S_3(q_n)$, $q_n^* = [Q_n^*(z), f_n^*(z), g_n^*(z)] = S(q_n)$, and observe that we have

$$h_n = Q_n(z)\Pi h_n + Q_n(z) , \tag{2.10}$$

$$f_n^* = F(z,w_n,\Pi f_n^*) - F(z,w_n,0) + A_1(z,w_n)(Tf_n^* - T_1 f_n^*) + A_2(z,w_n)\overline{(Tf_n^* - T_1 f_n^*)} + A_3(z,w_n) , \tag{2.11}$$

$$W_n g_n^* = F(z,w_n,W_n \Pi g_n^* + \Pi f_n^*) - F(z,w_n,\Pi f_n^*) + A_1(z,w_n)W_n + A_2(z,w_n)\overline{W}_n , \tag{2.12}$$

$$S_n'(\chi)h_n^* e^{\varphi_n} = F[z,w_n,S_n'(\chi)(1+\Pi h_n^*)e^{\varphi_n} + W_n \Pi g_n^* + \Pi f_n^*] - F[z,w_n,W_n \Pi g_n^* + \Pi f_n^*] , \tag{2.13}$$

$$Q_n^* = h_n^*/(1+\Pi h_n^*) , \quad n = 0,1,2,\ldots . \tag{2.14}$$

From the condition (2.9) and the integral equations (2.10), it follows that $L_{p_0}[h_n - h_0, \overline{D}] \to 0$, when $n \to \infty$. Using a similar method to that in the proof of Theorem 2.6 in Chapter 2, we can conclude that $L_{p_0}[f_n^* - f_0^*, \overline{D}] \to 0$, when $n \to \infty$. If $\Phi(\zeta) \equiv 0$, and $W_n(z) \equiv 0$, we may take $g_n^*(z) = 0$, $n = 0,1,\ldots$. If $\Phi(\zeta) \not\equiv 0$, by the same reason as that given above, we have that $L_{p_0}[g_n^* - g_0^*, \overline{D}] \to 0$, when $n \to \infty$. If $\Phi(\zeta) \equiv$ constant, and $\Phi_n'(\chi) \equiv 0$, we take $h_n^*(z) = Q_n^*(z) = 0$, $n = 0,1,\ldots$, while for $\Phi'(\zeta) \not\equiv 0$ and $\Phi_n'(\chi) \not\equiv 0$, we can also obtain that $L_{p_0}[h_n^* - h_0^*, \overline{D}] \to 0$, when $n \to \infty$. Consequently, $\|\omega_n^* - \omega_0^*\| = L_{p_0}[f_n^* - f_0^*, \overline{D}] + L_{p_0}[g_n^* - g_0^*, \overline{D}] + L_{p_0}[h_n^* - h_0^*, \overline{D}] \to 0$, when $n \to \infty$. This shows that S_3 is a continuous operator.

(2) $q^* = S(q)$ continuously maps B_1 onto a compact set in B_1. Under the condition in (1), we may verify that

$$L_{p_0}[Q_n^* - Q_0^*, \overline{D}] \to 0 , \text{ when } n \to \infty . \tag{2.15}$$

Indeed, from (2.14), we have

$$Q_n^* - Q_0^* = \frac{h_n^*}{1+\Pi h_n^*} - \frac{h_0^*}{1+\Pi h_0^*} = \frac{h_n^*-h_0^*-Q_n^*\Pi(h_n^*-h_0^*)}{1+\Pi h_0^*} .$$

Supposing that (2.15) is not true, there exists a subsequence $\{Q_{n_k}^*\}$, which again is denoted by $\{Q_n^*\}$, so that $L_{p_0}[Q_n^*-Q_0^*,\overline{D}] \to d > 0$. Due to

$$L_{p_0}[h_n^*-h_0^*,\overline{D}] \to 0 , \; L_{p_0}[\Pi(h_n^*-h_0^*,\overline{D})] \to 0 , \text{ when } n \to \infty ,$$

from $\{h_n^*-h_0^*\}$ and $\{\Pi(h_n^*-h_0^*)\}$, we can select subsequences $\{h_{n_k}^*-h_0^*\}$ and $\{\Pi h_{n_k}^*-h_0^*)\}$ converging to 0 for almost every point in D. By virtue of $|Q_n^*| \le q_0 < 1(n=0,1,2,...)$ and $\|1+\Pi h_0^*\| > 0$ (cf. [15]1) and [103]25)), it can be derived that $L_{p_0}[Q_{n_k}^*-Q_0^*,\overline{D}] \to 0$, when $n \to \infty$. This contradiction shows that $q^* = S(q)$ is a continuous mapping.

In order to prove that $q^* = S(q)$ maps B_1 onto a compact set in itself, we are free to choose $q_n = [Q_n(z), f_n(z), h_n(z)] \in B_1 (n=1,2,...)$ and denote $w_n(z) = S_2(q_n)$, $\omega_n^* = [f_n^*(z), g_n^*(z), h_n^*(z)] = S_3(q_n) \in B_2$, $q_n^* = [Q_n^*(z), f_n^*(z), g_n^*(z)] = S(q_n) \in B_1$, $n = 1,2,...$. Because $\psi_n(z)$, $\varphi_n(z)$, $\chi_n(z)$ and its inverse function $z_n(\chi)$ satisfy the estimates (2.9), (2.18), (2.19) in Chapter 2, we can choose subsequences of $\{\psi_n(z)\}$, $\{\varphi_n(z)\}$, $\{\chi_n(z)\}$, $\{w_n(z)\}$, and $\{W_n(z)\}$ converging uniformly to $\psi_0(z)$, $\phi_0(z)$, $\chi_0(z)$, $w_0(z)$, and $W_0(z)$ on any closed set in $D\setminus\{0\}$. Let $f_0^*(z)$, $g_0^*(z)$ and $h_0^*(z)$ denote solutions of the integral equations (2.11) - (2.13) with $n = 0$. Using the same method as in (1), we can prove that

$$L_{p_0}[f_n^*-f_0^*,\overline{D}] \to 0 , L_{p_0}[g_n^*-g_0^*,\overline{D}] \to 0 , L_{p_0}[h_n^*-h_0^*,\overline{D}] \to 0 , L_{p_0}[Q_n^*-Q_0^*,\overline{D}] \to 0.$$

So $q^* = S(q)$ is a compact mapping in B_1.

According to the Schauder fixed point theorem, there exists a function vector $q = [Q(z),f(z),g(z)] \in B_1$, so that $q = S(q)$. Thus $\omega = [f(z),g(z),h(z)] = [f^*(z),g^*(z),h^*(z)] = S_3(q) \in B_2$ satisfies the integral equations (2.4) - (2.7). We add equations (2.5) - (2.7) and noting

$$w_{\bar{z}} = S'(\chi)he^{\varphi(z)} + Wg + f, w_z = S'(\chi)\ (1+h)e^{\varphi(z)} + W\Pi g + \Pi f , \qquad (2.16)$$

the complex equation (21.) is obtained. This shows that

$$w(z) = S[\chi(z)]e^{\varphi(z)} + \psi(z) = \Phi[\zeta(z)]e^{\varphi(z)} + \psi(z)$$

is a solution of (2.1). □

Theorem 2.2 Under the hypotheses in Theorem 2.1, the nonlinear complex equation (2.1) possesses a solution of the type

$$w(z) = \Phi[\eta(z)]e^{\phi(z)} + \psi(z) , \eta(z) = \Delta[\chi(z)] , \qquad (2.17)$$

where $\psi(z)$, $\phi(z)$, $\chi(z)$ are as stated in Theorem 2.1 and $\Delta(\chi)$ is an analytic function which maps the (N+1)-connected domain $\chi(D)$ onto the (N+1)-sheeted unit disc $|\Delta| < 1$.

Proof. By [45], we see that there exists an analytic function $\Delta(\chi)$ as stated in this theorem. Applying a method similar to that used in the proof in Theorem 2.1, we can prove this theorem. (cf. [103]5)). □

2. The Riemann-Hilbert problem with piecewise continuous coefficients for complex equations of first order

Next, we consider the Riemann-Hilbert problem with the boundary condition (1.1), i.e.

$$Re[\overline{\lambda(t)}w(t)] = c(t) , t \in \Gamma \qquad (2.18)$$

for the complex equation (2.1) in the (N+1)-connected circular domain

D , where $\lambda(t)$, $c(t) = \prod_{j=1}^{m} |t-t_j|^{-\beta_j} c_0(t)$ in (2.18) have discontinuities at $t_1,\ldots,t_m$ on Γ and satisfy the conditions in (1.27). In addition, we assume that β_j satisfies

$$\beta_j/\beta = p_0\beta_j/(p_0-2) < 1 \ , \ 2 < p_0 < p \ , \ j=1,\ldots,m \ . \tag{2.19}$$

The above problem is called Problem A for (2.1), the constant $K = \sum_{j=1}^{m} K_j/2$ is called the index of Problem A, where $K_j (j=1,\ldots,m)$ are stated in (1.28).

The boundary condition of the corresponding modified Problem B for (2.1) is as follows:

$$Re[\overline{\lambda(t)}w(t)] = c(t) + |Y[\zeta(t)]|h(t) \ , \ t \in \Gamma \ , \tag{2.20}$$

$$Im[\overline{\lambda(a_j)}w(a_j)] = |Y[\zeta(a_j)]|b_j \ , \ j \in \{j\} \ , \tag{2.21}$$

in which $h(t)$, a_j , $b_j (j\in\{j\})$ are the same as stated in (1.30), (1.32), $\zeta(z)$ is a homeomorphic solution of the Beltrami equation similar to that in Theorem 2.1 and

$$Y(\zeta) = \prod_{j=1}^{m_1} (\zeta-\tau_j)^{\gamma_j} \prod_{j=m_1+1}^{m_2} \left(\frac{\zeta-\tau_j}{\zeta-\zeta_1}\right)^{\gamma_j} \cdots \prod_{j=m_N+1}^{m} \left(\frac{\zeta-\tau_j}{\zeta-\zeta_N}\right)^{\gamma_j} \ , \tag{2.22}$$

where $\gamma_j (j=1,\ldots,m)$ are stated in (1.28), $\gamma_j + \beta_j/\beta < 0$, $\tau_j = \zeta(t_j)$, $j = 1,\ldots,m$, ζ_j is the centre of the circle $L_j = \zeta(\Gamma_j)$, $j = 0,1,\ldots,N$. If the index $K < 0$, we allow the solution $w(z)$ to have a pole of order $\le [|K|+1/2]-1 = K' - 1$ at $z = 0$. If $K \ge 0$, we may require that the solution $w(z)$ of the above problem satisfies the conditions

$$Im[\overline{\lambda(a_j)}w(a_j)] = |Y[\zeta(a_j)]|b_j \ , \ j \in \{j\} \ , \tag{2.23}$$

and

$$w(z) = O(|z-t_j|^{-\eta_j}) \tag{2.24}$$

in the neighbourhood of t_j in D^* , $j = 1,\dots,m$, where

$$\eta_j = \begin{cases} \beta_j/\beta^2+\delta/\beta , & \gamma_j \geq 0 , \\ \beta_j/\beta^2+\delta/\beta , & \beta_j/\beta \geq |\gamma_j| , \\ (|\gamma_j|+\delta)/\beta , & \beta_j/\beta < |\gamma_j| , \end{cases} \quad \gamma_j < 0 ,$$

$\beta = 1-2/p_0 \leq \alpha$, δ is a small positive number.

In the following, we first give some estimates for solutions of Problem B^* for (2.1) with Condition C , and then by a method similar to that in the proof of Theorem 2.1, we obtain an existence theorem for Problem B^* and a result on solvability of Problem A^*. As an application of the above boundary value problems, we shall derive an existence theorem of nonlinear quasiconformal mappings from D onto an (N+1)-connected strip domain.

Theorem 2.3 Let the nonlinear complex equation (2.1) satisfy Condition C. Then the solutions of Problem B^* for (2.1) satisfy the estimates

$$C_*(W) = C_\beta[W,D_n] + L_{p_0}[|W_{\bar z}|+|W_z|,D_n] \leq M_1 , \tag{2.25}$$

$$C^*(W) = C_{\beta_0\beta}[XW,\bar D] \leq M_2 , \quad 0 < \beta_0 < \delta , \tag{2.26}$$

where

$$W(z) = \begin{cases} w(z) , & \text{for } K \geq 0 , \\ w(z)[\zeta(z)]^{K'-1} , & \text{for } K < 0 , \end{cases}$$

$\zeta(z)$ is a homeomorphic solution of the Beltrami equation, D_n is as stated in (1.62), $\beta = 1-2/p_0$, $X(z) = \prod_{j=1}^{m} (z-t_j)^{\eta_j}$, η_j is as stated in (2.24), and $M_1 = M_1(q_0,p_0,k_0,\lambda,c,D,D_n)$, $M_2 = M_2(q_0,p_0,k_0,\lambda,c,D,\delta,\beta_0)$.

Proof. We first discuss the case $K \geq -1$. By Theorem 2.4 in Chapter2, the solution $w(z)$ of Problem B^* can be expressed as

$$w(z) = \Phi[\zeta(z)]e^{\phi(z)} + \psi(z) , \tag{2.27}$$

where $\phi(z)$, $\psi(z)$, $\zeta(z)$ and the inverse function $z(\zeta)$ satisfy the estimates

$$C_\beta[\sigma,\overline{D}] + L_{p_0}[|\sigma_{\bar{z}}|+|\sigma_z|,\overline{D}] \le M_3 , \; \sigma(z) \in \{\varphi(z),\psi(z),\zeta(z)\} , \tag{2.28}$$

$$C_\beta[z(\zeta),\overline{G}] + L_{p_0}[|z_{\bar{\zeta}}|+|z_\zeta|,\overline{G}] \le M_4 , \; G = \zeta(D) , \tag{2.29}$$

in which $M_j = M_j(q_0,p_0,k_0,D)$, $j = 3,4$. On the basis of (2.27), the solution $w(z)$ of Problem B^* for (2.1) can be transformed into the solution $\Phi(\zeta)$ of Problem B_* for analytic functions, and satisfies the boundary condition

$$Re[\overline{\Lambda(\zeta)}\Phi(\zeta)] = R(\zeta) + |Y(\zeta)|H(\zeta) , \; \zeta \in L = \zeta(\Gamma) , \tag{2.30}$$

$$Im[\overline{\Lambda(a_j)}\Phi(a_j)] = |Y(a_j)|b_j - Im[\overline{\lambda(a_j)}\psi(a_j)] , \; j=1,\ldots,2K-N+1 , \tag{2.31}$$

where $R(\zeta) = c[z(\zeta)] - Re\{\overline{\lambda[z(\zeta)]}\psi[z(\zeta)]\}$, $\overline{\Lambda(\zeta)} = \overline{\lambda[z(\zeta)]}e^{\phi[z(\zeta)]}$, $H(\zeta) = h[z(\zeta)]$, and $\Lambda(\zeta)$, $R(\zeta)$ are discontinuous at the points $\tau_j = \zeta(t_j)$, $j = 1,\ldots,m$, on $L = \zeta(\Gamma)$, $R(\zeta) = c_0[z(\zeta)]\prod_{j=1}^{m}[z(\zeta)-z(\tau_j)]^{-\beta_j}$, $R_0(\zeta) = c_0[z(\zeta)]$ and $\Lambda(\zeta)$ satisfy

$$C_{\alpha\beta}[\Lambda,L^j] \le \ell_1 , \; C_{\alpha\beta}[R_0,L^j] \le \ell_1 , \tag{2.32}$$

where $L^j = \zeta(\Gamma^j)$, $\ell_1 = \ell_1(q_0,p_0,k_0,\lambda,c,D)$. By Theorem 1.7 and Theorem 1.8, the solution $\Phi(\zeta)$ of Problem B_* for analytic functions can be expressed as

$$\Phi(\zeta) = \frac{1}{2\pi}\int_\Gamma T(z,\zeta)R(\zeta)d\theta + \Phi_0(\zeta) , \tag{2.33}$$

and $\Phi(\zeta)$ satisfies the estimates

$$C[\Phi^*(\zeta),G_n] + C[\Phi^{*\prime}(\zeta),G_n] \le M_5 \tag{2.34}$$

$$C_{\beta_0}[X^*(\zeta)\Phi^*(\zeta),\overline{G}] \le M_6 \ , \tag{2.35}$$

where

$$\Phi^*(\zeta) = \begin{cases}\Phi(\zeta) \ , & \text{for } K \ge 0 \\ \Phi(\zeta)\zeta^{K'-1} \ , & \text{for } K < 0 \ ,\end{cases} \qquad G_n = \zeta(D_n) \ , \ X^*(\zeta) = \prod_{j=1}^{m}(\zeta-\tau_j)^{\eta\beta} \ ,$$

$0 < \beta_0 < \delta$, $M_5 = M_5(q_0,p_0,k_0,\lambda,c,D,D_n)$, $M_6 = M_6(q_0,p_0,k_0,\lambda,c,D,\delta,\beta_0)$. Combining (2.27) - (2.29), (2.34) - (2.35), we obtain the estimates (2.25) and (2.26).

If $K < -1$, similarly we have (2.27) - (2.32), but $\Phi(\zeta)$ has a pole of order $\le K' - 1 = [|K|+\frac{1}{2}]-1$ at $\zeta = 0$, the function $\Psi(\zeta) = \Phi(\zeta)\zeta^{K'-1}$ is analytic in D , and satisfies the boundary condition

$$Re[\overline{\Lambda_*(\zeta)}\Phi(\zeta)] = R(\zeta) + |Y(\zeta)|H(\zeta) \ , \ \zeta \in L = \zeta(\Gamma) \ , \tag{2.36}$$

where the index of $\Lambda_*(\zeta) = \Lambda(\zeta)\overline{\zeta}^{1-K'}$ is -1 or $-1/2$. Hence $\Psi(\zeta)$ satisfies the estimates (2.34), (2.35), and

$$w(z) = \Psi[\zeta(z)][\zeta(z)]^{1-K'}e^{\phi[z(\zeta)]} + \psi[z(\zeta)]$$

satisfies the estimates (2.25), (2.26). □

Secondly, we prove the existence of solutions for Problem B^* of (2.1).

<u>Theorem 2.4</u> Under the hypthesis in Theorem 2.3, Problem B^* of (2.1) with Condition C is solvable.

<u>Proof.</u> Let us introduce a bounded, closed and convex set B_1 in the Banach space B , which is defined as at the beginning of this section. We arbitrarily choose $q = [Q(z),f(z),g(z)] \in B_1$, and find a solution $h(z) \in L_{p_0}(\overline{D})$ of the following integral equation

$$h(z) = Q(z)\Pi h + Q(z) \ , \ h = -\frac{1}{\pi}\iint_D \frac{h(\zeta)}{(\zeta-z)^2}\, d\sigma_\zeta \ . \tag{2.37}$$

Let $\chi(z) = z + Th - T_0 h$, $\zeta(z) = \Psi[\chi(z)]$. We only point out the differences from the proof of Theorem 2.1. We need to find an analytic function $\Phi(\zeta)$ in $G = \zeta(D)$ satisfying the boundary conditions

$$Re\{\overline{\lambda[z(\zeta)]}\Phi(\zeta)\} = R[z(\zeta)] + |Y(\zeta)|h[z(\zeta)] \ , \ \zeta \in L = \zeta(\Gamma) \ , \tag{2.38}$$

$$Im\{\overline{\lambda(a_j)}\Phi[\zeta(a_j)]\} = |Y[\zeta(a_j)]|b_j \ , \ j \in \{j\} \ . \tag{2.39}$$

We denote that

$$w(z) = \Phi[\zeta(z)]e^{\varphi(z)} + \psi(z) \ , \tilde{w}(z) = \tilde{\Phi}[\zeta(z)]e^{\varphi(z)} + \psi(z) = \tilde{W}(z) + \psi(z) \tag{2.40}$$

where $\tilde{\Phi}(\zeta) = \Phi(\zeta) + 1/n\ \zeta^{-[|K|]-1}$,n is a positive integer, and find solutions $f^*(z)$, $g^*(z)$, $h^*(z)$ of the following integral equations

$$f^*(z) = F(z,w,\Pi f^*) - F(z,w,0) + A_1(z,w)Tf^* + A_2(z,w)\overline{Tf^*} + A_3(z,w) \ , \tag{2.41}$$

$$\tilde{W}g^*(z) = F(z,w,\tilde{W}\Pi g^*+\Pi f^*) - F(z,w,\Pi f^*) + A_1(z,w)\tilde{W} + A_2(z,w)\overline{\tilde{W}} \ , \tag{2.42}$$

$$\Phi^{*\prime}(\chi)h^*(z)e^{\varphi(z)} = F(z,w,\Phi^{*\prime}(\chi)(1+\Pi h^*)e^{\varphi} + \tilde{W}\Pi g^*+\Pi f^*) - F(z,w,\tilde{W}\Pi g^*+\Pi f^*) \ , \ \Phi'(\chi) = [\Phi^*(\Psi(\chi))]'_\chi \tag{2.43}$$

and

$$Q^*(z) = h^*/(1+\Pi h^*) \ . \tag{2.44}$$

Set $\psi^*(z) = Tf^*$, $\phi^*(z) = Tg^*$, $w^*(z) = \Phi^*[\zeta^*(z)]e^{\phi^*(z)} + \psi^*(z)$, $\tilde{w}^*(z) = \tilde{\Phi}^*[\zeta^*(z)]e^{\phi^*(z)} + \psi^*(z)$, $\tilde{W}^*(z) = \Phi^*[\zeta^*(z)]e^{\phi^*(z)}$, and denote by $q^* = [Q^*(z), f^*(z), g^*(z)] = S(q)$ the mappings from q onto q^* . By analogy with the proof of Theorem 2.1, we can verify that $q^* = S(q)$

continuously maps B_1 onto a compact set in B_1. Hence, by the Schauder fixed point theorem, there exists a function vector $q = [Q(z), f(z), g(z)]$, so that $q = S(q)$, and we have the corresponding function $w(z)$, $w(z)$ as stated in (2.40). From (2.41) - (2.43) with $f^* = f$, $g^* = g$, $h^* = h$, we can conclude the complex equation

$$\tilde{w}_{\bar{z}} = F(z,w,\tilde{w}_z) + A_1(z,w)(\tilde{w}-w) + A_2(z,w)(\overline{\tilde{w}-w}) . \qquad (2.45)$$

Using Theorem 2.3 and the compactness principle in Theorem 2.7 of Chapter 2, we may choose a subsequence of solutions

$$\tilde{w}_n(z) = \tilde{\Phi}_n[\zeta(z)]e^{\varphi(z)} + \psi(z) , \quad \tilde{\Phi}_n(\zeta) = \Phi(\zeta) + \frac{1}{n}\zeta^{-[|K|]-1} ,$$

the limit function $w(z)$ of which is a solution to Problem B^* of (2.1). □

Similarly as in the case of Theorem 1.6, we can obtain the following results for Problem A^* of (2.1).

Theorem 2.5 Let the nonlinear complex equation (2.1) satisfy Condition C. Then Problem A^* has the same results with respect to solvability as is stated in Theorem 1.6 (cf. [103]23)).

3. Some existence theorems for nonlinear quasiconformal mappings

Finally, we discuss the complex equation (2.1) with the conditions $A_j(z,w) = 0$, $j = 1,2,3$, i.e.

$$w_{\bar{z}} = Q_1(z,w,w_z)w_z + Q_2(z,w,w_z)\bar{w}_{\bar{z}} . \qquad (2.46)$$

From Theorem 2.1 and Theorem 2.4, we can obtain some existence theorems for nonlinear quasiconformal mappings for (2.46).

Theorem 2.6 Let the nonlinear complex equation (2.46) satisfy Condition C. There exists a homeomorphic solution $w(z)$ of (2.46), which quasiconformally maps D onto an (N+1)-connected strip domain H, where a so-called (N+1)-connected strip domain means a domain with the boundary $Im w = 0$, $Im w = 1$ and $N-1$ level rectilinear slits in $0 < Im w < 1$.

Proof. We set

$$\overline{\lambda(t)} = -i \ , \ t \in \Gamma \ , \ c(t) = \begin{cases} 0 \ , & -\pi < \theta < 0, \\ 1 \ , & 0 < \theta < \pi, \end{cases} \Big\} t = e^{i\theta} \in \Gamma_0 \\ \ h_j \ , \ t \in \Gamma_j \ , \ j = 1,\ldots,N \ . \tag{2.47}$$

By Theorem 2.4, Problem B^* for (2.46) with the index $K = 0$ has a solution $w(z)$. Now, we select any point w_0 so that $0 < Im\, w_0 < 1$ and $w_0 \notin w(\Gamma)$. It is easy to see that the function $W(z) = w(z) - w_0$ is a solution of the complex equation

$$W_{\bar{z}} = Q_1(z,w,w_z)W_z + Q_2(z,w,w_z)\overline{W_{\bar{z}}} \tag{2.48}$$

and $W(z)$ can be expressed as

$$W(z) = w(z) - w_0 = \Phi[\zeta(z)] \tag{2.49}$$

where $\zeta(z)$ is a homeomorphism stated in Theorem 2.4 and $\Phi(\zeta)$ is an analytic function in $G = \zeta(D)$. According to the argument principle and the boundary condition (2.47), we have

$$\Delta_\Gamma arg W(t) = \Delta_\Gamma arg[w(t)-w_0] = 2\pi \ . \tag{2.50}$$

Hence, there exists a point z_0 in D so that $w(z_0) = w_0$. Furthermore, choosing an arbitrary point w_1 with the conditions $Im w_1 < 0$ or $Im w_1 > 1$ and $w_1 \notin w(\Gamma)$, in a similar way to that above, we obtain $\Delta_\Gamma arg[w(t)-w_0] = 0$. In addition, we can prove $w(D) \cap w(\Gamma) = \emptyset$. Thus, the function $w(z)$ is a homeomorphic solution as required in this theorem. □

Theorem 2.7 Let the nonlinear complex equation (2.46) satisfy Condition C. Then there exists a homeomorphic solution $w(z)$ of (2.46), which quasiconformally maps D onto an $(N+1)$-connected circular domain G in the w-plane.

Proof. Let $\Phi(\zeta) = \zeta$ and $A_j(z,w) = 0$, $j = 1,2,3$ in Theorem 2.1. We obtain an existence theorem on nonlinear quasiconformal mappings for (2.46) of the domain D onto an (N+1)-connected circular domain G . □

By Theorem 2.2 and applying the same method as for the above theorem, we can obtain the following result for nonschlicht mappings.

Theorem 2.8 Under the hypothesis of Theorem 2.7, the complex equation (2.46) has a solution $w(z)$, which maps D onto the (N+1)-sheeted unit disc $|w| < 1$.

§ 3. The Riemann boundary value problem with piecewise continuous coefficients for complex equations of first order

1. The Riemann boundary value problem with piecewise continuous coefficients on closed curves

The domains D^+ , D^- are stated in § 3 of Chapter 2, i.e. $D^+ = D$ with the boundary

$$\Gamma = \bigcup_{j=0}^{N} \Gamma_j \in C^1_\mu (0 < \mu < 1) \ , \ D^- = \bigcup_{j=0}^{N} D_j^- = \mathbb{C} \setminus \overline{D} \ ,$$

D_j is the domain surrounded by $\Gamma_j (j=1,\ldots,N)$ and D_0^- is unbounded with the boundary Γ_0 . Moreover, we assume $0 \in D^+$.

Problem R. The Riemann boundary value problem R with piecewise continuous coefficients for (2.1) may be formulated as follows: Find a solution

$$w(z) = \begin{cases} w^+(z), z \in D^+ \\ w^-(z), z \in D^- \end{cases}$$

of (2.1) satisfying the boundary condition

$$w^+(t) = G(t)w^-(t) + g(t) \ , \ t \in \Gamma \ , \tag{3.1}$$

where $t_1,\ldots,t_m$ are all discontinuities of the first kind of $G(t)$

on Γ ,

$$g(t) = \prod_{j=1}^{m} |t-t_j|^{-\beta_j} g_0(t) ,$$

$\beta_j (j=1,\dots,m)$ are sufficiently small nonnegative constants, and $G(t)$, $g_0(t)$ satisfy the conditions

$$C_\alpha[G,\Gamma^j] \le \ell < \infty , \ |G(t)| \ge \ell^{-1} , \ C_\alpha[g_0,\Gamma^j] \le \ell , \tag{3.2}$$

in which $\alpha (0<\alpha<1)$, ℓ are positive constants, and $\Gamma^j (j=1,\dots,m)$ are stated in § 1. We require that the solution is bounded in a neighbourhood of $z = \infty$ and $w(z) = O(\ln|z-t_j|)$ in the neighbourhood of $z = t_j$, $j = 1,\dots,m$.

Putting

$$K_j = [\frac{\varphi_j}{2\pi}] + J_j , J_j = 0 \text{ or } 1 , \rho_j e^{i\varphi_j} = \frac{G(t_j-0)}{G(t_j+0)} , \gamma_j = \frac{\varphi_j}{2\pi} - K_j - i\frac{\ln\rho_j}{2\pi} , \tag{3.3}$$

we denote by $K = \sum_{j=1}^{m} K_j$ the index of Problem R. For the sake of convenience, we only discuss the case of $J_j = 0$ and $\beta_j = 0 (j=1,\dots,m)$. Otherwise, we need to add more restrictions on (2.1). The general case can be discussed, too.

The corresponding modified boundary value problem R^* for (2.1) is to find a solution

$$w(z) = \begin{cases} w^+(z) , & z \in D^+ \\ w^-(z) , & z \in D^- \end{cases} \quad \text{of (2.1)}$$

satisfying the boundary condition (3.1), and, if $K < 0$, we allow the solution $w(z)$ of (2.1) to have a pole of order $\le |K|-1$ at $z = \infty$.

To prove the solvability of Problem R^* for (2.1), we use a similar method to that in § 3 of Chapter2, and write an integral form of Problem R^* for analytic functions, i.e.

$$\left.\begin{aligned}
&\varphi(z) = X(z)\psi(z) \;,\; \psi(z) = \frac{1}{2\pi i}\int_\Gamma \frac{g(t)dt}{(t-z)X^+(t)} + P(z) \;, \\
&X(z) = \begin{cases} X^+(z) = \Omega^+(z)e^{\Gamma^+(z)}/\Pi(z) \;,\; z \in D^+ \;, \\ X^-(z) = \Omega^-(z)z^{-K}e^{\Gamma^-(z)} \;,\; z \in D^- \;, \end{cases} \\
&\Gamma(z) = \frac{1}{2\pi i}\int_\Gamma \frac{\ell n[t^{-K}\Pi(t)G(t)\Omega^-(t)/\Omega^+(t)]}{t-z}\,dt \;,\; P(z) = \begin{cases} c_0+\ldots+c_K z^K \;,\; K \ge 0 \;, \\ 0 \;,\; K < 0 \;, \end{cases}
\end{aligned}\right\} \tag{3.4}$$

where

$$\left.\begin{aligned}
&\Omega(z) = \begin{cases} \Omega^+(z) = \prod_{j=1}^{m_1}(z-t_j)^{\gamma_j}\prod_{j=m_1+1}^{m_2}\left(\frac{z-t_j}{z-s_1}\right)^{\gamma_j}\ldots\prod_{j=m_N+1}^{m}\left(\frac{z-t_j}{z-s_N}\right)^{\gamma_j} \;,\; z\in D^+ \;, \\ \Omega^-(z) = \prod_{j=1}^{m_1}\left(\frac{z-t_j}{z}\right)^{\gamma_j}\prod_{j=m_1+1}^{m}(z-t_j)^{\gamma_j} \;,\; z \in D^- \;, \end{cases} \\
&\Pi(z) = \prod_{i=1}^{N}\prod_{j=m_i+1}^{m_{i+1}}(z-s_i)^{K_j} \;,\; s_i \text{ is a point in } D_j \;,\; j = 1,\ldots,N \;.
\end{aligned}\right\} \tag{3.5}$$

In fact, through the transformation $\phi_1(z) = \phi(z)/\Omega(z)$, $z \in D^\pm$, the boundary condition

$$\phi^+(t) = G(t)\phi^-(t) + g(t) \;,\; t \in \Gamma \tag{3.6}$$

for ϕ can be reduced to the boundary condition

$$\begin{aligned}
&\phi_1^+(t) = G_1(t)\phi_1^-(t) + g_1(t) \;,\; t \in \Gamma \;, \\
&G_1(t) = G(t)\frac{\Omega^-(t)}{\Omega^+(t)} , g_1(t) = \frac{g(t)}{\Omega^+(t)} \;,\; t \in \Gamma
\end{aligned} \tag{3.7}$$

for ϕ_1 . Hence there exists the standard solution

$$X_1(z) = \begin{cases} X_1^+(z) = e^{\Gamma^+(z)}/\Pi(z) \ , \ z \in D^+ \ , \\ X_1^-(z) = z^{-K} e^{\Gamma^-(z)} \ , \ z \in D^- \ , \end{cases} \quad \Gamma(z) = \frac{1}{2\pi i} \int_\Gamma \frac{\ln t^{-K}\Pi(t)G_1(t)}{t-z} dt \ , \tag{3.8}$$

and

$$\phi_1(z) = X_1(z)\psi(z) \quad , \quad \psi(z) = \frac{1}{2\pi i} \int_\Gamma \frac{g_1(t)dt}{(t-z)X_1^+(t)} + P(z) \ , \tag{3.9}$$

where $P(z)$ is stated in (3.4). So we have (3.4).

Theorem 3.1 Suppose that

$$L_{p,2}\left[\frac{Q}{z\prod_{j=1}^{m}(z-t_j)}, \mathbb{C}\right] \le \ell < \infty \ ,$$

$$L_{p,2}\left[\frac{A_k}{\prod_{j=1}^{m}(z-t_j)^{\gamma_j}}, \mathbb{C}\right] \le \ell \ , \ k=1,2,3 \ , \ L_{p,2}[A_3 z^K, \mathbb{C}] \le \ell \ . \tag{3.10}$$

Then

$$w(z) = \begin{cases} w^+(z) \ , \ z \in D^+ \\ w^-(z) \ , \ z \in D^- \end{cases}$$

is a solution of Problem R^* for (2.1) satisfying Condition C^* if and only if

$$V(z) = [w(z)-\phi(z)]/X(z) \tag{3.11}$$

is a continuous solution of the equation

$$V_{\bar{z}} = F[z,\phi(z)+X(z)V(z),\phi'(z)+X'(z)V(z)+X(z)V_z]/X(z) \tag{3.12}$$

on $\mathbb{C}$ and $V(\infty) = 0 \ (K < 0)$, where $\phi(z)$ and $X(z)$ are as stated in (3.4).

Proof Because $\phi(z)$ and $X(z)$ satisfy the boundary conditions

$$\left.\begin{aligned} &\varphi^+(t) = G(t)\varphi^-(t) + g(t)\ , t \in \Gamma^* = \Gamma\setminus\{t_1,\dots,t_m\}\ , \\ &X^+(t) = G(t)X^-(t)\ ,\ t \in \Gamma^*\ , \end{aligned}\right\} \tag{3.13}$$

we know that

$$w(z) = \phi(z) + X(z)V(z) \tag{3.14}$$

satisfies the boundary condition

$$\begin{aligned} w^+(z) &= G(t)[\phi^-(t)+X^-(t)V^+(t)] + g(t) = \\ &= G(t)[\phi^-(t)+X^-(t)V^-(t)] + g(t)\ ,\ t \in \Gamma\ . \end{aligned} \tag{3.15}$$

From $X^-(t) \neq 0$, $t \in \Gamma$, it follows that $V^+(t) = V^-(t)$, $t \in \Gamma$, and so

$$V(z) = \begin{cases} V^+(z)\ , & z \in D^+ \\ V^-(z)\ , & z \in D^- \end{cases}$$

is a continuous solution of (3.12) on $\mathbb{C}$. □

The converse statement is true, too.

Theorem 3.2 Let the nonlinear complex equation (2.1) satisfy Condition C^* and (3.10). Then Problem R^* is solvable, and if $K \geq 0$, the general solution $w(z)$ of Problem R for (2.1) includes $2K+2$ arbitrary real constants and, if $K < 0$, Problem R has at least $-2K-2$ solvability conditions as in Theorem 3.8 of Chapter 2.

Proof. According to Condition C^* and (3.10) the corresponding nonlinear complex equation (3.12) satisfies a similar Condition C^*. By Theorem 3.4 in Chapter 2, the equation (3.12) has a solution $V(z) = T\omega$ ($\omega \in L_{p_0,2}(\mathbb{C})$) . If $K \geq 0$, the general solution $w(z)$ of Problem R^* is a solution of Problem R for (2.1). Therefore, the general solution $w(z)$ contains $2K+2$ arbitrary real constants.

If $K < 0$, and the solution of Problem R^* is bounded in a neighbourhood of ∞, then it is also a solution of Problem R. So Problem R has at least $-2K - 2$ solvability conditions. □

2. The Riemann boundary value problem for complex equations of first order on open arcs

Let L consist of n disjoint open arcs $L_1,\ldots,L_n$, $L_j \in C^1_\mu (0 < \mu < 1)$ with starting point a_j and the terminal point $b_j (j=1,\ldots,n)$. We denote by D the complement of L on the whole z-plane, and $L^* = L \setminus \{a_1,\ldots,a_n,b_1,\ldots,b_n\}$.

Problem R_1 Find a solution $w(z)$ of the complex equation (2.1) in D, continuous on both sides of L and satisfying the boundary condition

$$w^+(t) = G_1(t)w^-(t) + g_1(t) \ , \ t \in L^* \ , \tag{3.16}$$

where $w^+(t)$ and $w^-(t)$ are the left limit and right limit of $w(t)$ on L respectively, $G_1(t)$, $g_1(t)$ satisfy the following conditions:

$$C_\alpha[G,L] < \ell < \infty \ , \ |G_1(t)| > \ell^{-1} \ , \ C_\alpha[g_1,L] < \ell \ . \tag{3.17}$$

In order to apply Theorem 3.2, we introduce m arcs $L'_1,\ldots,L'_n$ ($L' = \bigcup_{j=1}^{n} L'_j \setminus \{a_1,\ldots,a_n,b_1,\ldots,b_n\}$) with the initial point b_j and the terminal point $a_{j+1} (j=1,\ldots,n, a_{n+1}=a_1)$, so that $\bigcup_{j=1}^{n} (L_j \cup L'_j)$ consists of a simply closed curve $\Gamma \in C^1_\mu (0 < \mu < 1)$. Let D^+, D^- denote the bounded domain and the unbounded domain with boundary Γ. Now, we may rewrite Problem R_1 in the following form:

$$w^+(t) = G(t)w^-(t) + g(t) \ , \ t \in \Gamma \ , \tag{3.18}$$

where

$$G(t) = \begin{cases} 1 \ , \ t \in L' \ , \\ G_1(t) \ , \ t \in L \ , \end{cases} \qquad g(t) = \begin{cases} 0 \ , \ t \in L' \ , \\ g(t) \ , \ t \in L \ . \end{cases}$$

The boundary value problem with the boundary condition (3.18) is a

Problem R. Denote $G_1(a_j) = \rho_j e^{i\phi_j}$, $G_1(b_j) = \rho_j' e^{i(\phi_j+\Delta_j)}$; $\Delta_j = \Delta_{L_j} \, arg\, G(t)$. It is easy to see that

$$\left.\begin{aligned} &G_1(a_j-0) = 1 \, , \, G_1(a_j+0) = G(a_j) = \rho_j e^{i\varphi_j} \, , \\ &G_1(b_j-0) = G_1(b_j) = \rho_j' e^{i(\varphi_j+\Delta_j)} \, , \, G_1(b_j+0) = 1 \, , \\ &\frac{G_1(a_j-0)}{G_1(a_j+0)} = \frac{1}{\rho_j} e^{-i\varphi_j} \; \frac{G_1(b_j-0)}{G_1(b_j+0)} = \rho_j' e^{i(\varphi_j+\Delta_j)} \, , \, j = 1,\ldots,m \, . \end{aligned}\right\} \tag{3.19}$$

If ϕ_j satisfies $-2\pi < \phi_j \le 0$ or $0 < \phi_j < 2\pi$, i.e. $0 \le -\frac{\varphi_j}{2\pi} < 1$ or $-1 < -\frac{\varphi_j}{2\pi} < 0$, then

$$\left.\begin{aligned} &\gamma_j = \frac{1}{2\pi i} \ln(\frac{1}{\rho_j} e^{-i\varphi_j}) = -\frac{\varphi_j}{2\pi} + i \frac{\ln \rho_j}{2\pi} \\ &\gamma_j' = \frac{1}{2\pi i} \ell \;\; \rho_j' e^{i(\varphi_j+\Delta_j)} = \frac{\varphi_j+\Delta_j}{2\pi} - K_j - i \frac{\ln \rho_j}{2\pi} \, , \\ &K_j = [\frac{\varphi_j+\Delta_j}{2\pi}] + J_j \, , \, J_j = 0 \text{ or } 1 \, . \end{aligned}\right\} \tag{3.20}$$

The number $K = \sum_{j=1}^{n} K_j$ is called the index of Problem R_1 and Problem R. If $K < 0$, we permit the solution $w(z)$ to have a pole of order $\le |K| - 1$. The modified boundary value problems are called Problem R_1^* and Problem R^*. Next, there is no harm in discussing the case $0 \le \phi_j < 2\pi$ and $J_j = 0$, $j = 1,\ldots,n$. Since Problem R_1^* is a special case of Problem R^*, and $\phi^+(t) = \phi^-(t)$, $X^+(t) = X^-(t)$, $t \in L'$, where $\phi(z)$ and $X(z)$ are sectionally analytic functions similar to the types as stated in (3.4), we can verify that $\phi(z)$ and $X(z)$ are analytic on L', and by Theorem 3.1, it follows that $w(z)$ in (3.14) is continuous on L'. Thus, we obtain the following theorem.

Theorem 3.3 Let the complex equation (2.1) satisfy the Condition C^* on $\mathbb{C}$ and

$$L_{p,2}\left[\frac{Q}{z\prod_{j=1}^{n}(z-a_j)(z-b_j)},\mathbb{C}\right] \le \ell\ ,\ L_{p,2}\left[\frac{A_k}{\prod_{j=1}^{n}(z-a_j)^{\gamma_j}(z-b_j)^{\gamma_j}},\mathbb{C}\right] \le \ell\ , \qquad (3.21)$$

$$k=1,2,3,\ L_{p,2}[A_3 z^K,\mathbb{C}] \le \ell(K \ge 0)\ .$$

Then (1) Problem R_1^* is solvable, the solution $w(z)$ can be expressed as

$$w(z) = \phi(z) + X(z)V(z)\ , \qquad (3.22)$$

where $V(z)$ is a solution of the nonlinear complex equation (3.12), $\phi(z)$, $X(z)$ possess the following representations

$$\varphi(z) = X(z)\psi(z)\ ,\ \psi(z) = \frac{1}{2\pi i}\int_L \frac{g(t)dt}{(t-z)X^+(t)} + P(z)\ ,$$

$$X(z) = \begin{cases} X^+(z) = \prod_{j=1}^{n}(z-a_j)^{\gamma_j}(z-b_j)^{\gamma_j'}e^{\Gamma^+(z)}\ , & z \in D^+\ , \\ X^-(z) = z^{-K}\prod_{j=1}^{n}\left(\frac{z-a_j}{z}\right)^{\gamma_j}\left(\frac{z-b_j}{z}\right)^{\gamma_j'}e^{\Gamma^-(z)}\ , & z \in D^-\ , \end{cases}$$

$$\Gamma(z) = \frac{1}{2\pi i}\int_L \frac{\ell n[t^{-K-\sum_{j=1}^{n}(\gamma_j+\gamma_j')}G(t)]}{(t-z)t^{\sum_{j=1}^{n}(\gamma_j+\gamma_j')}}dt, P(z) = \begin{cases} c_0+\dots+c_K z^K, K \ge 0, \\ 0\ ,\ K < 0\ . \end{cases} \qquad (3.23)$$

(2) If $K \ge 0$, the general solution of Problem R_1 includes $2K + 2$ arbitrary real constants, and if $K < 0$, Problem R_1 for (2.1) has at least $-2K - 2$ solvability conditions.

Finally, we mention that if the functions $G_1(t)$ and $g_1(t)$ in (3.16) have discontinuities of the first kind on $L_j(j=1,\dots,n)$, then corresponding results can be obtained.

§ 4. The Poincaré problem with piecewise continuous coefficients for elliptic equations of second order

In this section, we shall discuss the Poincaré problem with the boundary condition

$$\left.\begin{aligned}&\frac{\partial u}{\partial \mu} + 2\varepsilon\sigma(t)u(t) = 2\tau(t) \ , \ t \in \Gamma \ , \\ &\text{i.e. } Re[\overline{\lambda(t)}u_t] + \varepsilon\sigma(t)u(t) = \tau(t) \ , \ t \in \Gamma\end{aligned}\right\} \tag{4.1}$$

for the nonlinear elliptic equation of second order

$$\begin{aligned}u_{z\bar{z}} &= F(z,u,u_z,u_{zz}) \ , \ F(z,u,u_z,u_{zz}) = Re[Q(z,u,u_z,u_{zz})u_{zz}+A_1u_z] \\ &+ \varepsilon A_2 u + A_3 \ , \ A_j = A_j(z,u,u_z) \ , \ j=1,2,3 \ ,\end{aligned} \tag{4.2}$$

where (4.2) satisfies Condition C stated in § 1 of Chapter 3, $\vec{\mu}$ is an arbitrary vector on Γ , $\overline{\lambda(t)} = cos(\mu,x) + icos(\mu,y)$, and $\lambda(t)$, $\sigma(t)$, $\tau(t) = \prod_{j=1}^{m} |t-t_j|^{-\beta_j}\tau_0(t)$ satisfy the conditions

$$|\lambda(t)| = 1 \ , \ C_\alpha[\lambda,\Gamma^j] \le \ell \ , \ C_\alpha[\tau_0,\Gamma^j] \le \ell \ , \tag{4.3}$$

in which $\alpha(0<\alpha<1)$, ℓ $(0<\ell<\infty)$ are real constants, $\varepsilon(0<\varepsilon<1)$ is a positive constant, $\Gamma^j(j=1,\ldots,m)$ are as stated in § 1, and β_j , γ_j in § 1 are chosen so that $\gamma_j \ge 0$, $\gamma_j + \beta_j/\beta < 1$, $\beta_j/\beta^2 < 1$, $j=1,\ldots,m$. The above boundary value problem is called <u>Problem P^*</u>. The constant $K = Ind\lambda(t)$ is called the index of Problem P^*. We may assume that $K_j = Ind\lambda(t)|_{\Gamma_j}$ $(j=1,\ldots,N_0)$ are integers and $K_j = Ind\lambda(t)|_{\Gamma_j}$ $(j=N_0+1,\ldots,N)$ are not integers.

Next, the solvability of Problem P^* for (4.2) will be studied, in particular, the result of solvability for the mixed boundary value problem M of (4.2) will be derived.

1. A priori estimates of solutions to the modified Poincaré problem with piecewise continuous coefficients for (4.2)

Problem Q^*. The so-called modified Problem Q^* with piecewise continuous coefficients for the first order complex equation

$$w_{\bar z} = Re[Qw_z+A_1w] + \varepsilon A_2U + A_3 \tag{4.4}$$

is to find a solution $[w(z),U(z)]$, where $w(z)$ is continuous on $\hat{D}=\bar{D}\setminus\{t_1,\ldots,t_m,0\}$, and satisfies the boundary condition and the relation

$$Re[\overline{\lambda(t)}w(t)] + \varepsilon\sigma(t)U(t) = \tau(t) + |Y[\zeta(t)]|h(t)\ ,\ t\in\Gamma\ , \tag{4.5}$$

$$U(z) = 2Re\int_{a_0}^{z}[W(z)+\sum_{j=1}^{N}\frac{id_j}{z-z_j}]dz + b_0\ , \tag{4.6}$$

where $Y(t)$, $\zeta(z)$ are as stated in (2.22), (2.26) respectively, $w(z) = \Phi[\zeta(z)]e^{\phi(z)} + \psi(z)$ is a representation similar to (2.3),

$$W(z) = \begin{cases} w(z)\ , \text{ for } K\ge 0\ , \\ \hat{\Phi}[\zeta(z)]e^{\varphi(z)} + \psi(z)\ , \text{ for } K<0\ , \end{cases}$$

$\hat{\Phi}(\zeta) = \Phi(\zeta) - \tilde{\Phi}(\zeta)$, $\tilde{\Phi}(\zeta)$ is the main part of the Laurent series of $\Phi(\zeta)$ in the neighbourhood of $\zeta=0$, $d_j(j=1,\ldots,N)$ are appropriate real constants so that the function determined by the integral in (4.6) is single valued in D, and

$$h(t) = \begin{cases} 0\ ,\ t\in\Gamma\ ,\ \text{for } K>N-1\ , \\ \left.\begin{matrix} h_j\ ,\ t\in\Gamma_j\ ,\ j=1,\ldots,N-K'\ ,\ K'=[|K|+\frac{1}{2}], \\ 0\ ,\ t\in\Gamma_j\ ,\ j=N-K'+1,\ldots,N+1\ , \end{matrix}\right\}\text{for } 0\le K\le N-1\ , \\ h_j\ ,\ t\in\Gamma_j\ ,\ j=1,\ldots,N+1\ ,\ \text{for } K<0\ , \end{cases} \tag{4.7}$$

$h_j(j=1,\ldots,N+1)$ are as stated in (1.30). Besides, if $K\ge 0$, we may require that the solution $w(z)$ satisfies the point conditions

$$Im[\overline{\lambda(a_j)}w(a_j)] = |Y(\zeta(a_j))|b_j \ , \ t \in \{j\} \ , \tag{4.8}$$

where a_j , b_j $(j \in \{j\})$ are as stated in (1.32).

Next, we first give the estimates for solutions of Problem R^*, i.e. the conditions (4.2), (4.1) on

$$A = \varepsilon A_2 u + A_3 \ , \ \Delta_0(t) = -\varepsilon\sigma(t)u(t) \prod_{j=1}^{m} |t-t_j|^{\beta_j} + \tau_0(t)$$

$$= \Delta(t) \prod_{j=1}^{m} |t-t_j|^{\beta_j}$$

and b_j are replaced by the conditions

$$L_p[A,\overline{D}] \le k_1 \ , \ C_\alpha[\Delta_0,\Gamma^j] \le \ell_1 \ , \ |b_j| \le \ell_1 \ , \tag{4.9}$$

where k_1 , ℓ_1 are positive constants.

Theorem 4.1 Let $[w(z),U(z)]$ be a solution for Problem R^* of (4.4) with Condition C and (4.9). Then $[w(z),U(z)]$ satisfies the estimates

$$C[W,D_n] + C[|W_{\overline{z}}|+|W_z|,D_n] \le (k_1+\ell_1)M_1 \ , \tag{4.10}$$

$$C_{\beta\beta_0}[XW,\overline{D}] \le (k_1+\ell_1)M_2 \ , \ C_{\beta\beta_0}[U,\overline{D}] \le (k_1+\ell_1)M_3 \ , \tag{4.11}$$

where $X(z) = \prod_{j=1}^{m} |z-t_j|^{\eta_j}$, $\beta = 1-2/p_0$, $0 < \beta_0 < \delta$, D_n , η_j , δ are as stated in (2.24) - (2.26), $M_1 = M_1(q_0,p_0,k_0,\alpha,\ell,D,D_n)$, $M_j = M_j(q_0,p_0,k_0,\alpha,\ell,D)$, $j = 2,3$.

Proof. Put $w^*(z) = w(z)/(k_1+\ell_1)$, $U^*(z) = U(z)/(k_1+\ell_1)$, and $w^*(z)$, $U^*(z)$ satisfy the corresponding equation

$$w^*_{\overline{z}} = Re[Qw^*_z + A_1 w^*] + A_2 U^* + A/(k_1+\ell_1) \ , \tag{4.12}$$

the boundary conditions

$$Re[\overline{\lambda(t)}w^*(t)] = \Delta(t)/(k_1+\ell_1) + |Y[\lambda(t)]|h(t) \ , \ t \in \Gamma \ , \tag{4.13}$$

$$Im[\overline{\lambda(a_j)}w^*(a_j)] = |Y(\zeta(a_j))|b_j/(k_1+\ell_1) \ , \ j \in \{j\} \ , \tag{4.14}$$

and the relation

$$U^*(z) = 2Re\int_{a_0}^{z}[W^*(z) + \sum_{j=1}^{N}\frac{id_j}{z-z_j}]dz + b_0 \ , \tag{4.15}$$

where the relation between $W^*(z)$ and $w^*(z)$ is the same as that between $W(z)$ and $w(z)$. It is easy to see that

$$L_p[A/(k_1+\ell_1),\overline{D}] \le 1 \ , \ C_\alpha[\Delta_0/(k_1+\ell_1),\Gamma^j] \le 1 \ , \ |b_j/(k_1+\ell_1)| \le 1 \ , \ j \in \{j\} \ .$$

Analogously to (2.3), $w^*(z)$ can be expressed as

$$w^*(z) = \Phi^*[\zeta^*(z)]e^{\phi^*(z)} + \psi^*(z) \ , \tag{4.16}$$

in which $\phi^*(z)$, $\psi^*(z)$, $\zeta^*(z)$ and its inverse function $z^*(\zeta)$ satisfy the estimations (2.28), (2.29). On the basis of (4.16), the boundary conditions (4.13) and (4.14) can be reduced to the boundary conditions for the analytic function $\Phi^*(\zeta)$

$$Re[\overline{\Lambda(\zeta)}\Phi^*(\zeta)] = R(\zeta) + |Y(\zeta)|h[z(\zeta)] \ , \ \zeta \in L = \zeta(\Gamma) \ , \tag{4.17}$$

$$Im[\overline{\Lambda(a_j')}\Phi^*(a_j')] = b_j^* = |Y(a_j')|b_j/(k_1+\ell_1) - Im[\overline{\lambda(a_j)}\psi^*(a_j)] \ , \ j \in \{j\} \ , \tag{4.18}$$

where $\overline{\Lambda(\zeta)} = \overline{\lambda[z(\zeta)]}e^{\phi[z(\zeta)]}$, $R(\zeta) = \Delta[z(\zeta)]/(k_1+\ell_1) - Re\{\overline{\lambda[z(\zeta)]}\psi[z(\zeta)]\}$, $a_j' = \zeta(a_j)$, $j \in \{j\}$, $t_j' = \zeta(t_j)$, $j = 1,\dots,m$,

and $\Lambda(\zeta)$, $R(\zeta) = R_0(\zeta)\prod_{j=1}^{m}|\zeta-t_j'|^{-\beta_j/\beta}$ satisfy the conditions

$$C_{\alpha\beta}[\Lambda,L^j] \le \ell_2 \ , \ C_{\alpha\beta}[R_0,L^j] \le \ell_2 \ , \ j = 1,\dots,m \ , \ |b_j^*| \le \ell_2 \ , \ j \in \{j\} \ , \tag{4.19}$$

in which $L^j = \zeta(\Gamma^j)$, $\ell_2 = \ell_2(q_0,p_0,k_0,\alpha,\ell,D)$. According to Theorem 1.8, it is known that the function $\Phi^*(\zeta)$ satisfies the estimates

$$C[\Phi^*,G_n] + C[\Phi^{*\prime},G_n] \le M_4 \ , \tag{4.20}$$

$$C_{\beta_0}[X^*\Phi^*,\overline{G}] \le M_5 \ , \tag{4.21}$$

where $G_n = \zeta(D_n)$, $X^*(\zeta) = \prod_{j=1}^{m} |\zeta-t_j'|^{\eta_j\beta}$, $\beta = 1-2/p_0$, $0 < \beta_0 < \delta$, $M_4 = M_4(q_0,p_0,k_0,\alpha,\ell,D,D_n)$, $M_5 = M_5(q_0,p_0,k_0,\alpha,\ell,D)$. Combining (4.15), (4.16), (2.28), (2.29), (4.20) and (4.21), we conclude the estimates

$$C[W^*,D_n] + C[|W^*_{\bar z}|+|W^*_z|,D_n] \le M_1 \ , \tag{4.22}$$

$$C_{\beta\beta_0}[XW^*,\overline{D}] \le M_2 \ , \ C_{\beta\beta_0}[U^*,\overline{D}] \le M_3 \ . \tag{4.23}$$

From (4.22) and (4.23), (4.10) and (4.11) follow. □

Theorem 4.2 Let the equation (4.4) satisfy Condition C and the constant ε in (4.4), (4.5) be small enough. Then the solution $[w(z),U(z)]$ for Ptoblem Q^* of (4.4) satisfies the estimates

$$\Delta_0 = C[W,D_n] + C[|W_{\bar z}|+|W_z|,D_n] \le M_6 \ , \tag{4.24}$$

$$\Delta_1 = C_{\beta\beta_0}[XW,\overline{D}] \le M_7 \ , \ \Delta_2 = C_{\beta\beta_0}[U,\overline{D}] \le M_8 \ , \tag{4.25}$$

where $W(z)$, $X(z)$, $U(z)$, D_n , β_0 , β are as stated in Theorem 4.1, and $M_6 = M_6(q_0,p_0,k_0,\alpha,\ell,D,D_n)$, $M_j = M_j(q_0,p_0,k_0,\alpha,\ell,D)$, $j = 7,8$.

Proof. Let the solution $[w(z),U(z)]$ be inserted into the equation (4.4), the boundary condition (4.5), (4.8) and the relation (4.6). Denoting $A = \varepsilon A_2 U + A_3$ and $\prod_{j=1}^{m} |t-t_j|^{\beta_j}\Delta(t) = \prod_{j=1}^{m} |t-t_j|^{\beta_j}\sigma(t)u(t) + \tau_0(t) = \Delta_0(t)$, we have

$$L_p[A,\overline{D}] \le (\varepsilon\Delta_2+1)k_0 = k_1 \, , \, C_\alpha[\Delta_0,\Gamma^j] \le (\varepsilon\Delta_2+1)\ell = \ell_1 \, . \tag{4.26}$$

By Theorem 4.1, the estimates

$$\Delta_0 \le (k_1+\ell_1)\; M_1 = (\varepsilon\Delta_2+1)(k_0+\ell)M_1 \, , \tag{4.27}$$

$$\Delta_1 \le (\varepsilon\Delta_2+1)(k_0+\ell)M_2 \, , \, \Delta_2 \le (\varepsilon\Delta_2+1)(k_0+\ell)M_3 \tag{4.28}$$

can be concluded. Choosing the constant sufficiently small such that $1-\varepsilon(k_0+\ell_2)M_3 > 0$, it is clear that

$$\Delta_2 \le (k_0+\ell)M_3/[1-\varepsilon(k_0+\ell)M_3] = M_8 \, , \tag{4.29}$$

i.e. the second estimate in (4.25) holds. From this estimate and (4.27), (4.28), the estimates of Δ_0 and Δ_1 in (4.24) and (4.25) follow. □

2. Solvability of Problem Q^* and Problem P^* for (4.1)

First of all, we prove that under the hypothesis of Theorem 4.2, Problem Q^* has a solution. Then, we derive a result on solvability of Problem P^*.

Theorem 4.3 Suppose that the second order equation (4.2) satisfies Condition C and the constant ε in (4.2) and (4.5) is sufficiently small. Than Problem Q^* for (4.4) is solvable.

Proof. Let us introduce a closed and convex set B_M in the Banach space $L_\infty(D) \times L_{p_0}(\overline{D}) \times L_{p_0}(\overline{D}) \times C^1(\overline{D})$, in which the elements are vectors of functions $\omega = [Q(z),f(z),g(z),U(z)]$ with norms $|\omega| = C[U,\overline{D}] + L_\infty[Q,\overline{D}] + L_{p_0}[f,\overline{D}] + L_{p_0}[g,\overline{D}]$ satisfying the following condition

$$|Q(z)| \le q_0 \, , \, L_{p_0}[f,\overline{D}] \le k_1 \, , \, L_{p_0}[g,\overline{D}] \le k_1 \, , \, C[U,\overline{D}] + C[XW,\overline{D}] \le M_7 + M_8 \, , \tag{4.30}$$

where q_0 , M_7 , M_8 are nonnegative constants as stated in

Condition C and (4.25), k_1 is an appropriate constant similar to M_3 in (2.28). We arbitrarily select $\omega = [Q(z), f(z), g(z), U(z)] \in B_M$ and find a solution $\rho \in L_{p_0}(\overline{D})$ of the integral equation

$$\rho(z) - Q(z)\Pi\rho = Q(z) \quad , \quad \Pi\rho = -\frac{1}{\pi}\iint_D \frac{\rho(\zeta)}{(\zeta-z)^2}\, d\sigma_\zeta \ . \tag{4.31}$$

As in the proof of Theorem 2.4, we have $\chi(z)$, $\Psi(\chi)$ and $\zeta(z) = \Psi[\chi(z)]$, and its inverse function $z(\zeta)$. Set $\psi(z) = Tf$, $\phi(z) = Tg$. We insert the above function $U(z)$ into the boundary condition (4.5) - (4.8), and consider the boundary value problem Q^* for the analytic function $\Phi(\zeta)$ with the boundary condition

$$\begin{aligned} &Re\{\overline{\lambda[z(\zeta)]}Z(\zeta)\} + \tau[z(\zeta)]U[z(\zeta)] = \\ &\qquad = \tau[z(\zeta)] + |Y(\zeta)|h[z(\zeta)] \ , \zeta \in L = \zeta(\Gamma) \ , \end{aligned} \tag{4.32}$$

$$Im\{\overline{\lambda[z(\zeta)]}Z(\zeta)\}\big|_{\zeta=a_j^*=\zeta(a_j)} = |Y(a_j)|b_j \ , \ j \in \{j\} \ , \tag{4.33}$$

where $Z(\zeta) = \Phi(\zeta)e^{\phi[z(\zeta)]} + \psi[z(\zeta)]$. By Theorem 1.4 and Theorem 1.5, the above Problem Q^* has a unique solution $\Phi(\zeta)$ in $G = \zeta(D)$. Let $w(z) = \Phi[\zeta(z)]e^{\phi(z)} + \psi(z)$ and $\tilde{W}(z) = \tilde{\Phi}[\zeta(z)]e^{\phi(z)}$ with $\tilde{\Phi}(\zeta) = \Phi(\zeta) + 1/(n\zeta^{[|K|]+1})$ $(n > 0)$, where $\tilde{\Phi}(\zeta) \not\equiv 0$ and $\tilde{\Phi}'(\zeta) \not\equiv 0$. Using continuity method, we can obtain that the integral equations

$$f^*(z) = F(z,U,w,\Pi f^*) - F(z,U,w,0) + Re[A_1(z,U,w)Tf^*] + \varepsilon A_2 U + A_3 \tag{4.34}$$

$$\tilde{W}g^*(z) = F(z,U,w,\tilde{W}\Pi g^* + \Pi f^*) - F(z,U,w,\Pi f^*) + Re\ A_1(z,U,w)\tilde{W} \tag{4.35}$$

have unique solutions $f^*(z)$, $g^*(z) \in L_{p_0}(\overline{D})$, respectively. Afterwards, we seek a unique solution $\rho^*(z) \in L_{p_0}(\overline{D})$ of the integral equation

$$\begin{aligned} \tilde{\Phi}'\rho^* e^{\varphi(z)} &= F[z,U,w,\tilde{\Phi}'(1+\Pi\rho^*)e^{\varphi} + \tilde{W}\Pi g^* + \Pi f^*] - \\ &\quad - F(z,U,w,\tilde{W}\Pi g^* + \Pi f^*) \ , \ \tilde{\Phi}' = [\tilde{\Phi}(\zeta(\chi))]'_\chi \ . \end{aligned} \tag{4.36}$$

Let $Q^*(z) = \rho^*(z)/(1+\Pi\rho^*)$, $\psi^*(z) = Tf^*$, $\phi^*(z) = Tg^*$, $\chi^*(z) = z + T\rho^*$. There are analytic functions $\Psi^*(\chi^*)$ in $\chi^*(D)$ and $\Phi^*(\zeta^*)$ in $G^* = \Psi^*[\chi^*(D)]$ similar to $\Psi(\chi)$ and $\Phi(\zeta)$. We set $w^*(z) = \Phi^*[\zeta^*(z)]e^{\phi^*(z)} + \psi^*(z)$, with $\zeta^*(z) = \Psi^*[\chi^*(z)]$ on $\overline{D}$. Then there exists a unique real function

$$U^*(z) = 2Re\int_{a_0}^{z}\left[W^*(z)+\sum_{j=1}^{N}\frac{id_j}{z-z_j}\right]dz + b_0 , \tag{4.37}$$

where $W^*(z)$ is as stated in (4.15). Denote by $\omega^* = [Q^*(z),f^*(z),g^*(z),U^*(z)] = S(\omega)$ a mapping from ω onto ω^* . According to the method used in the proof of Theorem 2.4, we can verify that $\omega^* = S(\omega)$ continuously maps B_M onto a compact set in B_M . Hence, by the Schauder fixed-point theorem, the mapping $\omega^* = S(\omega)$ possesses a fixed-point $\omega = [Q(z),f(z),g(z),U(z)] \in B_M$, and from (4.34) - (4.36) with $\omega = S(\omega)$ and noting that

$$\tilde{w}_{\bar{z}} = \tilde{\Phi}'e^{\phi(z)}\rho + \tilde{W}g + f , \quad \tilde{w}_z = \tilde{\Phi}'e^{\phi(z)}(1+\Pi\rho) + \tilde{W}\Pi g + \Pi f ,$$

it follows that $\tilde{w}(z) = \tilde{W}(z) + \psi(z) = \tilde{\Phi}[\zeta(z)]e^{\phi(z)} + \psi(z)$ satisfies the complex equation

$$\tilde{w}_{\bar{z}} = F(z,U,w,\tilde{w}_z) + Re[A_1(z,U,w)(\tilde{w}-w)] \tag{4.38}$$

for almost every point $z \in D$. Applying the above result and the principle of compactness for a sequence of solutions $\{w_n(z)\}$ of (4.38), where $\tilde{w}_n(z) = \tilde{\Phi}_n[\zeta(z)]e^{\phi(z)} + \psi(z)$ with $\tilde{\Phi}_n(\zeta) = \Phi(\zeta) + 1/n\ \zeta^{-[|K|]-1}$ and choosing a subsequence $\{n_k\}$ with $n_k \to \infty$, we can conclude that Problem Q^* of the equation (4.4) has a solution $[w(z),U(z)]$. □

Theorem 4.4 Under the hypothesis of Theorem 4.3, the following statements hold.

(1) If the index $K > N-1$, Problem P^* of (2.1) has N solvability conditions.

(2) If $0 \le K \le N-1$, the total number of solvability conditions of Problem P^* is not greater than $2N - K' = 2N - [K+1/2]$.

(3) If $K < 0$, the Problem P^* has $2N - 2K - 1$ solvability conditions.

Proof. Using the methods of Theorem 6.4 in Chapter 3 and Theorem 1.6, we can prove this theorem. □

3. The mixed boundary value problem of elliptic equations of second order

Finally, as an example of the Poincaré problem with piecewise continuous coefficients, we shall prove the solvability for the mixed boundary value problem of the nonlinear uniformly elliptic equation of second order (4.2) with $\varepsilon = 1$.

Problem M. The mixed problem M of (4.2) ($\varepsilon = 1$) is to find a continuously differentiable solution $u(z)$ of (4.2) in $D^* = \bar{D} \setminus \{t_1, \ldots, t_m\}$, $m = 2n$, continuous on $\bar{D}$ and satisfying the boundary condition

$$u(t) = 2r(t),\ t \in \Gamma',\ \frac{\partial u}{\partial n} = 2\tau(t),\ t \in \Gamma'', \tag{4.39}$$

where Γ' and Γ'' are arcs with end points in $\{t_1, \ldots, t_m\}$, and $\Gamma'' = \Gamma \setminus \Gamma'$, $r(t) \in C^1_\alpha(\Gamma')$, $\tau(t) \in C_\alpha(\Gamma'')$, $0 < \alpha <$, $\vec{n}$ is the outward normal vector on the boundary Γ''.

Without loss of generality, we assume that all points with even index in $\{t_1, \ldots, t_{2n}\}$ are on the Γ_j, $j=0,\ldots,N$. If there is not a point of $\{t_1, \ldots, t_{2n}\}$ on $\Gamma_j (0 \le j \le N)$, then the index of $\lambda(t)$ on Γ_j is $K_j = \frac{1}{2\pi} \Delta_{\Gamma_j} \arg\lambda(t) = \begin{cases} -1, & j=0, \\ 1, & j=1,\ldots,N \end{cases}$.

The above problem M is equivalent to the following special Problem P^* with the boundary condition

$$\left.\begin{aligned} &\frac{\partial u}{\partial s} = 2\tau(t) = 2\frac{\partial r}{\partial s}, \quad t \in \Gamma' \\ & \qquad\qquad\qquad\qquad\qquad \text{i.e. } Re[\overline{\lambda(t)u_t}] = \tau(t), \quad t \in \Gamma, \\ &\frac{\partial u}{\partial n} = 2\tau(t), \quad t \in \Gamma'', \end{aligned}\right\} \tag{4.40}$$

where $\vec{s}$ is the tangent vector on the boundary Γ', $\lambda(t)$ is as stated in (4.3). In fact

$$u(z) = 2Re\int_{a_1}^{z} u_t dt + b_1 \tag{4.41}$$

is a solution for Problem P^* of (4.2) $(\varepsilon = 1)$.

We take $J_{2k-1} = 1$, $J_{2k} = 0 (k=1,\ldots,n)$ in the formula

$$\gamma_j = \frac{\phi_j}{\pi} - K_j, \quad e^{i\phi_j} = \frac{\lambda(t_j-0)}{\lambda(t_j-0)}, \quad K_j = [\frac{\phi_j}{\pi}] + J_j, \quad j = 1,\ldots,2n, \tag{4.42}$$

which is similar to Corollary 1.3(3). Hence, the index of Problem P^* is $N - 1 + \frac{n}{2}$.

Now, we prove the uniqueness of the solution for Problem M of (4.2) $(\varepsilon = 1)$ and give the estimates for solutions to the problem.

Theorem 4.5 Let (4.2) with $\varepsilon = 1$ satisfy Condition C^* and (4.2) in Chapter 3. Then the continuous solution $u(z)$ of Problem M is unique.

Proof. Let $u_1(z)$, $u_2(z)$ be two solutions to Problem M of (4.2) $(\varepsilon = 1)$ and $u(z) = u_1(z) - u_2(z)$. Obviously, $u(z)$ is a solution of

$$u_{z\bar{z}} = Re[q(z)u_{zz}+a_1u_z+a_2u], \tag{4.43}$$

and satisfies the boundary condition

$$u(t) = 0, \quad t \in \Gamma', \quad \frac{\partial u}{\partial n} = 0, \quad t \in \Gamma'', \tag{4.44}$$

where $q(z)$, $a_1(z)$, $a_2(z)$ are as stated in (4.3) of Chapter 3. If

$u(z) \not\equiv 0$, then there exist t' , $t'' \in \Gamma$, so that

$$M = u(t') = \max_{z\in\overline{D}} u(z) \geq 0 \quad \text{and} \quad m = u(t'') = \min_{z\in\overline{D}} u(z) \leq 0 \ .$$

It is easy to see $M = u(t') > 0$ or $m = u(t'') < 0$. We assume $t' \in \Gamma''$. However, by Corollary 2.11 in Chapter 3, $\partial u/\partial n > 0$. This contradiction proves $u(z) \equiv 0$, i.e. $u_1(z) = u_2(z)$ on $\overline{D}$. □

Theorem 4.6 Suppose that (4.2) with $\varepsilon = 1$ satisfies Condition C^* and the coefficient $Q(z,u,u_z,u_{zz}) = Q(z) \in W^1_p(D^\varepsilon_j)$; D^ε_j is a neighbourhood of the point t_j in $\overline{D}$, $p > 2$, $j = 1,\ldots,2n$. Then the continuous solution $u(z)$ for Problem M of (4.2) $(\varepsilon = 1)$ in D satisfies the estimates

$$C[u_z,D_m] + C[|u_{z\bar{z}}|+|u_{zz}|,D_m] \leq M_9 \ , \tag{4.45}$$

$$C_{\beta\beta_0}[Xu_z,\overline{D}] < M_{10} \ , \ C_{\beta\beta_0}[u,\overline{D}] < M_{11} \ , \tag{4.46}$$

where $X(z) = \prod_{k=1}^{n} |z-t_{2k-1}|^{\frac{1}{2}+\delta}$, $\beta = 1-2/p_0$, $0 < \beta_0 < \delta$, D_m is stated in (4.10), $M_9 = M_9(q_0,p_0,k_0,\alpha,\ell,D,D_m)$, $M_j = M_j(q_0,p_0,k_0,\alpha,\ell,D)$, $j = 10,11$.

Proof. We substitute the solution $u(z)$ into the equation (4.2), and then (4.2) reduces to the linear equation

$$u_{z\bar{z}} = Re[Q(z)u_{zz}+A_1(z)u_z] + A_2(z)u + A_3(z) \ . \tag{4.47}$$

Similarly to the proof of Theorem 5.2 in Chapter 3, the equation (4.47) can be transformed into the following equation

$$U_{z\bar{z}} = Re[Q(z)U_{zz}+A(z)U_z] \ , \ A(z) = -2(\ell n\Psi)_{\bar{z}} + 2Q(z)(\ell n\Psi)_z + A_1(z) \ , \tag{4.48}$$

where $U(Z) = [u(z)-\psi(z)]/\Psi(z)$, $\psi(z)$ and $\Psi(z)$ are stated there, and $U(z)$ satisfies the boundary condition

$$\left.\begin{array}{l}\dfrac{\partial U}{\partial s} = 2\,\dfrac{\partial r}{\partial s}\,,\ t \in \Gamma'\,,\\[2ex] \dfrac{\partial U}{\partial n} + \dfrac{\partial \Psi}{\partial n}\,U = 2\tau(t) - \dfrac{\partial \psi}{\partial n}\,,\ t \in \Gamma''\,,\end{array}\quad \text{i.e. } Re[\overline{\lambda(t)}u_t] + \sigma^*(t)U(t) = \tau^*(t)\,,\ t \in \Gamma\,,\right\} \tag{4.49}$$

where

$$\sigma^*(t) = \begin{cases} 0\,, & t \in \Gamma'\,,\\ \dfrac{1}{2}\dfrac{\partial \Psi}{\partial n}\,, & t \in \Gamma''\,,\end{cases} \qquad \tau^*(t) = \begin{cases} \dfrac{\partial r}{\partial s}\,, & t \in \Gamma'\,,\\ \tau(t) - \dfrac{\partial \psi}{\partial n}/2\,, & t \in \Gamma''\,.\end{cases} \tag{4.50}$$

As in (4.42), the index of $\lambda(t)$ equals $N-1+n/2$. Using the method of proof in Theorem 5.2 of Chapter 3, Theorem 4.2 and Theorem 4.5, we can give estimates of the solution $U(z)$ of (4.48) with the boundary condition (4.49), (4.50), i.e.

$$C[U_z,D_m] + C[|U_{z\bar{z}}|+|U_{zz}|,D_m] \le M_{12}\,, \tag{4.51}$$

$$C_{\beta\beta_0}[XU_z,\bar{D}] \le M_{13}\,,\ C_{\beta\beta_0}[U,\bar{D}] \le M_{14}\,, \tag{4.52}$$

where $X(z)$, β_0, β, D_m are stated in (4.45), (4.46), $M_{12} = M_{12}(q_0,p_0,k_0,\alpha,\ell,D,D_m)$, $M_j = M_j(q_0,p_0,k_0,\alpha,\ell,D)$, $j = 13,14$. Hence $u(z) = U(z)\Psi(z) + \psi(z)$ satisfies the estimates (4.45), (4.46). □

In the following, we first prove the solvability of Problem M for harmonic functions, and then derive the solvability for Problem M of (4.2) $(\varepsilon = 1)$.

Theorem 4.7 Problem M for harmonic functions has a solution.

Proof. We first prove the solvability of Problem B^* for the analytic function $\Phi(z)$ with the boundary conditions

$$Re[\overline{\lambda(t)}\Phi(t)] = \tau(t)\,,\ t \in \Gamma\,, \tag{4.53}$$

$$Im[\overline{\lambda(a_j)}\Phi(a_j)] = |Y(a_j)|b_j\,,\ j = 1,\dots,N+n-1\,, \tag{4.54}$$

where $\lambda(t)$ is stated in (4.40),

a_j , $b_j(j=1,\ldots,N+n-1$, $a_j \neq t_k$, $k=1,\ldots,2n)$ are stated in (1.32),

$$Y(z) = \left[\prod_{j=1}^{m_1}(z-t_j) \prod_{j=m_1+1}^{m_2}\left(\frac{z-t_j}{z-z_1}\right) \cdots \prod_{j=m_N+1}^{2n}\left(\frac{z-t_j}{z-z_N}\right)\right]^{-1/2} ,$$

in which $t_{m_k+1},\ldots,t_{m_{k+1}}$ are points of discontinuity of $\lambda(t)$ on Γ_k, $k = 0,1,\ldots,N$, $m_0 = 0, m_{N+1} = 2n$. By Theorem 1.4 and Theorem 1.5, Problem B^* for analytic functions has a unique solution $\Phi_0(z)$. If the function

$$u_0(z) = 2Re\int_{t_1}^{z}\Phi_0(\zeta)d\zeta + r(t_1) \tag{4.55}$$

is single-valued in D and $u(t_{2k-1}) = r(t_{2k-1})$, $k = 2,\ldots,n$, then $u(z)$ is a solution of Problem P^* for harmonic functions. Ohterwise, we find $N+n-1$ linearly independent analytic functions $\Phi_k(z)(k=1,\ldots,N+n-1)$ in D satisfying the boundary conditions

$$Re[\overline{\lambda(t)}\Phi_k(t)] = 0 \ , \ t \in \Gamma \ , \tag{4.56}$$

$$Im[\overline{\lambda(a_j)}\Phi_k(a_j)] = |Y(a_j)|\delta_{jk} \ , \tag{4.57}$$

where $\delta_{jk} = \begin{cases} 1 \ , \ j=k \ , \\ 0 \ , \ j\neq k \ , \end{cases}$ $k = 1,\ldots,N+n-1$. We can prove

$$J = \begin{vmatrix} I_{11} & \cdots & I_{1N} & u_1(t_3) & \cdots & u_1(t_{2n-1}) \\ \vdots & & \vdots & \vdots & & \vdots \\ I_{N+n-1\,1} & \cdots & I_{N+n-1\,N} & u_{N+n-1}(t_3) & \cdots & u_{N+n-1}(t_{2n-1}) \end{vmatrix} \neq 0 ,$$

where

$$I_{kj} = 2Re \int_{\Gamma_j} \Phi_k(z)dz \ , \ j=1,\dots,N \ ,$$

$$u_k(t_{2j-1}) = 2Re \int_{t_1}^{t_{2j-1}} \Phi_k(z)dz \ , \ j=2,\dots,n \ , \ k=1,\dots,N+n-1 \ .$$

In fact, if $J = 0$, then there exist real constants $c_1,\dots,c_{N+n-1}$, which are not all equal to zero, so that

$$\Phi_*(z) = \sum_{j=1}^{N+n-1} c_j \Phi_j(z)$$

is analytic in D, and satisfies the boundary conditions

$$\left.\begin{aligned} &Re \int_{\Gamma_j} \Phi_*(z)dz = 0 \ , \ j = 1,\dots,N, \\ &u(t_{2j-1}) = 2Re \int_{t_1}^{t_{2j-1}} \Phi_*(z)dz = 0 \ , \ j=1,\dots,n \ . \end{aligned}\right. \tag{4.58}$$

Hence, the function

$$u_*(z) = 2Re \int_{t_1}^{z} \Phi_*(\zeta)d\zeta \tag{4.59}$$

is harmonic in D, and satisfies the boundary condition

$$\left.\begin{aligned} &\frac{\partial u_*}{\partial s} = 0 \ , \ t \in \Gamma' \ , \frac{\partial u_*}{\partial n} = 0 \ , \ t \in \Gamma'' \ , u(t_{2j-1}) = 0 \ , \ j = 1,\dots,n \ , \\ &\text{i.e. } u_*(t) = 0 \ , \ t \in \Gamma' \ , \frac{\partial u_*}{\partial n} = 0 \ , \ t \in \Gamma'' \ . \end{aligned}\right\} \tag{4.60}$$

By Theorem 4.5, the homogeneous problem M_0 has a unique trivial solution $u_*(z) = 0$, and then $u_{*z} = \Phi_*(z) = 0$ in D. This contradiction proves $J \neq 0$. Thus, the algebraic system of equations

$$\left.\begin{aligned}\sum_{k=1}^{N+n-1} c_k I_{kj} &= -I_{0j} = -Re\int_{\Gamma_j} \Phi_0(z)dz\ ,\ j=1,\dots,N\ ,\\ \sum_{k=1}^{N+n-1} c_k u_k(a_{2j-1}) &= -u_0(a_{2j-1}) + r(a_{2j-1})\ ,\ j=2,\dots,n,\end{aligned}\right\} \tag{4.61}$$

has a unique solution $c_1,\dots,c_{N+n-1}$, and the function

$$u(z) = 2Re\int_{a_1}^{z} [\Phi_0(\zeta) + \sum_{k=1}^{N+n-1} c_k\Phi_k(\zeta)]d\zeta \tag{4.62}$$

is a solution of Problem M for harmonic functions. □

Theorem 4.8 Let the equation (4.2) ($\varepsilon = 1$) satisfy the same condition as in Theorem 4.6. Then its Problem M is solvable.

Proof. Just as in the proof of Lemma 5.5 in Chapter 3, we first prove that Problem M of the equation

$$u_{z\bar{z}} = Re[Q(z,u_{zz})u_{zz}+A_1(z)u_z] + A_3(z) \tag{4.63}$$

satisfying Condition C^* has a solution $u(z)$, where (5.44) and (5.46) in Chapter 3 are replaced by

$$Re\{\overline{\lambda[z(\zeta)]}\Phi(\zeta)e^{\phi[z(\zeta)]}\} = \tau[z(\zeta)] - \tag{4.64}$$

$$-Re\ \{\overline{\lambda[z(\zeta)]}\psi[z(\zeta)]\}\ ,\ \zeta \in L = \zeta(\Gamma)\ ,$$

$$u(a_{2j-1}) = r(a_{2j-1})\ ,\ j = 1,\dots,n\ , \tag{4.65}$$

where $\lambda(t)$, $\tau(t)$ are as stated in (4.40) .

Secondly, according to the proof of Theorem 5.6 in Chpater 3, we introduce a bounded and open set B_M in the Banach space $B = C(\bar{D}) \times L_\infty(D) \times L_{p_0}(\bar{D}) \times L_{p_0}(\bar{D})$, where the elements are all vectors of four functions $\omega = [u(z),Q(z),f(z),g(z)]$ with norms

$$\|\omega\| = C[u,\bar{D}] + L_\infty[Q,D] + L_{p_0}[f,\bar{D}] + L_{p_0}[g,\bar{D}]$$

satisfying the conditions

$$C[u,\overline{D}] + C[V,\overline{D}] < M_{10} + M_{11} , \tag{4.66}$$

$$L_{\infty}[Q,D] < (1+q_0)/2 , \quad L_{p_0}[f,\overline{D}] < K_0 , \; L_{p_0}[g,\overline{D}] < K_0 ,$$

where $V(z) = u_z X(z)$, $X(z)$ and the constants M_{10} , M_{11} are as stated in (4.46), K_0 is an unknown constant to be determined appropriately. We replace (5.56) and (5.58) in Chapter 3 by (4.64) and (4.65), and use the Leray-Schauder theorem to prove the existence of solutions for Problem M of (4.2) ($\varepsilon = 1$). □

V Boundary value problems for elliptic systems of two second order equations

In this chapter, we first transform the uniformly elliptic system of two second order equations into a complex equation of second order, and then discuss several boundary value problems for the complex equations in a multiply connected domain, i.e. the Dirichlet problem, the Neumann problem, the oblique derivative problem and the Riemann-Hilbert problem. We give some a priori estimates and integral representations of solutions, the total number of solutions and solvability conditions for the above boundary value problems for the complex equation.

§ 1. Reduction and conditions for elliptic systems of two second order equations

1. Complex form of elliptic systems of two linear equations of second order

Let a system of two linear elliptic equations of second order

$$\left.\begin{aligned} &a_{11}u_{xx}+a_{12}v_{xx}+2b_{11}u_{xy}+2b_{12}v_{xy}+c_{11}u_{yy}+c_{12}v_{yy}+\\ &\qquad + d_{11}u_x+d_{12}v_x+e_{11}u_y+e_{12}v_y+f_{11}u+f_{12}v = g_1 ,\\ &a_{21}u_{xx}+a_{22}v_{xx}+2b_{21}u_{xy}+2b_{22}v_{xy}+c_{21}u_{yy}+c_{22}v_{yy}+\\ &\qquad + d_{21}u_x+d_{22}v_x+e_{21}u_y+e_{22}v_y+f_{21}u+f_{22}v = g_2 ,\end{aligned}\right\} \qquad (1.1)$$

be given where the coefficients

a_{jk} , b_{jk} , c_{jk} , d_{jk} , e_{jk} , f_{jk} , $g_j (j,k=1,2)$ are real functions of $(x,y) \in D$, D is a bounded domain in the plane. The system (1.1) is called elliptic, if the determinant of $A\lambda^2 + 2B\lambda + C$ for any real number λ is positive in D , i.e.

$$K(\lambda) = |A\lambda^2+2B\lambda+C| > 0 , \tag{1.2}$$

where

$$A = \begin{pmatrix} a_{11} a_{12} \\ a_{21} a_{22} \end{pmatrix} , \; B = \begin{pmatrix} b_{11} b_{12} \\ b_{21} b_{22} \end{pmatrix} \quad C = \begin{pmatrix} c_{11} c_{12} \\ c_{21} c_{22} \end{pmatrix} .$$

Setting $w = u + iv$, we have

$$\left.\begin{aligned}
u_{xx} &= w_{z\bar{z}} + \bar{w}_{z\bar{z}} + \frac{1}{2}(w_{\bar{z}\bar{z}}+\bar{w}_{zz}+w_{zz}+\bar{w}_{\bar{z}\bar{z}}) , \\
v_{xx} &= i[-w_{z\bar{z}}+\bar{w}_{z\bar{z}}+\frac{1}{2}(-w_{\bar{z}\bar{z}}+\bar{w}_{zz}-w_{zz}+\bar{w}_{\bar{z}\bar{z}})] , \\
u_{yy} &= w_{z\bar{z}} + \bar{w}_{z\bar{z}}-\frac{1}{2}(w_{\bar{z}\bar{z}}+\bar{w}_{zz}+w_{zz}+\bar{w}_{\bar{z}\bar{z}}) , \\
v_{yy} &= i[-w_{z\bar{z}}+\bar{w}_{z\bar{z}}+\frac{1}{2}(w_{\bar{z}\bar{z}}-\bar{w}_{zz}+w_{zz}-\bar{w}_{\bar{z}\bar{z}})] , \\
u_{xy} &= \frac{i}{2}(-w_{\bar{z}\bar{z}}-\bar{w}_{zz}+w_{zz}+\bar{w}_{\bar{z}\bar{z}}) , \\
v_{xy} &= \frac{1}{2}(-w_{\bar{z}\bar{z}}+\bar{w}_{zz}+w_{zz}+\bar{w}_{\bar{z}\bar{z}}) .
\end{aligned}\right\} \tag{1.3}$$

Substituting u_{xx} , v_{xx} , u_{yy} , v_{yy} , u_{xy} , v_{xy} in (1.3) and $x = 1/2(z+\bar{z})$, $y = 1/2i(z-\bar{z})$, $u = 1/2(w+\bar{w})$, $v = 1/2i(w-\bar{w})$ into (1.1),

$$\left.\begin{aligned}
&(a_{11}-ia_{12}+c_{11}-ic_{12})w_{z\bar{z}} + (a_{11}+ia_{12}+c_{11}+ic_{12})\bar{w}_{z\bar{z}} = \\
&\qquad = f_1(z,w,w_z,\bar{w}_z,w_{zz},\bar{w}_{zz}) , \\
&(a_{21}-ia_{22}+c_{21}-ic_{22})w_{z\bar{z}} + (a_{21}+ia_{22}+c_{21}+ic_{22})\bar{w}_{z\bar{z}} = \\
&\qquad = f_2(z,w,w_z,\bar{w}_z,w_{zz},\bar{w}_{zz})
\end{aligned}\right\} \tag{1.4}$$

can be obtained. If

$$\begin{vmatrix} a_{11}-ia_{12}+c_{11}-ic_{12} & a_{11}+ia_{12}+c_{11}+ic_{12} \\ a_{21}-ia_{22}+c_{21}-ic_{22} & a_{21}+ia_{22}+c_{21}+ic_{22} \end{vmatrix} = 2i|A+C| \neq 0 \tag{1.5}$$

in D, then (1.4) can be solved for $w_{z\bar{z}}$, namely

$$\left.\begin{aligned} w_{z\bar{z}} &= F(z,w,w_z,\bar{w}_z,w_{zz},\bar{w}_{zz}) , \\ F &= Q_1(z)w_{zz} + Q_2(z)\bar{w}_{\bar{z}\bar{z}} + Q_3(z)\bar{w}_{zz} + Q_4(z)w_{\bar{z}\bar{z}} + \\ &+ A_1(z)w_z + A_2(z)\bar{w}_{\bar{z}} + A_3(z)\bar{w}_z + A_4(z)w_{\bar{z}} + \\ &+ A_5(z)w + A_6(z)\bar{w} + A_7(z) . \end{aligned}\right\} \tag{1.6}$$

Following [31] the system (1.1) with constant coefficients is called strongly elliptic if, for any real numbers β and $\gamma(\beta^2 < \gamma)$,

$$|A + 2\beta B + \gamma C| \neq 0 . \tag{1.7}$$

(There is no harm in assuming $|A+2\beta B+\gamma C| > 0$.) In [54], the equivalence of the above strong ellipticity and M.I. Vishik's strong ellipticity is proved in a similar way. We can give the definition of strong ellipticity for the general linear system (1.1). The condition (1.5) is a special case of the condition of strong ellipticity with $\beta = 0$ and $\gamma = 1$.

2. Complex form of elliptic systems of two nonlinear equations of second order

Let

$$\Phi_j(x,y,u,v,u_x,u_y,v_x,v_y,u_{xx},u_{xy},u_{yy},v_{xx},v_{xy},v_{yy})\,(j=1,2)$$

be continuous real functions in $(x,y) \in D$ and the real variables $u,v,u_x,u_y,v_x,v_y,u_{xx},u_{xy},u_{yy},v_{xx},v_{xy},v_{yy}$, and possess continuous partial derivatives with respect to $u_{xx},u_{xy},u_{yy},v_{xx},v_{xy},v_{yy}$. The nonlinear system of second order equations

$$\Phi_j(x,y,u,v,u_x,u_y,v_x,v_y,u_{xx},u_{xy},u_{yy},v_{xx},v_{xy},v_{yy}) = 0 , \; j=1,2 \tag{1.8}$$

is called elliptic, if the following inequality for any real number λ holds in D :

$$K(\lambda) = |A\lambda^2 + 2B\lambda + C| > 0 , \tag{1.9}$$

where

$$A = \begin{pmatrix} \Phi_{1u_{xx}} & \Phi_{1v_{xx}} \\ \Phi_{2u_{xx}} & \Phi_{2v_{xx}} \end{pmatrix} , \quad 2B = \begin{pmatrix} \Phi_{1u_{xy}} & \Phi_{1v_{xy}} \\ \Phi_{2u_{xy}} & \Phi_{2v_{xy}} \end{pmatrix} , \quad C = \begin{pmatrix} \Phi_{1u_{yy}} & \Phi_{1v_{yy}} \\ \Phi_{2u_{yy}} & \Phi_{2v_{yy}} \end{pmatrix} .$$

Similarly, we can give the definition of strong ellipticity for the nonlinear system (1.8). After that, under the conditions of ellipticity or strong ellipticity, system (1.8) is reduced to a complex form.

Theorem 1.1 (1) If (1.8) satisfies the condition of strong ellipticity in a bounded domain D, then it can be solved for $w_{z\bar{z}}$ so that a complex equation of the type

$$w_{z\bar{z}} = F(z,w,w_z,\bar{w}_z,w_{zz},\bar{w}_{zz}) , \tag{1.10}$$

is obtained or it can be solved for $u_{z\bar{z}}$, $v_{z\bar{z}}$ so that a system of complex equations

$$\left.\begin{aligned} u_{z\bar{z}} &= F_1(z,u,v,u_z,v_z,u_{zz},v_{zz}) , \\ v_{z\bar{z}} &= F_2(z,u,v,u_z,v_z,u_{zz},v_{zz}) \end{aligned}\right\} \tag{1.11}$$

results.

(2) If the system (1.8) satisfies the conditions

$$\sup_{j=1,2} \{|\Phi_{ju_{xx}}|,|\Phi_{ju_{xy}}|,|\Phi_{ju_{yy}}|,|\Phi_{jv_{xx}}|,|\Phi_{jv_{xy}}|,|\Phi_{jv_{yy}}|\} \tag{1.12}$$

$$\le \delta^{-1} , \quad |A| \ge \delta > 0 ,$$

where δ is a positive constant, then through a linear transformation of the independent variables, a complex equation of new variables (ξ,η)

$$w_{\zeta\bar{\zeta}} = F(\zeta,w,w_\zeta,\bar{w}_\zeta,w_{\zeta\zeta},\bar{w}_{\zeta\zeta}) \quad (\zeta=\xi+i\eta) \tag{1.13}$$

can be obtained.

<u>Proof.</u> It is clear that

$$J = \frac{D(\Phi_1,\Phi_2)}{D(w_{z\bar{z}},\bar{w}_{z\bar{z}})} \tag{1.14}$$

$$= \begin{vmatrix} \Phi_{1u_{xx}}-i\Phi_{1v_{xx}}+\Phi_{1u_{yy}}-i\Phi_{1v_{yy}} & \Phi_{1u_{xx}}+i\Phi_{1v_{xx}}+\Phi_{1u_{yy}}+i\Phi_{1v_{yy}} \\ \Phi_{2u_{xx}}-i\Phi_{2v_{xx}}+\Phi_{2u_{yy}}-i\Phi_{2v_{yy}} & \Phi_{2u_{xx}}+i\Phi_{2v_{xx}}+\Phi_{2u_{yy}}+i\Phi_{2v_{yy}} \end{vmatrix}$$

$$= 2i|A+C| \ .$$

If the system (1.8) satisfies the strong ellipticity condition (1.7) and $\beta = 0$, $\gamma = 1$, then $|A+C| \neq 0$, $J \neq 0$ in D . Thus, we can obtain the complex equation (1.10).

Moreover, noting that

$$(\ )_{xx} = 2(\ )_{z\bar{z}} + (\ )_{zz} + (\ )_{\bar{z}\bar{z}} \ , \ (\ )_{yy} = 2(\ )_{z\bar{z}} - (\ )_{zz} - (\ )_{\bar{z}\bar{z}} \ ,$$

we have

$$\Phi_{ju_{z\bar{z}}} = 2(\Phi_{ju_{xx}}+\Phi_{ju_{yy}}) \ , \ \Phi_{jv_{z\bar{z}}} = 2(\Phi_{jv_{xx}}+\Phi_{jv_{yy}}) \ , \ j = 1,2 \ .$$

Hence, (1.14) and the condition of strong ellipticity (1.7) give

$$\frac{D(\Phi_1,\Phi_2)}{D(u_{z\bar{z}},v_{z\bar{z}})} = 4|A+C| = -2iJ \neq 0 \ . \tag{1.15}$$

So the system (1.11) can be derived.

(2) Through the linear transformation

$$\xi = x - ty \ , \ \eta = x + ty \ , \tag{1.16}$$

where t is an unknown positive constant to be determined

appropriately, we obtain

$$u_{xx} = u_{\xi\xi} + 2u_{\xi\eta} + u_{\eta\eta} \,,\; v_{xx} = v_{\xi\xi} + 2v_{\xi\eta} + v_{\eta\eta} \,,\; u_{xy} = (-u_{\xi\xi}+u_{\eta\eta})t \;,$$

$$u_{yy} = (u_{\xi\xi}-2u_{\xi\eta}+u_{\eta\eta})t^2 \,,\; v_{yy} = (v_{\xi\xi}-2v_{\xi\eta}+v_{\eta\eta})t^2 \,,\; v_{xy} = (-v_{\xi\xi}+v_{\eta\eta})t \;.$$

Hence

$$\Phi_{ju_{\xi\xi}} = \Phi_{ju_{xx}} - \Phi_{ju_{xy}} t + \Phi_{ju_{yy}} t^2 \,,\; \Phi_{ju_{yy}} = \Phi_{ju_{xx}} + \Phi_{ju_{xy}} t + \Phi_{ju_{yy}} t^2 \;,$$

$$\Phi_{jv_{\xi\xi}} = \Phi_{jv_{xx}} - \Phi_{jv_{xy}} t + \Phi_{jv_{yy}} t^2 \,,\; \Phi_{jv_{yy}} = \Phi_{jv_{xx}} + \Phi_{jv_{xy}} t + \Phi_{jv_{yy}} t^2 \;,$$

$$j = 1,2 \;.$$

Thus

$$\begin{pmatrix} \Phi_{1u_{\xi\xi}} & \Phi_{1v_{\xi\xi}} \\ \Phi_{2u_{\xi\xi}} & \Phi_{2v_{\xi\xi}} \end{pmatrix} + \begin{pmatrix} \Phi_{1u_{\eta\eta}} & \Phi_{1v_{\eta\eta}} \\ \Phi_{2u_{\eta\eta}} & \Phi_{2v_{\eta\eta}} \end{pmatrix} = 2(A+Ct^2) \;, \tag{1.17}$$

$$|A+Ct^2| + |A| + (|D_1|+|D_2|)t^2 + |C|t^4 \;, \tag{1.18}$$

where $D_1 = \begin{pmatrix} \Phi_{1u_{xx}} & \Phi_{1v_{yy}} \\ \Phi_{2u_{xx}} & \Phi_{2v_{yy}} \end{pmatrix}$, $D_2 = \begin{pmatrix} \Phi_{1u_{yy}} & \Phi_{1v_{xx}} \\ \Phi_{2u_{yy}} & \Phi_{2v_{xx}} \end{pmatrix}$.

Provided the positive constant t is small enough, we must have

$$|A+Ct^2| \geq \frac{1}{2}\delta > 0 \;. \tag{1.19}$$

From (1.17) - (1.19), the complex equation (1.13) can be derived. □

In the following, we shall discuss (1.10) or (1.11) satisfying the following uniform ellipticity condition

$$|F(z,w,w_z,\bar{w}_z,U_1,V_1) - F(z,w,w_z,\bar{w}_z,U_2,V_2)| \leq \tag{1.20}$$
$$\leq q_0|U_1-U_2| + q_0'|V_1-V_2| \;,$$

or

$$|F_j(z,u,v,u_z,v_z,U_1,V_1) - F_j(z,u,v,u_z,v_z,U_2,V_2)| \le \tag{1.21}$$
$$\le q_{j0}|U_1-U_2| + q'_{j0}|V_1-V_2| \ , \ j = 1,2 \ ,$$

where

$$q_0 + q'_0 < 1 \ , \ q_{j0} + q'_{j0} < \frac{1}{2} \ , \ j = 1,2 \ ,$$

in which q_0 , q'_0 , q_{j0} , $q'_{j0}(j=1,2)$ are nonnegative constants.

3. Conditions for uniformly elliptic complex equations of second order

Let D be a bounded (N+1)-connected domain with the boundary $\Gamma \in C^2_\mu(0 < \mu < 1)$. Without loss of generality, we may assume that D is an (N+1)-connected circular domain as stated in § 1 of Chapter 1.

We consider the uniformly elliptic complex equation of the type

$$\left.\begin{aligned}
&w_{z\bar{z}} = F(z,w,w_z,\bar{w}_z,w_{zz},\bar{w}_{zz}) \ ,\\
&F = Q_1 w_{zz} + Q_2\bar{w}_{\bar{z}\bar{z}} + Q_3\bar{w}_{zz} + Q_4 w_{\bar{z}\bar{z}} +\\
&\quad + A_1 w_z + A_2\bar{w}_{\bar{z}} + A_3\bar{w}_z + A_4 w_{\bar{z}} + A_5 w + A_6\bar{w} + A_7 \ ,\\
&Q_j = Q_j(z,w,w_z,\bar{w}_z,w_{zz},\bar{w}_{zz}) \ , \ j = 1,\ldots,4 \ ,\\
&A_j = A_j(z,w,w_z,\bar{w}_z) \ , \ j = 1,\ldots,7
\end{aligned}\right\} \tag{1.22}$$

and suppose that (1.22) satisfies Condition C:

(1) $Q_j(z,w,w_z,\bar{w}_z,U,V)(j=1,\ldots,4)$, $A_j(z,w,w_z,\bar{w}_z)(j=1,\ldots,7)$ are measurable in $z \in D$ for all continuously differentiable functions $w(z)$ and all measurable functions $U(z)$, $V(z)$ in D , satisfying the conditions

$$L_p[A_j,\bar{D}] \le k_0 \ , \ j = 1,\ldots,7 \ , \ L_p[A_j,\bar{D}] \le \varepsilon k_0 \ , \ j = 3,\ldots,6 \ , \tag{1.23}$$

where $p(2 < p < \infty)$, $k_0(0 \le k_0 < \infty)$, $\varepsilon(0 < \varepsilon \le 1)$ are real constants.

In addition, we assume $A_j = 0$, for $z \notin D$, $j = 1,\ldots,7$.

(2) The above functions are continuous in w , w_z , $\bar{w}_z \in \mathbb{C}$ for almost every point $z \in D$ and $U,V \in \mathbb{C}$.

(3) The complex equation (1.22) satisfies the uniform ellipticity condition (1.20) for almost every point $z \in D$ and w , w_z , $\bar{w}_z$, U_1 , U_2 , V_1 , $V_2 \in \mathbb{C}$, where q_0 , q_0' are nonnegative constants satisfying the condition $q_0 + q_0' < 1$, $q_0' \le \varepsilon$.

We will discuss also a uniformly elliptic complex system of the following type:

$$\left.\begin{aligned} & u_{jz\bar{z}} = F_j(z,u_1,u_2,u_{1z},u_{2z},u_{1zz},u_{2zz}) , \\ & F_j = Re[Q_{j1}u_{1zz}+Q_{j2}u_{2zz}+A_{j1}u_{1z}+A_{j2}u_{2z}] +A_{j3}u_1+A_{j4}u_2+A_{j5} , \\ & Q_{jk} = Q_{jk}(z,u_1,u_2,u_{1z},u_{2z},u_{1zz},u_{2zz}) , \ k = 1,2 , \\ & A_{jk} = A_{jk}(z,u_1,u_2,u_{1z},u_{2z}) , \ k = 1,\ldots,5 , \ j = 1,2 , \end{aligned}\right\} \qquad (1.24)$$

and suppose that (1.24) satisfies the following Condition C:

(1) $Q_{jk}(z,u_1,u_2,u_{1z},u_{2z},U,V)(k=1,2)$, $A_{jk}(z,u_1,u_2,u_{1z},u_{2z})$ $(k=1,\ldots,5)$, $(j=1,2)$ are measurable in $z \in D$ for all continuously diffenrentiable functions $u_j(z)(j=1,2)$, and all measurable functions $U(z)$, $V(z)$ in D , satisfying

$$\begin{aligned} & L_p[A_{jk},\bar{D}] \le k_0(k=1,\ldots,5,j=1,2) , \\ & L_p[A_{1k},\bar{D}] \le \varepsilon k_0(k=2,3,4) , \ L_p[A_{24},\bar{D}] \le \varepsilon k_0 \end{aligned} \qquad (1.25)$$

where $p(>2)$, $k_0(>0)$, $\varepsilon(0 < \varepsilon < 1)$ are constants.

(2) The above functions are continuous in $u_j \in \mathbb{R}$ (the whole real axis) , $u_{jz} \in \mathbb{C}$ $(j=1,2)$ for almost every point $z \in D$ and $U,V \in \mathbb{C}$.

(3) The complex system (1.24) satisfies the uniform ellipticity condition (1.21) for almost every point $z \in D$ and $u_j \in \mathbb{R}$, u_{jz} , $U,V \in \mathbb{C}(j=1,2)$, where q_{j0} , $q_{j0}'(j=1,2)$ are

nonnegative constants satisfying $q_{j0} + q'_{j0} < 1/2$ $(j=1,2)$, $q'_{10} \le \varepsilon$ and ε is as stated in (1.25).

A so-called solution $w(z)$ for (1.22) and $w(z) = u_1(z) + iu_2(z)$ for (1.24) in D are continuously differentiable functions on $\overline{D}$ satisfying (1.22) and (1.24), respectively for almost every point $z \in D$ and the condition $w(z) \in W^1_{p_0}(D)$, $2 < p_0 < p$.

§ 2. The Dirichlet problem and the Neumann problem for elliptic complex equations of second order

1. Formulations of the Dirichlet problem and the Neumann problem

Problem D. The Dirichlet boundary value problem for the second order complex equation (1.22) in an (N+1)-connected circular domain D is to find a continuously differentiable solution $w(z) = u_1(z) + iu_2(z)$ of (1.22) on $\overline{D}$ satisfying the boundary condition

$$w(t) = r(t) \quad \text{or} \quad u_j(t) = \gamma_j(t) \ (j=1,2) \ , \ t \in \Gamma \ , \tag{2.1}$$

where $r(t) = r_1(t) + ir_2(t)$, $C^1_\alpha[r_j,\Gamma] \le \ell < \infty$, $1/2 < \alpha < 1$. Problem D with $r(t) \equiv 0$ is called Problem D_0.

Problem N. The Neumann boundary value problem of the complex equation (1.22) may be formulated as follows: Find a continuously differentiable solution $w(z) = u_1(z) + iu_2(z)$ of (1.22) on $\overline{D}$ satisfying the boundary condition

$$\left.\begin{aligned} &\frac{\partial w}{\partial n} = \tau(t) + h(t) \ , \ t \in \Gamma \ , \ w(0) = w_0 \ , \\ &\text{or} \quad Re[\overline{\lambda_j(t)} u_{jt}] = \tau_j(t) + h_j(t) \ , \ t \in \Gamma \ , \ u_j(0) = u_{j0} \ , \ j=1,2 \ , \end{aligned}\right\} \tag{2.2}$$

where $\vec{n}$ is the outer normal vector on Γ , and $C_\alpha[\tau,\Gamma] \le \ell$, $2\tau(t) = \tau_1(t) + i\tau_2(t)$, $2h(t) = h_1(t) + ih_2(t)$,

$$h_j(t) = \begin{cases} h_{j0} \ , \ t \in \Gamma_0 \ , \\ 0 \ , \ t \in \Gamma \setminus \Gamma_0 \ , \end{cases} \quad \overline{\lambda_j(t)} = (-1)^m (t - z_k) \ , \ m = \begin{cases} 0 \ , \ t \in \Gamma_k \ , \ k=0 \ , \\ 1 \ , \ t \in \Gamma_k \ , \ k=1,\dots,N \ , \end{cases}$$

in which $\alpha(1/2<\alpha<1)$, $\ell(0\le\ell<\infty)$, $w_0 = u_{10} + iu_{20}$ are constants, satisfying $|u_{j0}| \le \ell$, $j = 1,2$, $h_{j0}(j=1,2)$ are unknown constants to be determined appropriately. Problem N with $\tau(t) \equiv 0$ is denoted by Problem N_0.

Problem O. This is a special regular oblique derivative boundary value problem. Its boundary condition is as follows:

$$\left.\begin{aligned} &\frac{\partial u_j}{\partial \nu_j} = 2[\tau_j(t)+h_j(t)] , \ t\in\Gamma , \ u_j(0) = u_{j0} , \ j=1,2, \\ &\text{or } \ Re[\overline{\lambda_j(t)}u_{jt}] = \tau_j(t) + h_j(t) , \\ &\overline{\lambda(t)} = \begin{cases} e^{-i\theta_{j0}} t^{K_j-N} , & t\in\Gamma_0 , \ K_j \ge N-1 , \ j=1,2 , \\ e^{-i\theta_{jk}}(t-z_k) , & t\in\Gamma_k , \ k=1,\dots,N , \end{cases} \end{aligned}\right\} \quad (2.3)$$

where

$$h_j(t) = 0 , \text{ for } K_j \ge N , \ h_j(t) = \begin{cases} h_{j0} , & t\in\Gamma_0 , \\ 0 , & t\in\Gamma\setminus\Gamma_0 , \end{cases} \quad \text{for } K_j = N ,$$

h_{j0} is an appropriate constant, $1\le j\le 2$, $\theta_{jk}(k=0,\dots,N)$, $u_{j0}(j=1,2)$ are all constants satisfying the conditions $|u_{j0}| \le \ell_0$, and $C_\alpha[\tau_j,\Gamma] \le \ell$, $j=1,2$. We may require that the solution $w(z) = u_1(z) + iu_2(z)$ satisfies the following point conditions

$$Im[\overline{\lambda_j(t)}u_{jt}]\big|_{t=a_k} = b_{jk} , \ k=N+1,\dots,2K_j - N+1 , \ j=1,2 , \quad (2.4)$$

in which a_k are distinct points on Γ_0 , b_{jk} are constants satisfying $|b_{jk}| \le \ell$, $|\theta_{jk}| \le \pi$ $(k=N+1,\dots,2K_j-N+1 , j=1,2)$. If $\theta_{jk} = \pm\frac{\pi}{2}$ $(1\le k\le N , 1\le j\le 2)$, we assume

$$\int_{\Gamma_k} \tau_j(t)ds = 0 , \ u_j(a_k) = b_{jk} , \ |b_{jk}| \le \ell , \ 1\le k\le N , \ 1\le j\le 2 . \quad (2.5)$$

There is no harm in setting $\theta_{jk} = \pm\frac{\pi}{2}$ $(k=1,\ldots,N_{j0})$, and $\theta_{jk} \neq \pm\frac{\pi}{2}$ $(k=N_{j0}+1,\ldots,N)$. Problem O with $\tau_j(t) = 0$, $b_{jk} = 0$ $(k=1,\ldots,N_{j0},N+1,\ldots,2K_j-N+1,j=1,2)$ is called Problem O_0.

It is clear that if $\vec{\nu}_j = \vec{n}$, then $\theta_{j0} = 0$, $\theta_{jk} = \pi$ $(k=1,\ldots,N,j=1,2)$, $K_j = N-1$. Thus, Problem O with these conditions is Problem N. If $\vec{\nu}_j = \vec{s}_j$, then $\theta_{j0} = -\frac{\pi}{2}$, $\theta_{jk} = \frac{\pi}{2}$ $(k=1,\ldots,N,j=1,2)$, $K_j = N-1$. Provided we take $b_{jk} = r_j(a_k)$, $k=1,\ldots,N$, $j=1,2$ in (2.5), Problem O with these conditions is Problem D. Hence, we may consider only Problem O.

2. Solvability of Problem O for the linear elliptic complex equation

We discuss the linear uniformly elliptic complex equation of second order with the parameter $\varepsilon(-\infty < \varepsilon < \infty)$:

$$\begin{aligned} &w_{z\bar{z}} - Q_1(z)w_{zz} - Q_2(z)\bar{w}_{\bar{z}\bar{z}} - Q_3(z)\bar{w}_{zz} - Q_4(z)w_{\bar{z}\bar{z}} = \\ &= \varepsilon\, f(z,w,w_z,\bar{w}_z) + A_7(z) \ , \ f = A_1(z)w_z + A_2(z)\bar{w}_z + A_3(z)\bar{w}_z \\ &+ A_4(z)w_{\bar{z}} + A_5(z)w + A_6(z)\bar{w} \ , \end{aligned} \tag{2.6}$$

where the coefficients are measurable functions in $z \in D$ satisfying

$$\begin{aligned} &|Q_1(z)| + |Q_2(z)| \le q_0 \ , \ |Q_3(z)| + |Q_4(z)| \le q_0' \ , \\ &L_p[A_j,\bar{D}] \le k_0 \ , \end{aligned} \tag{2.7}$$

where q_0 , $q_0'(q_0+q_0' < 1)$, $p(>2)$, k_0 are nonnegative constants.

From § 4 in Chapter 1, we can give an integral representation for solutions $w(z) = u_1(z) + iu_2(z)$ to Problem O for the complex equation, namely

$$w(z) = U(z) + H\rho \quad \text{or} \quad u_j(z) = U_j(z) + H_j\rho_j \ (j=1,2) \ , \tag{2.8}$$

where $U_j(z)$ is a harmonic function, which is similar to (4.20) in Chapter 1 and can be expressed as

$$U_j(z) = V_j(z) + V_{j0}(z) \ , \ V_j(z) = \frac{2}{\pi}\int_\Gamma P_j(z,t)\ \tau_j(t)d\theta \ , \tag{2.9}$$

in which $P_j(z,t)$ is the Poisson kernel corresponding to the boundary conditions (2.3) - (2.5), and

$$H_j\rho_j = \frac{2}{\pi}\iint_D G_j(z,\zeta)\rho_j(\zeta)d\sigma_\zeta = H_{j0}\rho_j + \sum_{k=0}^{N}\tilde{H}_{jk}\rho_j + H_j^*\rho_j \ , \ j=1,2 \ , \tag{2.10}$$

in which $\rho_j(z) = U_{jz\bar{z}}$, $G_j(z,\zeta)$ is an integral kernel similar to (4.23) in Chapter 1, and $H_j\rho_j$ possesses properties as stated in Theorem 4.5 of Chapter 1.

We choose sufficiently small constants q_0, q_0' , so that

$$(q_0+q_0')\ \Lambda_{p_0} < 1 \ , \ 2 < p_0 < min\ (p,\frac{1}{1-\alpha}) \ , \tag{2.11}$$

where Λ_p is the least constant so that $L_p[S\rho,\bar{D}] \le \Lambda_p L_p[\rho,\bar{D}]$ holds. If $N = 0$, i.e. D is the unit disc, from Condition C, we can derive the inequality (2.11) for Problem D and Problem N without assuming that q_0 and q_0' are small enough.

Let $w(z)$ in (2.8) be substituted into the linear complex equation (2.6), then

$$\rho(z) - Q_1(z)(H\rho)_{zz} - Q_2(z)(\overline{H\rho})_{\bar{z}\bar{z}} - Q_3(z)(\overline{H\rho})_{zz} - Q_4(z)(H\rho)_{\bar{z}\bar{z}} \tag{2.12}$$
$$= \varepsilon\ f[z,H\rho,(H\rho)_z,(\overline{H\rho})_z] + g(z,\varepsilon) \ ,$$

where $g(z,\varepsilon) = Q_1U_{zz} + Q_2\bar{U}_{\bar{z}\bar{z}} + Q_3\bar{U}_{zz} + Q_4U_{\bar{z}\bar{z}} + \varepsilon\ f(z,U,U_z,\bar{U}_z) + A_7(z)$. From (2.11), we see that the above equation has an inverse operator R and obtain the integral equation

$$\rho(z) = \varepsilon\ R\{f[z,H\rho,(H\rho)_z,(H\rho)_{zz}]\} + R[g(z,\varepsilon)] \ . \tag{2.13}$$

Because $H\rho$ is a completely continuous operator from $L_{p_0}(\bar{D})$ into $C_\beta^1(\bar{D})(\beta=1-2/p_0)$, the inverse operator is also completely continuous.

By the Fredholm theorem for integral equations, the homogeneous integral equation

$$\rho(z) = \varepsilon R\{f[z,H\rho,(H\rho)_z,(H\rho)_{zz}]\} \tag{2.14}$$

possesses discrete eigenvalues

$$\varepsilon_j \;,\; 0 < |\varepsilon_j| \le |\varepsilon_{j+1}| \; (j=1,2,\ldots) \;, \tag{2.15}$$

where $|\varepsilon_1| > 0$ because for $\varepsilon = 0$, Problem 0 for the complex equation (2.6) is solvable, and then (2.13) is also solvable. This shows that when $\varepsilon \neq \varepsilon_j$ $(j=1,1,\ldots)$, the nonhomogeneous integral equation (2.12) has a unique solution $\rho(z) \in L_{p_0}(\bar{D})$. Hence, Problem 0 for the complex equation (2.6) is solvable. The case $\varepsilon = \varepsilon_j$ can be discussed in a similar way to that of Theorem 6.5, Theorem 6.6 in Chpater 3. Thus one obtains

<u>Theorem 2.1</u> Suppose that the linear complex equation (2.6) satisfies the condition (2.7) and (2.11). If $\varepsilon \neq \varepsilon_j$ $(j=1,2,\ldots)$, where $\varepsilon_j (j=1,2,\ldots)$ are the eigenvalues of (2.14), then Problem D, Problem N and Problem 0 of (2.6) are solvable, and if ε is an eigenvalue of rank q as stated in (2.15), then Problem 0 of (2.6) has q solvability conditions.

3. Solvability of Problem 0 for the nonlinear elliptic complex equation

Now we will discuss Problem 0 for the nonlinear elliptic complex equaiton (1.22) with Condition C and give a result for Problem 0 for (1.22).

<u>Theorem 2.2</u> Let the nonlinear complex equation (1.22) satisfy Condition C, and q_0, q_0' and k_0 in Condition C satisfy

$$L = (q_0+q_0')\Lambda_{p_0} + 2k_0 C^1[H\rho,\bar{D}] < 1 \;, \tag{2.16}$$

where Λ_{p_0} is the least constant so that

$$L_{p_0}[(H\rho)_{zz},\overline{D}] \le \Lambda_{p_0} L_{p_0}[\rho,\overline{D}] \ , \ 2 < p_0 < min\,(p_0,\frac{1}{1-\alpha}) \ .$$

Then Problem 0 of (1.22) possesses a solution of the type (2.8).

Proof. If the function $w(z) = U(z) + H\rho$ in (2.8) is a solution of (1.22), and is substituted into (1.22), we obtain the integral equation

$$\left.\begin{aligned} \rho &= Q_1(H\rho)_{zz} + Q_2(\overline{H\rho})_{\overline{z}\overline{z}} + Q_3(\overline{H\rho})_{zz} + Q_4(H\rho)_{\overline{z}\overline{z}} \\ &\quad + A_1(H\rho)_z + A_2(\overline{H\rho})_{\overline{z}} + A_3(\overline{H\rho})_z + A_4(H\rho)_{\overline{z}} + A_5 H\rho + A_6\overline{H\rho} + A \ , \\ A &= Q_1U_{zz} + Q_2\overline{U}_{\overline{z}\overline{z}} + Q_3\overline{U}_{zz} + Q_4U_{\overline{z}\overline{z}} + A_1U_z + A_2\overline{U}_{\overline{z}} + A_3\overline{U}_z \\ &\quad + A_4U_{\overline{z}} + A_5U + A_6\overline{U} + A_7 \ . \end{aligned}\right\} \qquad (2.17)$$

By Condition C and Theorem 4.2 in Chapter 1, we see that $A \in L_{p_0}(\overline{D})$. From (2.16) it follows

$$L_{p_0}[\rho,\overline{D}] \le \frac{1}{1-L} L_{p_0}[A,\overline{D}] \ , \text{ i.e. } L_{p_0}[\rho,\overline{D}] \le M_1 \ , \qquad (2.18)$$

where M_1 is a constant.

Denote by B_M the set of measurable functions $\rho(z)$ satisfying the second inequality in (2.18). We arbitrarily choose a function $\rho(z) \in B_M$, and then construct a function of the type (2.8), i.e. $w(z) = U(z) + H\rho$, where $U(z)$ is a solution for Problem 0 of complex harmonic functions, where a complex harmonic function denotes a complex function, the real and imaginary parts of which are harmonic. Inserting w , w_z , $\overline{w}_z$ into the coefficients of (1.22) and replacing $w_{z\overline{z}}$, w_{zz} , $\overline{w}_{zz}$ by $\rho^*(z)$, $U_{zz} + (H\rho^*)_{zz}$, $\overline{U}_{zz} + (\overline{H\rho^*})_{zz}$ respectively, one obtains the integral equation

$$\rho^* = Q_1(H\rho^*)_{zz} + Q_2(\overline{H\rho^*})_{\bar{z}\bar{z}} + Q_3(\overline{H\rho^*})_{zz} + Q_4(H\rho^*)_{\bar{z}\bar{z}} \tag{2.19}$$
$$+ A_1(H\rho^*)_z + A_2(\overline{H\rho^*})_{\bar{z}} + A_3(\overline{H\rho^*})_z + A_4(H\rho^*)_{\bar{z}} + A_5 H\rho^* +$$
$$+ A_6\overline{H\rho}^* + A ,$$

where $Q_j = Q_j[z,w,w_z,\bar{w}_z,U_{zz}+(H\rho^*)_{zz},\overline{U}_{zz}+(\overline{H\rho^*})_{zz}]$, $j = 1,\ldots,4$, $A_j = A_j(z,w,w_z,\bar{w}_z)$, $j=1,\ldots,7$. For the sake of convenience, we write the above equation in the form

$$\rho^* = F^*[z,w,w_z,\bar{w}_z,\rho^*] . \tag{2.20}$$

By the principle of contracting mappings we can find a solution $\rho^*(z)$ of (2.20). Denote by $\rho^*(z) = S[\rho(z)]$ and $\rho^*(z) = S_1[w(z)]$ the mappings from $\rho(z)$ onto $\rho^*(z)$ and $w(z)$ onto $\rho(z)$ respectively.

In order to prove that $\rho^* = S(\rho)$ maps B_M onto a compact set of itself we are free to select a sequence of functions $\rho_n(z) \in B_M (n=1,2,\ldots)$ and set $\rho_n^*(z) = S[\rho_n(z)]$. Evidently $\rho_n^*(z) \in B_M$. By Theorem 4.2 in Chapter 1, $w_n(z) = U_n(z) + H\rho_n$ satisfies the estimate

$$C_\beta^1[w_n,\overline{D}] \le M_2 = M_2(p_0,\alpha,\ell,M,D) , \tag{2.21}$$

where $\beta = 1-2/p_0$. Hence, we can choose subsequences $\{w_{n_k}(z)\}$, $\{w_{n_k z}\}$ which uniformly converge to $w_0(z)$, w_{0z} on $\overline{D}$, and from $\rho_{n_k}^*(z) = S_1[w_{n_k}(z)]$, and $\rho_0^*(z) = S_1[w_0(z)]$, we have

$$\rho_{n_k}^* - \rho_0^* = F^*(z,w_{n_k},w_{n_k z},\overline{w_{n_k z}},\rho_n^*) - F^*(z,w_{n_k},w_{n_k z},\overline{w_{n_k z}},\rho_0^*) \tag{2.22}$$
$$+ c_n(z) , c_n(z) = F^*(z,w_{n_k},w_{n_k z},\overline{w_{n_k z}},\rho_0^*) - F^*(z,w_0,w_{0z},\overline{w_{0z}},\rho_0) .$$

It is easy to see that $c_n(z) \to 0\ (n\to\infty)$ for almost every point

$z \in D$. Using the method in (2.43) of Chapter 2, $L_{p_0}[c_n,\overline{D}] \to 0$ $(n \to \infty)$ can be proved. From $L_{p_0}[\rho_n^* - \rho_0, \overline{D}] \leq \frac{1}{1-L} L_{p_0}[c_n, \overline{D}]$, it follows that $L_{p_0}[\rho_n^* - \rho_0^*, \overline{D}] \to 0$ for $n \to \infty$. Similarly, we can verify that $\rho^* = S(\rho)$ continuously maps B_M into itself. By the Schauder fixed-point theorem there exists a solution $\rho(z) \in B_M$ so that $\rho = S(\rho)$. Consequently, $w(z) = U(z) + H\rho$ is just a solution to Problem 0 of (1.22). □

The condition (2.16) in Theorem 2.2 is very strong. In the following, we shall weaken it.

4. Solvability of Problem 0 for the nonlinear complex system of second order

We divide the complex equation (1.22) into real and imaginary parts to obtain the system of complex equations (1.24), and suppose that (1.24) satisfies Condition C in the (N+1)-connected circular domain D . We first give the estimates for solutions to Problem 0 for (1.24).

Theorem 2.3 Let (1.24) satisfy Condition C in D and the constant ε be small enough. Then the solution $[u_1(z), u_2(z)]$ to Problem 0 for (1.24) satisfies the estimates

$$L_j = L_{p_0}[\rho_j, \overline{D}] = L_{p_0}[u_{jz\bar{z}}, \overline{D}] < M_3 \,,\; S_j = C_\beta^1[u_j, \overline{D}] < M_4 \,,\; j=1,2 \,, \qquad (2.23)$$

where

$$\rho_j(z) = u_{jz\bar{z}} \,,\; j=1,2 \,,\; 2 < p_0 < \mathit{min}(p, \frac{1}{1-\alpha}) \,,\; \beta = 1 - p_0/2 \,,$$

and $M_k = M_k(q_0, p_0, k_0, \alpha, \ell, D)$, $q_0 = (q_{10}, q_{10}', q_{20}, q_{20}')$, $k=3,4$.

Proof. From the representation (2.8) of solutions to Problem 0 for (1.24), and Theorem 4.2, Theorem 4.5 in Chapter 1, we have

$$L_{p_0}[(H_j\rho_j)_{zz}, \overline{D}] \leq M_5 L_j \,,\; C_\beta^1[H_j\rho_j, \overline{D}] \leq M_6 L_j \,,\; j=1,2 \,, \qquad (2.24)$$

$$L_{p_0}[U_{jzz},\overline{D}] \le M_7 \quad , \quad C^1_\beta[U_j,\overline{D}] \le M_8 \ , \ j=1,2 \ , \tag{2.25}$$

in which $M_k = M_k(p_0,\alpha,\ell,D)$, $k = 5,\ldots,8$. Substituting $u_j(z) = U_j(z) + H_j\rho_j(j=1,2)$ into the first equation of (1.24), and rewriting it in the following form

$$u_{1z\bar{z}} = Re[Q_{11}u_{1zz}+A_{11}u_{1z}] + A_1 \ , \tag{2.26}$$

$$A_1 = Re[Q_{12}u_{2zz}+A_{12}u_{2z}] + A_{13}u_1 + A_{14}u_2 + A_{15}$$

it is easy to see that if the constant ε is small enough, then

$$L_{p_0}[A_1,\overline{D}] \le \varepsilon\, L_{p_0}[u_{2zz},\overline{D}] + \varepsilon k_0 \sum_{j=1}^{2} C^1_\beta[u_j,\overline{D}] + k_0 \tag{2.27}$$

$$\le \varepsilon\, M_5L_2 + \varepsilon k_0\ (S_1+S_2) + \varepsilon\, M_7 + k_0 = H_3 \ .$$

The function $u_1^* = u_1/H_3$ is a solution of the equation

$$u^*_{1z\bar{z}} = Re[Q_{11}u^*_{1zz}+A_{11}u^*_{1z}] + A_1/H_3 \tag{2.28}$$

and satisfies the boundary condition

$$\frac{\partial u_1^*}{\partial \nu_1} = 2[\tau_1(t)+h_1(t)]/H_3 \ , \tag{2.29}$$

i.e. $Re[\overline{\lambda_1(t)}u^*_{1t}] = [\tau_1(t)+h_1(t)]/H_3$, $t \in \Gamma$, $u_1(0) = u_{10}$.

Using Theorem 4.2 in Chapter1 and the method of proof for Theorem 6.2 in Chapter 3, we obtain the estimates

$$L_{p_0}[u^*_{1z\bar{z}},\overline{D}] \le M_9 \ , \ C^1_\beta[u_1^*,\overline{D}] \le M_{10} \ , \tag{2.30}$$

where $M_k = M_k(q_0,p_0,k_0,\alpha,\ell,D)$, $k = 9,10$. Thus, $u_1(z) = H_3u_1^*(z)$ satisfies the estimate

$$L_1 \le H_3M_9 = \varepsilon\, M_9[M_5L_2+k_0(S_1+S_2)] + M_9(k_0+\varepsilon M_7) \ , \tag{2.31}$$

$$S_1 \le H_3M_{10} = \varepsilon\, M_{10}[M_5L_2+k_0(S_1+S_2)] + M_{10}(k_0+\varepsilon M_7) \; . \tag{2.32}$$

Next, we consider the second equation in (1.24) and rewrite it in the form

$$u_{2z\bar{z}} = \mathit{Re}[Q_{22}u_{2zz}+A_{22}u_{2z}] + A_2 \; , \tag{2.33}$$

$$A_2 = \mathit{Re}[Q_{21}u_{1zz}+A_{21}u_{1z}] + A_{23}u_1 + A_{24}u_2 + A_{25} \; ,$$

where A_2 satisfies

$$L_{p_0}[A_2,\bar{D}] \le M_5L_1 + k_0(S_1+\varepsilon S_2) + M_7 + k_0 = H_4 \; . \tag{2.34}$$

Similarly to (2.31), we can find the estimate

$$L_2 \le H_4M_9 = M_9[M_5L_1+k_0(S_1+\varepsilon S_2)] + M_9(k_0+M_7) \; , \tag{2.35}$$

$$S_2 \le H_4M_{10} = M_{10}[M_5L_1+k_0(S_1+\varepsilon S_2)] + M_{10}(k_0+M_7) \; . \tag{2.36}$$

From (2.24) and (2.25), the estimate

$$S_j \le M_6L_j + M_8 \; , \; j = 1,2 \; , \tag{2.37}$$

can be obtained. Combining (2.31), (2.35), and (2.37), we have

$$L_2 \le M_9\{M_5L_1+k_0[M_6(L_1+\varepsilon L_2)+2M_8]\} + M_9(k_0+M_7) \; , \tag{2.38}$$

$$\text{i.e. } L_2 \le \{M_9[(M_5+k_0M_6)L_1+2k_0M_8+k_0+M_7]\}/[1-\varepsilon k_0M_6M_9]$$

$$= M_{11}L_1 + M_{12} \; ,$$

$$L_1 \le \varepsilon\, M_9[k_0M_6L_1+(M_5+k_0M_6)L_2+2k_0M_8] + M_9(k_0+\varepsilon M_7) \tag{2.39}$$

$$\le \varepsilon\, M_9[k_0M_6+M_{11}(M_5+k_0M_6)]L_1 + \varepsilon\, M_9[2k_0M_8+(M_5+k_0M_6)M_{12}] +$$

$$+ M_9(k_0+M_7) \; ,$$

$$\text{i.e. } L_1 = \frac{\varepsilon M_9[2k_0M_8+(M_5+k_0M_6)M_{12}]+M_9(k_0+M_7)}{1-\varepsilon M_9[k_0M_6(1+M_{11})+M_5M_{11}]} \; ,$$

where the constant ε is so small, that $1 - \varepsilon k_0 M_6 M_9 > 0$ and $1 - \varepsilon M_9[k_0 M_6(1+M_{11})+M_5 M_{11}] > 0$. Hence, (2.23) is derived. □

Secondly, using the above result and the Leray-Schauder theorem, we prove the solvability for Problem 0 of (1.24).

Theorem 2.4 Under the hypotheses in Theorem 2.3 Problem 0 for (1.24) is solvable.

Proof. First of all, we consider the system

$$u_{jz\bar{z}} = F_j^{(n)}(z,u_1,u_2,u_{1z},u_{2z},u_{1zz},u_{2zz}) , \ j=1,2 \tag{2.40}$$

where $F_j^{(n)} = \sigma_n(z)F_j$, $Q_{jk}^{(n)} = \sigma_n(z)Q_{jk}$, $A_{jk}^{(n)} = \sigma_n(z)A_{jk}$, $\sigma_n(z)$, D_n and F_j are stated in (2.34) in Chapter 2 and (1.24), respectively. Let B_M be a bounded and open set in the Banach space $B = L_{p_0}(D_n) \times L_{p_0}(D_n)$, the elements of which are all system of measurable functions $\rho = [\rho_1(z),\rho_2(z)]$ satisfying the inequality

$$L_{p_0}[\rho_j,\overline{D}_n] \le M_3 , \ j = 1,2 , \ 2 < p_0 < min(p,\frac{1}{1-\alpha}) , \tag{2.41}$$

where M_3 is the constant in (2.23). We arbitrarily choose $\rho = [\rho_1(z),\rho_2(z)] \in \overline{B}_M$ and consider the functions

$$u_j(z) = U_j(z) + H_j\rho_j = V_j(z) + H_0\rho_j , \ j=1,2 , \tag{2.42}$$

in which $U_j(z)$, $H_j\rho_j(j=1,2)$ are as stated in (2.8) - (2.10),

$$H_0\rho_j = \frac{2}{\pi}\iint_{D_n} \ell n|1-\frac{z}{\zeta}| \ \rho_j(\zeta)d\sigma_\zeta ,$$

and $V_j(z)$ is a harmonic function, j=1,2 .

We substitute $u_j(z)$, $V_j(z)$ into suitable positions of (2.40), and observe the system of integral equations with the parameter $k(0 \le k \le 1)$

$$\rho_j^* = kF_j^{(n)}(z,u_1,u_2,u_{1z},u_{2z},V_{1zz}+\Pi\rho_1^*,V_{2zz}+\Pi\rho_2^*)\ (j=1,2)\ . \tag{2.43}$$

Noting $q_{j0}+q'_{j0}<\frac{1}{2}$ $(j=1,2)$ in Condition C by the principle of contracting mappings, the system (2.43) has a solution $\rho^*=[\rho_1^*(z),\rho_2^*(z)]\in B$. Denote by $\rho^*=S(\rho,k)$ $(0\le k\le 1)$ the mapping from $\rho=[\rho_1(z),\rho_2(z)]$ onto $\rho^*=[\rho_1^*(z),\rho_2^*(z)]$. We can verify that $\rho^*=S(\rho,k)$ satisfies the conditions of the Leray-Schauder theorem.

(1) For every $k\in[0,1]$, $\rho^*=S(\rho,k)$ maps the Banach space B continuously onto itself, and is completely continuous on B_M. Moreover, for $\rho\in\overline{B_M}$, $S(\rho,k)$ is uniformly continuous with respect to $k\in[0,1]$.

We are free to choose $\rho^{(m)}=[\rho_1^{(m)}(z),\rho_2^{(m)}(z)]\in\overline{B_M}$ and consider the functions

$$u_j^{(m)}(z) = V_j^{(m)}(z) + H_0\rho_j^{(m)}\ ,\ j=1,2\ ,\ m=1,2\ldots\ , \tag{2.44}$$

where $V_j^{(m)}(z)$ is a harmonic function satisfying the boundary conditions

$$Re[\overline{\lambda_j(t)V_{jt}^{(m)}}] = \tau_j(t) + h_j(t) - Re[\overline{\lambda_j(t)T\rho_j^{(m)}}]\ ,\ j=1,2\ , \tag{2.45}$$

in which

$$T\rho_j^{(m)} = (H_0\rho_j^{(m)})_z = -\frac{1}{\pi}\iint_{D_n}\frac{\rho_j^{(m)}(\zeta)}{\zeta-z}\,d\sigma_\zeta\ (j=1,2)\ .$$

Noting that $T\rho_j^{(m)}$ is analytic in $\mathbb{C}\setminus D_n$, and using (2.23) and Theorem 4.5 in Chapter 1, we obtain the estimates

$$C_\beta^1[u_j^{(m)},D_n]\le M_{13}\ ,\ L_{p_0}[u_{jzz}^{(m)},\overline{D_n}]\le M_{14}\ ,\ C_\beta^1[V_j^{(m)},\overline{D_n}]\le M_{15}\ , \tag{2.46}$$

in which the constants β, p_0 are the same as in (2.23), $M_k=M_k(q_0,p_0,k_0,\alpha,\ell,D,D_n)$, $k=13,14,15$. Hence, from

$\{u_j^{(m)}(z)\}$, $\{u_{jz}^{(m)}\}$, $\{V_{jzz}^{(m)}\}$, we can select subsequences which still will be denoted by $\{u_j^{(m)}(z)\}$, $\{u_{jz}^{(m)}\}$, $\{V_{jz}^{(m)}\}$ and which uniformly converge to $u_j^{(0)}(z)$, $u_{jz}^{(0)}$, $V_{jzz}^{(0)}$ $(j=1,2)$ on D_n , respectively. Set $\rho^{*(m)} = S[\rho^{(m)},k]$, $0 \le k \le 1$, $m = 1,2,\ldots,$ and let $\rho^{*(0)} = [\rho_1^{*(0)}(z),\rho_2^{*(0)}(z)]$ be a solution of the system of integral equations

$$\rho_j^{*(0)} = k\, F_j^{(n)}(z,u_1^{(0)},u_2^{(0)},u_{1z}^{(0)},u_{2z}^{(0)},V_{1zz}^{(0)} + \Pi\rho_1^{*(0)} , \quad V_{2zz}^{(0)} + \Pi\rho_2^{*(0)}) \; , \; j=1,2 \; , \tag{2.47}$$

and conclude

$$\begin{aligned}\rho_j^{*(m)} - \rho_j^{*(0)} &= k[F_j^{(n)}(z,u_1^{(m)},u_2^{(m)},u_{1z}^{(m)},u_{2z}^{(m)},V_{1zz}^{(m)} + \Pi\rho_1^{*(m)} , \\ &V_{2z}^{(m)} + \Pi\rho_2^{*(m)})-F_j^{(n)}(z,u_1^{(m)},u_2^{(m)},u_{1z}^{(m)},u_{2z}^{(m)},V_{1zz}^{(m)} + \Pi\rho_1^{*(0)} , \\ &V_{2zz}^{(m)} + \Pi\rho_z^{*(0)}) + c_j^{(m)}] \; , \\ c_j^{(m)} &= F_j^{(n)}(z,u_1^{(m)},u_2^{(m)},u_{1z}^{(m)},u_{2z}^{(m)},V_{1zz}^{(m)} + \Pi\rho_1^{*(0)},V_{2zz}^{(m)} + \Pi\rho_2^{*(0)}) - \\ &F_j^{(n)}(z,u_1^{(0)},u_2^{(0)},u_{1z}^{(0)},u_{2z}^{(0)},V_{1zz}^{(0)} + \Pi\rho_1^{*(0)},V_{2zz}^{(0)} + \Pi\rho_2^{*(0)}) \; .\end{aligned} \tag{2.48}$$

Due to Condition C and similarly to (2.43) in Chapter 2, we can prove that

$$L_{p_0}[c_j^{(m)},\overline{D}_n] \to 0 \; , \quad \text{as} \quad m \to \infty \; . \tag{2.49}$$

From $\sum_{j=1}^{2} L_{p_0}[\rho_j^{*(m)}-\rho_j^{*(0)},\overline{D_n}] \le \sum_{j=1}^{2} L_{p_0}[c_j^{(m)},\overline{D_n}]/[1-2 \max_{j=1,2} (q_{j0}+q'_{j0})\Lambda_{p_0}]$,

it follows

$$L_{p_0}[\rho_j^{*(m)}-\rho_j^{*(0)},\overline{D}_n] \to 0 \; , \quad \text{as} \quad m \to \infty \; .$$

So $\rho^* = S(\rho,k)$ is completely continuous. By similar methods, we can prove the other statements.

(2) If $k = 0$, then $\rho^* = [\rho_1^*(z), \rho_2^*(z)] = [0,0] \in B_M$.

(3) The solution $[\rho_1(z), \rho_2(z)]$ of the system of integral equations

$$\rho_j = k\, F_j^{(n)}(z, u_1, u_2, u_{1z}, u_{2z}, V_{1zz} + \Pi\rho_1, V_{2zz} + \Pi\rho_2)\ ,\ j=1,2\ ,\ 0 \le k \le 1\ , \tag{2.50}$$

satisfies the first estimate in (2.23), i.e. (2.41), so $\rho = S(\rho, k)$ $(0 \le k \le 1)$ has a solution on the boundary ∂B_M . Hence, by the Leray-Schauder theorem, the system (2.50) with $k = 1$ has a solution $\rho = [\rho_1(z), \rho_2(z)] \in B_M$, and then the functions $u_j(z)$ $(j=1,2)$ in (2.42) form a solution to Problem 0 for (2.40).

Finally, we eliminate the assumption of $Q_{jk} = A_{jk} = 0$ in the neighbourhood of the boundary Γ . By Theorem 2.3, the solution $[u_1^{(n)}(z), u_2^{(n)}(z)]$ to Problem 0 for (2.40) satisfies (2.23), for $n = 1,2,\ldots$. Therefore, from $\{u_j^{(n)}(z)\}$ $(j=1,2)$, we can choose subsequences which uniformly converge to $u_j^{(0)}(z)$ $(j=1,2)$ on $\bar{D}$, and $[u_1^{(0)}(z), u_2^{(0)}(z)]$ satisfies the boundary conditions (2.3) - (2.5). It remains to prove that $[u_1^{(0)}(z), u_2^{(0)}(z)]$ is a solution of the system (1.24). The following theorem solves this problem. □

<u>Theorem 2.5</u> Under the conditions in Theorem 2.4, and if the sequences of solutions $\{u_j^{(n)}(z)\}$ as stated above uniformly converge to $u_j^{(0)}(z)$ on $\bar{D}$, $j=1,2$, then $[u_1^{(0)}(z), u_2^{(0)}(z)]$ is a solution of (1.24) in D .

<u>Proof.</u> Substituting $[u_1^{(n)}(z), u_2^{(0)}(z)]$ into the system (2.40), we have

$$u_{jz\bar{z}}^{(n)} = F_j^{(n)}(z, u_1^{(n)}, u_2^{(n)}, u_{1z}^{(n)}, u_{2z}^{(n)}, u_{1zz}^{(n)}, u_{2zz}^{(n)})\ ,\ j=1,2\ . \tag{2.51}$$

The solution $[u_1^{(n)}(z), u_2^{(n)}(z)]$ can be expressed as

$$u_j^{(n)}(z) = V_j^{(n)}(z) + H_0\rho_j^{(n)} \ , \ H_0\rho_j^{(n)} = \frac{2}{\pi}\iint_D \ell n \left|1-\frac{2}{\zeta}\right|\rho_j^{(n)}(\zeta)d\sigma_\zeta \ , \qquad (2.52)$$

and $\rho_j^{(n)}(z) = u_{jz\bar{z}}^{(n)}$ satisfies

$$L_{p_0}[\rho_j^{(n)},\bar{D}] \le M_{16} = M_{16}(q_0,p_0,k_0,\alpha,\ell,D) \ , \ j=1,2 \ . \qquad (2.53)$$

Hence, from $\{\rho_j^{(n)}(z)\}$, we can choose a subsequences still denoted by $\{\rho_j^{(n)}(z)\}$, which weakly converges to $\rho_j^{(0)}(z)$ on $\bar{D}$, $j=1,2$. Similarly, we can select subsequences $\{H_0\rho_j^{(n)}\}$, $\{T\rho_j^{(n)}\}$, $\{V_j^{(n)}(z)\}$, and $\{V_{jz}^{(n)}\}$ which uniformly converge to $H_0\rho_j^{(0)}$, $T\rho_j^{(0)}$, $V_j^{(0)}(z)$, and $V_{jz}^{(0)}$ on $\bar{D}$ respectively, and a subsequence $\{V_{jzz}^{(n)}\}$ which uniformly converges to $V_{jzz}^{(0)}$ on any closed set D_* in D . It is not difficult to see that

$$u_j^{(0)}(z) = V_j^{(0)}(z) + H_0\rho_j^{(0)} \in W_{p_0}^2(D_*) \ , \ j=1,2 \ . \qquad (2.54)$$

We are free to choose a disc $\bar{S}_R = \{|z-z_*| \le R\} \subset D \ (R > 0)$. From (2.53), it can be seen that $u_j^{(n)}(z)$ possesses the representation

$$\left.\begin{aligned} u_j^{(n)}(z) &= U_j^{(n)}(z) + \tilde{H}\rho_j^{(n)} \ , \\ \tilde{H}\rho_j^{(n)} &= \frac{2}{\pi}\iint_S \ell n \left|\frac{R(\zeta-z)}{R^2-(\zeta-z_*)(z-z_*)}\right| \rho_j^{(n)}(\zeta)d\sigma_\zeta \ , \ j=1,2 \ , \end{aligned}\right\} \qquad (2.55)$$

and that $\{U_j^{(n)}(z)\}$, $\{U_{jz}^{(n)}\}$ uniformly converge to $U_j^{(0)}(z)$, $U_{jz}^{(0)}$ on $\bar{S}_R$, respectively and $\{U_{jzz}^{(n)}\}$ uniformly converges to $U_{jzz}^{(0)}$ on any closed set in S_R . We can prove that $\{U_{jzz}^{(0)}\}$ satisfies the estimate

$$L_{p_0}[U_{jzz}^{(n)},\bar{S}_R] \le M_{17} = M_{17}(q_0,p_0,k_0,\alpha,\ell,S_R) \ , \ n=0,1,2,\ldots \ . \qquad (2.56)$$

By the principle of contracting mapping the system of integral equations

$$\rho_j^* = F_j(z,u_1^{(0)},u_2^{(0)},u_{1z}^{(0)},u_{2z}^{(0)},U_{1zz}^{(0)} + \tilde{S}\rho_1^*,U_{2zz}^{(0)} + \tilde{S}\rho_2^*)\ ,\ j=1,2\ , \quad (2.57)$$

has a unique solution $[\rho_1^*(z),\rho_2^*(z)]$, where $\tilde{S}\rho_j^* = (\tilde{H}\rho_j^*)_{zz}$, $j=1,2$. From (2.51) and (2.57), we have

$$\rho_j^{*(n)} - \rho_j^* = F_j^{(n)}(z,u_1^{(n)},u_2^{(n)},u_{1z}^{(n)},u_{2z}^{(n)},U_{1zz}^{(n)}+\tilde{S}\rho_1^{*(n)},U_{2zz}^{(n)}+\tilde{S}\rho_2^{*(n)}) \quad (2.58)$$
$$- F_j^{(n)}(z,u_1^{(n)},u_2^{(n)},u_{1z}^{(n)},u_{2z}^{(n)},U_{1zz}^{(n)}+\tilde{S}\rho_1^*,U_{2zz}^{(n)}+\tilde{S}\rho_2^*) + c_n\ ,$$

where

$$c_n = F_j^{(n)}(z,u_1^{(n)},u_2^{(n)},u_{1z}^{(n)},u_{2z}^{(n)},U_{1zz}^{(n)} + S\rho_1^*,U_{2zz}^{(n)} + S\rho_2^*) -$$
$$- F_j(z,u_1^{(0)},u_2^{(0)},u_{1z}^{(0)},u_{2z}^{(0)},U_{1zz}^{(0)} + \tilde{S}\rho_1^*, U_{1zz}^{(0)} + \tilde{S}\rho_2^*)\ .$$

If n is sufficiently large, then $F_j^{(n)} = F_j$ on $\overline{S}_R$. Similarly to the proof of (2.49) we can prove that $L_{p_0}[c_n,\overline{S}_R] \to 0$, as $n \to \infty$, and $L_{p_0}[\rho_j^{*(n)}-\rho_j^*,\overline{S}_R] \to 0$ as $n \to \infty$. Hence $[u_1^{(0)}(z),u_2^{(0)}(z)]$ is a solution of (1.24) on $\overline{S}_R$. Because of the arbitrariness of S_R , $[u_1^{(0)}(z),u_2^{(0)}(z)]$ is a solution of (1.24) in D . This completes the proof. □

§ 3. The oblique derivative problem for elliptic complex equations of second order

In this section, we first discuss the irregular oblique derivative problem of (1.22) in a simply connected domain, and then consider the problem for (1.22) in a multiply connected domain. Finally, as an application of the above result, we shall derive the solvability for the oblique derivative problem for the complex equation of first order in a multiply connected domain.

Problem P. The irregular oblique derivative problem of (1.22) is to find a continuously differentiable solution $w(z)$ satisfying the boundary conditions

$$Re[\overline{\lambda_j(t)}V_j(t)+\varepsilon\beta_j(t)w(t)] = \tau_j(t) \ , \ t \in \Gamma \ , \ j=1,2 \ , \tag{3.1}$$

where $V_1(z) = w_z$, $V_2(z) = \bar{w}_z$, and $\lambda_j(t)$, $\beta_j(t)$, $\tau_j(t)$ satisfy the conditions

$$\left.\begin{array}{l} |\lambda_j(t)| = 1(t\in\Gamma) \ , \ C_\alpha[\lambda_j,\Gamma] \le \ell \ , \ C_\alpha[\beta_j,\Gamma] \le \ell \ , \\ C_\alpha[\tau_j,\Gamma] \le \ell \ , \ j=1,2 \ . \end{array}\right\} \tag{3.2}$$

in which $\alpha(\frac{1}{2}<\alpha<1)$, $\ell\ (0\le\ell<\infty)$, $\varepsilon(0<\varepsilon<1)$ are constants.

1. The oblique derivative problem for complex equations of second order in the unit disc

We discuss a special case of the boundary condition (3.1), namely

$$Re[\bar{t}^{K_j}V_j(t)] = \tau_j(t) \ , \ t \in \Gamma \ , \ j=1,2 \ , \ V_1(z) = w_z \ , \ V_2(z) = \bar{w}_z \ , \tag{3.3}$$

where K_1, K_2 are integers. The boundary value problem of (1.22) is called Problem P_1. If $K_j < 0$ (j=1 or 2), Problem P_1 is not solvable. Hence, we introduce Problem Q_1 for (1.22) with the modified boundary condition

$$Re[\bar{t}^{K_j}V_j(t)] = \tau_j(t)+h_j(t) \ , t \in \Gamma \ , \ j=1,2 \ , V_1 = w_z \ , \ V_2 = \bar{w}_z \ , \tag{3.4}$$

where

$$h_j(t) = \begin{cases} 0 \ , \ t \in \Gamma \ , \ \text{for} \ K_j \ge 0 \ , \ 1 \le j \le 2 \ , \\ h_{j0} + Re \displaystyle\sum_{k=1}^{-K_j-1} (h_{jk}^{+}+ih_{jk}^{-})t^k \ , \ t \in \Gamma \ , \text{for} \ K_j < 0 \ . \end{cases} \tag{3.5}$$

Similarly to §§ 2, 4 in Chapter 1, we can give an integral representation for the solutions to Problem Q_1 for (1.22), i.e.

$$
\left.\begin{aligned}
&w(z) = U(z) + H\rho \ , \ U(z) = \Phi_0(z) + \int_0^z \Phi_1(\zeta)d\zeta + \overline{\int_0^z \Phi_2(\zeta)d\zeta} \ , \\
&H\rho = \frac{2}{\pi}\iint_D \{\ln\left|1-\frac{z}{\zeta}\right| \rho(\zeta) + \frac{1}{2}[g_1(z,\zeta)+\overline{g_2(z,\zeta)}]\overline{\rho(\zeta)}\}d\sigma_\zeta \ ,
\end{aligned}\right\} \tag{3.6}
$$

where

$$
\rho(z) = w_{z\bar{z}} \in L_{p_0}(\bar{D}) \ , \ 2 < p_0 < \min\{p, \frac{1}{1-\alpha}\} \ , \ \Phi_0(z) = a + ib \ ,
$$

$$
\Phi_j(z) = \begin{cases} \dfrac{z^{K_j}}{2\pi i}\displaystyle\int_\Gamma \tau_j(t)\frac{t+z}{t-z}\frac{dt}{t} + \sum_{m=0}^{2K_j} c_m^{(j)} z^m \ , \\ \qquad c_{2K_j-m}^{(j)} = -\overline{c_m^{(j)}} \ , \ m=0,1,2,\ldots,K_j \ , \text{ for } K_j \geq 0 \ , \\ \dfrac{1}{\pi i}\displaystyle\int_\Gamma \frac{\tau_j(t)dt}{t^{-K_j}(t-z)} \ , \text{ for } K_j < 0 \ , \ 1 \leq j \leq 2 \ , \end{cases} \tag{3.7}
$$

and

$$
g_j(z,\zeta) = \begin{cases} \zeta^{-2K_j-2}[\ln(1-\bar{\zeta}z) + \displaystyle\sum_{m=1}^{2K_j+1} \frac{(\bar{\zeta}z)^m}{m}] \ , \text{ for } K_j \geq 0 \ , \\ \zeta^{-2K_j-2} \ln(1-\bar{\zeta}z) \ , \text{ for } K_j < 0 \ , \ 1 \leq j \leq 2 \ . \end{cases} \tag{3.8}
$$

Now, we give some estimates of solutions to Problem P_1 for (1.22).

Theorem 3.1 Let the complex equation (1.22) satisfy the Condition C and the constant ε in (1.20), (1.23) be sufficiently small. Then the solution $w(z)$ to Problem Q_1 for (1.22) satisfies the estimates

$$
S(w) = C_\beta^1[w,\bar{D}] + L_{p_0}[|w_{zz}|+|\bar{w}_{zz}|,\bar{D}] < M_1 \ , \tag{3.9}
$$

where

$$
\beta = 1 - 2/p_0 \ , \ 2 < p_0 < \min(p, \frac{1}{1-\alpha}) \ ,
$$

$$
M_1 = M_1(q_0, p_0, k_0, \alpha, \ell, c_m^{(j)}) \ ,
$$

$c_m^{(j)}$ is as stated in (3.7).

Proof. Insert the solution $w(z) = U(z) + H\rho(z) = U(z) + w^*(z)$ into (1.22), and rewrite this equation as

$$w^*_{z\bar{z}} = Q_1 w^*_{zz} + Q_2 \overline{w^*_{zz}} + A_1 w^*_z + A_2 \overline{w^*_z} + A \,, \tag{3.10}$$

$$A = Q_3 \overline{w}^*_{zz} + Q_4 w^*_{\bar{z}\bar{z}} + A_3 \overline{w}^*_z + A_4 w^*_{\bar{z}} + A_5 w^* * A_6 \overline{w}^* + A_7 + A_0 \,,$$

$$A_0 = Q_1 U_{zz} + Q_2 \overline{U}_{\bar{z}\bar{z}} + Q_3 \overline{U}_{zz} + Q_4 U_{\bar{z}\bar{z}} + A_1 U_z + A_2 \overline{U}_{\bar{z}} + A_3 \overline{U}_z + $$
$$+ A_4 U_{\bar{z}} + A_5 U + A_6 \overline{U} \,,$$

where $A_0(z)$, $A(z)$ satisfy the conditions

$$L_{p_0}[A_0,\overline{D}] \le k_1 = k_1(q_0,p_0,k_0,\alpha,\ell,D,c_m^{(j)}) \,, \tag{3.11}$$

$$L_{p_0}[A,\overline{D}] \le q_0' L_{p_0}[\overline{w}^*_{zz},\overline{D}] + 2\varepsilon k_0[C(\overline{w}^*_z,\overline{D})+C(w^*,\overline{D})] + k_0 + k_1$$
$$\le \varepsilon(q_0'+2k_0)S(w^*) + k_0 + k_1 = k \,.$$

Set $W^*(z) = w^*(z)/(\ell+k)$. It is clear that $W^*(z)$ satisfies the complex equation

$$W^*_{z\bar{z}} = Q_1 W^*_{zz} + Q_2 \overline{W^*_{zz}} + A_1 W^*_z + A_2 \overline{W^*_z} + A/(\ell+k) \tag{3.12}$$

and the boundary condition

$$Re[\bar{z}^{k_j} V_j^*(z)] = \tau_j(z)/(\ell+k) + h_j(z) \,, \; z \in \Gamma \,, \; j=1,2 \,, \tag{3.13}$$

where $V_1^*(z) = W^*_z$, $V_2^*(z) = \overline{W^*_z}$. Taking $L_{p_0}[A/(\ell+k),\overline{D}] \le 1$, $L_{p_0}[\tau_j/(\ell+k),\Gamma] \le 1$ $(j=1,2)$ into account and using the result of Theorem 4.3 in Chapter 2, we can obtain the estimate

$$S(W^*) \le M_2 = M_2(q_0,p_0,k_0,\alpha,\ell,D,c_m^{(j)}) \,. \tag{3.14}$$

Thus we have

$$S(w^*) \le M_2(\ell+k) = M_2[\varepsilon(q_0'+2k_0)S(w^*)+k_0+k_1+\ell] . \qquad (3.15)$$

Choosing ε so small that $\varepsilon(q_0'+2k_0)M_2 < 1$, from (3.15) it follows

$$S(w^*) \le M_2(k_0+k_1+\ell)/[1-\varepsilon M_2(q_0'+2k_0)] . \qquad (3.16)$$

Combining (3.11) with (3.16) the estimate (3.9) can be derived. □

By the above theorem and using the Leray-Schauder theorem, we can prove the solvability theorem for Problem Q_1 .

Theorem 3.2 Under the hypotheses in Theorem 3.1, Problem Q_1 is solvable and the solvability of Problem P_1 is as follows:

(1) If $K_1 \ge 0$. $K_2 \ge 0$, Problem P_1 for (1.22) possesses a solution of the type (3.6)

(2) If $K_1 < 0$, $K_2 \ge 0$ (or $K_1 \ge 0$,$K_2 < 0$) , Problem P_1 has $2|K_1|-1$ (or $2|K_2|-1$) solvability conditions

$$\left.\begin{aligned} &Re\Big[\frac{1}{2\pi i}\int_\Gamma \frac{\tau_j(t)}{t}\,dt - \frac{1}{\pi}\iint_D \zeta^{|K_j|-1}\omega_j(\zeta)d\sigma\Big] = 0 , \\ &\frac{1}{\pi i}\int_\Gamma \frac{\tau_j(t)}{t^{m+1}}\,dt - \frac{1}{\pi}\iint_D [\zeta^{|K_j|-m-1}\omega_j(\zeta)+\bar\zeta^{|K_j|+m-1}\overline{\omega_j(\zeta)}]\,d\sigma_\zeta = 0 , \\ &\qquad m=1,\dots,|K_j|-1 , \; j=1 \text{ (or } 2) , \end{aligned}\right\} \qquad (3.17)$$

where $\omega_1(z) = \rho(z) = w_{z\bar z}$, $\omega_2(z) = \overline{\rho(z)}$.

(3) If $K_1 < 0$, $K_2 < 0$, the Problem P_1 has $2(|K_1|+|K_2|-1)$ solvability conditions as stated in (3.17) with $j=1,2$. If these conditions are satisfied, the solution $w(z)$ to Problem P_1 for (1.22) can be expressed as (3.6).

Proof. Let us introduce the Banach space $B = L_{p_0}(\overline{D})$, and denote by B_M the set of measurable functions $\rho(z)$ satisfying the condition

$$L_{p_0}[\rho,\overline{D}] < M_1 , \tag{3.18}$$

where M_1 is the constant stated in (3.9). We arbitrarily choose a function $\rho(z) \in B_M$, and consider the double integral

$$H_0\rho = \frac{2}{\pi}\iint_{D_n} \ln\left|1-\frac{z}{\zeta}\right| \rho(\zeta)d\sigma_\zeta , \tag{3.19}$$

where D_n is as stated in (2.40). Putting

$$\Psi_j(z) = \Phi_j(z) - \begin{cases} \dfrac{1}{\pi}\displaystyle\iint_{D_n} \frac{z^{2K_j+1}\overline{\rho(\zeta)}}{1-\bar{\zeta}z} d\sigma_\zeta , K_j \geq 0 , \\ \\ \dfrac{1}{\pi}\displaystyle\iint_{D_n} \frac{\zeta^{-2K_j-1}\overline{\rho(\zeta)}}{1-\bar{\zeta}z} d\sigma_\zeta , K_j < 0 , \end{cases} \quad j=1,2 , \tag{3.20}$$

and

$$V(z) = U(z) + \frac{1}{\pi}\iint_{D_n} [g_1(z,\zeta)+\overline{g_2(z,\zeta)}]\overline{\rho(\zeta)}d\sigma_\zeta , \; w(z) = V(z)+H_0\rho , \tag{3.21}$$

in which $\Phi_j(z)$, $U(z)$, $g_j(z,\zeta)$ $(j=1,2)$ are as stated in (3.6). Substituting $w(z)$, $\Psi_j(z)$ $(j=1,2)$ into the appropriate positions of the complex equation

$$w_{z\bar{z}} = F_n(z,w,w_z,\bar{w}_z,w_{zz},\bar{w}_{zz}) , \; F_n = \sigma_n(z) F , \tag{3.22}$$

where $\sigma_n(z)$ is as stated in (2.40), and considering the integral equation with the parameter $k(0 \leq k \leq 1)$

$$\rho^* = kF_n[z,w,w_z,\bar{w}_z,\Psi_1'+\Pi\rho^*,\Psi_2'+\Pi\bar{\rho}^*] , \tag{3.23}$$

we can find a solution $\rho^*(z) \in B$. This mapping from ρ onto ρ^* is denoted by $\rho^* = S(\rho,k)$ $(0 \le k \le 1)$. Similarly to the proof of Theorem 2.4, we can prove that $\rho^* = S(\rho,k)$ satisfies the conditions of the Leray-Schauder theorem, so that there exists a solution $w(z) = U(z) + H\rho$ to Problem Q_1 for (1.22).

The result of the solvability for Problem P_1 can be derived, too. □

2. The oblique derivative problem for nonlinear complex equations of second order in a multiply connected domain

Let D be an (N+1)-connected circular domain as stated in § 1. We consider the oblique derivative problem P for the nonlinear complex equation (1.22) in D with the boundary condition (3.1), and introduce the corresponding modified boundary value Problem Q for the nonlinear system of first order complex equations

$$\left.\begin{aligned} &V_{1\bar{z}} = f(z,w,w_z,\bar{w}_z,V_1,V_2,V_{1z},V_{2z}) \ , \ f = Q_1V_{1z} + Q_2\bar{V}_{1\bar{z}} + Q_3V_{2z} + \\ &+ Q_4\bar{V}_{2\bar{z}} + A_1V_1 + A_2\overline{V_1} + A_3V_2 + A_4\bar{V}_2 + A_5w + A_6\bar{w} + A_7 \ , \ V_{2\bar{z}} = \bar{V}_{1z} = \overline{\rho(z)} \ , \end{aligned}\right\} \tag{3.24}$$

with the boundary conditions

$$\begin{aligned} &Re[\overline{\lambda_j(t)}V_j(t)] = \tilde{\tau}_j(t) + h_j(t) \ , \\ &\tilde{\tau}_j(t) = -\varepsilon\, Re[\beta_j(t)w(t)] + \tau_j(t) \ , \ t \in \Gamma \ , \end{aligned} \tag{3.25}$$

$$Im[\overline{\lambda_j(a_k)}V_j(a_k) + \varepsilon\, \beta_j(a_k)w(a_k)] = b_{jk} \ , \tag{3.26}$$

$$k \in \{k\} = \begin{cases} 1,\dots, 2K_j-N+1 \ , & K_j \ge N \ , \\ N-K_j+1,\dots,N+1 \ , & 0 \le K_j < N \ , \end{cases} \qquad j=1,2$$

and the relation

$$w(z) = w_0 + \int_0^z \left\{\left[V_1(\zeta) + \sum_{m=1}^{N} \frac{d_m}{\zeta - z_m}\right]d\zeta + \overline{V_2(\zeta)}\,\overline{d\zeta}\right\} \ , \tag{3.27}$$

where $Q_j = Q_j(z,w,w_z,\bar{w}_z,V_{1z},V_{2z})$, $j=1,\dots,4$, $A_j = A_j(z,w,w_z,\bar{w}_z)$,

$j=1,\ldots,7$, $\lambda_j(t)$, $\beta_j(t)$, $\tau_j(t)$ $(j=1,2)$ satisfy (3.2) , $a_k(k\in\{k\})$ are the same as those in (3.3) of Chapter 1, $b_{jk}(k\in\{k\}, j=1,2)$ are real constants satisfying $|b_{jk}|\le \ell$, w_0 is a complex constant satisfying $|w_0|\le \ell$, $d_m(m=1,\ldots,N)$ are appropriate complex constants so that the function determind by the integral in (3.27) is single-valued in D , and

$$h_j(t)=\begin{cases}\left.\begin{array}{l} h_0^{(j)}+Re\displaystyle\sum_{m=1}^{|K_j|-1}(H_m^{(j)}+iH_{-m}^{(j)})t^m\ , t\in\Gamma_0\ ,\\ h_m^{(j)}\ ,\ t\in\Gamma_m\ ,\ m=1,\ldots,N\ ,\end{array}\right\} K_j<0\ ,\\ \\ \left.\begin{array}{l} h_m^{(j)}\ ,\ t\in\Gamma_m\ ,\ m=1,\ldots,N-K_j\ ,\\ 0\ ,\ t\in\Gamma_m\ ,\ m=N-K_j+1,\ldots,N+1\ ,\end{array}\right\} 0\le K_j\le N\ ,\\ 0\ ,\ t\in\Gamma\ ,\ K_j\ge N\ ,\ j=1,2\ ,\end{cases} \tag{3.28}$$

in which $h_m^{(j)}$ $(m=1,\ldots,N)$, $H_{\pm m}^{(j)}$ $(m=1,\ldots,|K_j|-1\ , j=1,2)$ are unknown real constants to be determined appropriately.

Now, we give a representation of solutions to Problem Q of the first order system (3.24).

<u>Theorem 3.3</u> Let $[w(z),V_1(z),V_2(z)]$ be a solution to Problem Q for the first order system (3.24). Then $w(z)$ possesses the representation (3.27), where $V_j(z)$ $(j=1,2)$ can be expressed by

$$\left.\begin{array}{l} V_j(z)=\Phi_j(z)+\tilde{T}_j\omega_j\ ,\\ \Phi_j(z)=\dfrac{1}{2\pi}\displaystyle\int_\Gamma T_j(z,t)\tilde{\tau}_j(t)d\theta+\Phi_{j0}(z)\ ,\end{array}\right\} \tag{3.29}$$

$$\tilde{T}_j\omega_j=-\frac{1}{\pi}\iint_D[G_{j1}^*(z,\zeta)Re\omega_j(\zeta)+G_{j2}^*(z,\zeta)i\,Im\omega_j(\zeta)]d\sigma_\zeta\ ,\ j=1,2\ ,$$

in which $\omega_1(z)=\rho(z)=w_{z\bar{z}}=V_{1\bar{z}}$, $\omega_2(z)=\overline{\rho(z)}$, $T_j(z,t)$ are the Schwarz kernels for analytic functions as stated in (3.25) of Chapter 1, $\Phi_{j0}(z)$ is an analytic function in D satisfying the homogeneous boundary condition

$$Re[\overline{\lambda_j(t)}\Phi_{j0}(t)] = h_j(t) \ , \ t \in \Gamma \ , \ j=1,2 \ , \tag{3.30}$$

$G^*_{jk}(z,\zeta)$ $(j,k=1,2)$ are the Green functions of the corresponding boundary value problem of the type

$$\left.\begin{aligned} G^*_{j1}(z,\zeta) &= \Delta_j(z)[G_{j1}(z,\zeta)Re\,\Delta_j(\zeta) - G_{j2}(z,\zeta)\,i\,Im\,\Delta_j(\zeta)]/|\Delta_j(\zeta)|^2 \ , \\ G_{j2}(z,\zeta) &= \Delta_j(z)[-G_{j1}(z,\zeta)\,i\,Im\,\Delta_j(\zeta) + G_{j2}(z,\zeta)Re\Delta_j(\zeta)]/|\Delta_j(\zeta)|^2 \ , \end{aligned}\right\} \tag{3.31}$$

$G_{jk}(z,\zeta)$ are Green functions satisfying the homogeneous canonical boundary conditions, and $\Delta_j(z) = e^{iS_j(z)}/\Pi_j(z)$, $S_j(z)$, $\Pi_j(z)$ are as stated in (3.4) of Chapter 1 with the index K_j , $j=1,2$.

In particular, if D is the unit disc $\{|z| < \}$ the solution $w(z)$ to Problem Q_1 for the complex equation (1.22) can be expressed as (3.6).

Proof. From (3.27), it can be seen that $V_{1\bar{z}} = \overline{V}_{2z}$, i.e. $V_{2\bar{z}} = \overline{V}_{1z} = \overline{\rho(z)}$. Because $[V_1(z),V_2(z)]$ is a solution of the system (3.24) and satisfies the boundary conditions (3.25), (3.26), using Theorem 3.6 in Chapter 1, we can obtain (3.29), where $\omega_1(z) = \rho(z) = V_{1\bar{z}} = w_{z\bar{z}}$, $\omega_2(z) = \overline{\rho(z)}$.

As D is the unit disc $\{|z| < 1\}$, is not difficult to see that the solution $w(z)$ to Problem Q_1 for (1.22) possesses the representation (3.6). □

Next, we give a priori estimates of solutions to Problem Q for (3.24).

Theorem 3.4 Let the complex equation (1.22) satisfy Condition C and the constant ε in (1.23), (1.20), (3.1) be small enough. Then the solution $[w(z),V_1(z),V_2(z)]$ for the corresponding Problem Q of (3.24) satisfies the estimates

$$L_j = C_\beta[V_j,\bar{D}] + L_{p_0}[|V_{j\bar{z}}|+|V_{jz}|,\bar{D}] < M_3 , \qquad (3.32)$$

$$S = C_\beta^1[w,\bar{D}] < M_4 , \qquad (3.33)$$

where $\beta = 1 - 2/p_0$, $p_0(2 < p_0 < p)$ are as stated in (3.9), and $M_k = M_k(q_0,p_0,k_0,\alpha,\ell,D)$, $k=2,3$.

<u>Proof.</u> From (3.25) and (3.27), we have

$$C_\alpha[\tilde{\tau}_j,\Gamma] \le \ell[\varepsilon C_\beta^1(w,\Gamma)+1] \le \ell[\varepsilon S+1] = \ell_1 , \; j=1,2 \qquad (3.34)$$

$$S = C_\beta^1[w,\bar{D}] \le \ell[M_5 \sum_{j=1}^{2} C_\beta(V_j,\bar{D})+1] \le \ell[M_5(L_1+L_2)+1] , \qquad (3.35)$$

where $M_5 = M_5(q_0,p_0,k_0,\alpha,\ell,D)$. The first equation in (3.24) can be rewritten as

$$V_{1\bar{z}} = Q_1V_{1z} + Q_2\overline{V_{1\bar{z}}} + A_1V_1 + A_2\overline{V_1} + A , \qquad (3.36)$$

$$A = Q_3V_{2z} + Q_4\overline{V_{2\bar{z}}} + A_3V_2 + A_4\overline{V_2} + A_5w + A_6\bar{w} + A_7 ,$$

in which A satisfies the inequality

$$L_{p_0}[A,\bar{D}] \le \varepsilon[(1+2k_0)L_2+2k_0S] + k_0 = k_1 . \qquad (3.37)$$

From the second equation in (3.24) the inequality

$$L_{p_0}[\overline{V_{1z}},\bar{D}] = L_{p_0}[\rho,\bar{D}] \le L_1 \qquad (3.38)$$

can be derived. Because $V_1(z)$ is a solution of the complex equation (3.36) with the boundary conditions (3.25), (3.26) $(j=1)$, by Theorem 4.3 in Chapter 2, we can obtain the estimate

$$L_1 \le M_6(\ell_1+k_1) = \varepsilon[(1+2k_0)M_6L_2+(\ell+2k_0)M_6S]+M_6(\ell+k_0) . \qquad (3.39)$$

Similarly, from $V_{2\bar{z}} = \overline{V_{1z}}$ and (3.25), (3.26) $(j=2)$, we have the estimate

$$L_2 \le M_7(\ell_1+L_1) = \varepsilon\,\ell\,M_7 S + M_7 L_1 + M_7\,\ell\,, \tag{3.40}$$

where $M_k = M_k(q_0,p_0,k_0,\alpha,\ell,D)$, $k=6,7$. Combining (3.35), (3.39) and (3.40) and observing that the constant ε is sufficiently small, the estimates (3.32) and (3.33) are derived. □

Finally, we prove the solvability of Problem Q for (3.24).

Theorem 3.5 Under the hypothesis in Theorem 3.4, Problem Q for (3.24) is solvable, and the results on the solvability for Problem P of (1.22) are as follows:

(1) If $K_j = 1/2\pi\,\Delta_\Gamma\,\mathit{arg}\,\lambda_j(t) \ge N(j=1,2)$, Problem P for (1.22) has $2N$ solvability conditions.

(2) If $0 \le K_j < N(j=1,2)$, the total number of solvability conditions for Problem P is not greater than $4N - K_1 - K_2$.

(3) If $K_j < 0(j=1,2)$, under $4N - 2K_1 - 2K_2 - 2$ solvability conditions, Problem P is solvable.

Besides, we can give the total number of sovability conditions for Problem P for (1.22) in the other cases.

Proof. Using a similar method to that used in the proofs of Theorem 2.4 and Theorem 3.2, we introduce the Banach space $B = C^1(\overline{D}) \times L_p(D_n)$ and denote by B_M the set of systems $\Omega = [w(z),\rho(z)]$, satisfying the inequalities

$$C^1[w,\overline{D}] < M_4\,,\quad L_{p_0}[\rho,\overline{D_n}] < M_3\,, \tag{3.41}$$

where M_3 , M_4 are the constants in (3.32) and (3.33), and D_n is as stated in (2.40). We select arbitrarily $\Omega = [w(z),\rho(z)] \in \overline{B}_M$ and consider the double integrals

$$T\omega_j = -\frac{1}{\pi}\iint_{D_n} \frac{\omega_j(\zeta)}{\zeta - z}\,d\sigma_\zeta\,,\ j=1,2\,,\ \omega_1(z) = \rho(z)\,,\ \omega_2(z) = \overline{\rho(z)}\,. \tag{3.42}$$

Substituting $w(z)$, $T\omega_j (j=1,2)$ into suitable positions of the boundary conditions (3.25), (3.26), we have

$$Re\{\overline{\lambda_j(t)}[\Psi_j(t)+T\omega_j]\} = -\varepsilon\, Re[\beta_j(t)w(t)] + h_j(t)\ ,\ t\in\Gamma\ ,\ j=1,2\ , \tag{3.43}$$

$$Im\{\overline{\lambda_j(t)}[\Psi_j(t)+T\omega_j] + \varepsilon\,\beta_j(t)w(t)\}\Big|_{t=a_k} = b_{jk}\ ,\ k\in\{k\}\ . \tag{3.44}$$

By using Theorem 3.1 and Theorem 3.2 in Chapter 1, there exist analytic functions $\Psi_j(z)$ $(j=1,2)$ in D satisfying the boundary conditions (3.43), (3.44) and the estimates

$$C_\beta[\Psi_j',\overline{D_n}] \le M_8 = M_8(p_0,k_0,\ell,D,D_n,M_3,M_4)\ ,\ j=1,2\ . \tag{3.45}$$

Setting $V_j(z) = \Psi_j(z) + T\omega_j$, $j=1,2$ and substituting them into (3.27), it is not difficult to see that $w(z)$ satisfies the estimate

$$C_\beta^1[w,\overline{D}] \le M_9 = M_9(p_0,\alpha,\ell,D,D_n,M_3,M_4)\ . \tag{3.46}$$

By Condition C and the principle of contraction, the complex equation

$$\rho^*(z) = k\ f_n(z,w,w_z,\bar{w}_z,V_1,V_2,\Psi_1+\Pi\rho^*,\Psi_2+\Pi\bar{\rho}^*)\ ,\ f_n = \sigma_n f\ ,\ 0\le k\le 1 \tag{3.47}$$

has a unique solution $\rho^*(z)\in L_{p_0}(\overline{D_n})$. Set
$\omega_1^*(z) = \rho_1^*(z)$, $\omega_2^*(z) = \overline{\rho_2(z)}$. Then we find analytic functions $\Psi_j^*(z)$ $(j=1,2)$ in D , which satisfy the boundary conditions

$$Re[\overline{\lambda_j(t)}V_j^*(t) + \varepsilon\,\beta_j(t)w(t)] = \tau_j(t) + h_j(t)\ ,\ t\in\Gamma\ ,\ j=1,2\ , \tag{3.48}$$

$$Im[\overline{\lambda_j(t)}V_j^*(t) + \varepsilon\,\beta_j(t)w(t)]\Big|_{t=a_k} = b_{jk}\ ,\ k\in\{k\}\ ,\ j=1,2\ , \tag{3.49}$$

where $V_j^*(z) = \Psi_j^*(z) + T\omega_j^*$, $j=1,2$, and then obtain a single-valued function

$$w^*(z) = w_0 + \int_0^z \Big\{\Big[V_1^*(\zeta) + \sum_{m=1}^N \frac{d_m^*}{z-z_m}\Big]dz + \overline{V_2^*(\zeta)d\bar{\zeta}}\Big\} \tag{3.50}$$

in D . Denoting by $\Omega^* = S(\Omega,k)$ $(0 \le k \le 1)$ the mapping from Ω onto $\Omega^* = [w^*(z),\rho^*(z)]$, we can prove that $\Omega^* = S(\Omega,k)$ satisfies the assumptions in the Leray-Schauder theorem. Hence, there exists a solution $[w(z),V_1(z),V_2(z)]$ to Problem Q for the system

$$V_{1\bar{z}} = f_n(z,w,w_z,\bar{w}_z,V_1,V_2,V_{1z},V_{2z}) \ , \ V_{2\bar{z}} = \bar{V}_{1z} \ . \tag{3.51}$$

According to the proofs of Theorem 2.4 and Theorem 2.5, we can eliminate the assumption $Q_j = 0$, $j=1,\ldots,4$, $A_j = 0$, $j=1,\ldots,7$ on $D \setminus D_n$. Therefore, Problem Q for (3.24) has a solution $[w(z),V_1(z),V_2(z)]$.

Let us substitute the solution into the boundary conditions (3.25), (3.26) and the relation (3.27). The functions $h_j(t)$ $(j=1,2)$ and the constants $d_m (m=1,\ldots,N)$ are determined. If the functions and the constants are equal to zero, i.e.

$$h_j(t) = \begin{cases} h_0^{(j)} = 0 \ , \ m=1,\ldots,N-K_j & \text{if } 0 \le K_j < N \ , \\ h_m^{(j)} = 0 \ , \ m=0,1,\ldots,N \ , & \\ H_{\pm m}^{(j)} = 0 \ , \ m=1,\ldots,|K_j|-1 & \left.\right\} \text{if } K_j < 0 \ , \ j=1,2 \end{cases} \tag{3.52}$$

and

$$d_m = 0 \ , \ m=1,\ldots,N \ , \tag{3.53}$$

then the solution $[w(z),V_1(z),V_2(z)]$ to Problem Q for (3.24) is also a solution to Problem P for the second order complex equation (1.22). Thus, we have the results on the solvability for Problem P for (1.22) as stated in the theorem (cf. [103], 17)). □

In addition, we mention that if

$$\lambda(t) = \begin{cases} \bar{t}^{K_j} \ , \ t \in \Gamma_0 \ , \\ e^{-i\theta_{jm}} \ , \ t \in \Gamma_m \ , \ m=1,\ldots,N \ , \end{cases}$$

and $K_1 \le 0$, $K_2 \le 0$ in Theorem 3.5, due to $\lambda_2 \le 1$ in Theorem 3.5 of Chapter 1, we do not have to assume that q_0' in (1.20) is sufficiently small, but instead ask $L_p[A_j,\overline{D}] \le \varepsilon k_0$, j=1,2 in (1.23). In this case, Problem Q for (1.22) is also solvable.

3. The oblique derivative problem for linear complex equations of second order in a multiply connected domain

Let D be an (N+1)-connected circular domain as before. We consider the linear complex equation of second order

$$\left.\begin{aligned} w_{z\bar{z}} &= Q_1(z)w_{zz}+Q_2(z)\bar{w}_{\bar{z}\bar{z}}+Q_3(z)\bar{w}_{zz}+Q_4(z)w_{\bar{z}\bar{z}}+\varepsilon G(z,w,w_z,\bar{w}_z)+A_7(z), \\ G &= A_1(z)w_z+A_2(z)\bar{w}_{\bar{z}}+A_3(z)\bar{w}_z+A_4(z)w_{\bar{z}}+A_5(z)w+A_6(z)\bar{w}, \ z \in D, \end{aligned}\right\} \quad 3.54)$$

where $Q_j(z)$ (j=1,...,4) , $A_j(z)$ (j=1,...,7) satisfy the conditions

$$|Q_1(z)|+|Q_2(z)| \le q_0, \ |Q_3(z)|+|Q_4(z)| \le q_0', \ q_0+q_0' < 1, \ z \in D, \quad (3.55)$$

$$L_p[A_j(z),\overline{D}] \le \ell, \ j=1,\dots,7, \ 2 < p < \infty, \quad (3.56)$$

and $\varepsilon(-\infty < \varepsilon < \infty)$ is a parameter. We denote by Problem P the boundary value problem for (3.54) with the boundary condition (3.1), (3.2).

To obtain the results on the solvability of Problem P for (3.54), we introduce the corresponding modified Problem Q for the first order system

$$\left.\begin{aligned} LV &= V_{1\bar{z}} - Q_1(z)V_{1z} - Q_2(z)\overline{V}_{1\bar{z}} - Q_3(z)V_{2z} - Q_4(z)\overline{V}_{2\bar{z}} \\ &= \varepsilon G(z,w,w_z,\bar{w}_z) + A_7(z), \end{aligned}\right\} \quad (3.57)$$

with the boundary condition

$$\ell_jV_j = Re[\overline{\lambda_j(t)}V_j(t)] = -\varepsilon\, Re[\beta_j(t)w(t)] + \tau_j(t) + h_j(t), \ j=1,2, \quad (3.58)$$

where $h_j(t)$ (j=1,2) are as stated in (3.28), w(z) and $V_j(z)$ (j=1,2) satisfy the relation (3.27). Problem Q with $A_7(z) \equiv 0$, $\tau_j(t) \equiv 0$ (j=1,2) is called Problem Q_0. In the following, we mainly discuss the case of $K_1 \ge N$, $0 \le K_2 < N$. If the constant q_0' in (3.55) is sufficiently small, we substitute $w(z) \in C^1(\overline{D})$ into the corresponding function in (3.57), (3.58), and find the general solution

$$\left.\begin{aligned} &\hat{V}_j(z) = V_{j0}(z) + V_{j*}(z) \\ &V_{j*}(z) = \sum_{m=1}^{2K_1-N+1} c_{1m}V_{1jm}(z) + \sum_{m=N-K_2+1}^{N+1} c_{2m}V_{2jm}(z)\ ,\ j=1,2 \end{aligned}\right\} \tag{3.59}$$

of the first order complex equation

$$LV = A_7(z) \quad \text{in } D$$

with the boundary condition

$$\ell_j V_j = \tau_j(t) + h_j(t) \quad \text{on } \Gamma\ ,\ j=1,2\ ,$$

where $[V_{10}(z), V_{20}(z)]$ is a special solution of the above boundary value problem, and $V_{1jm}(z)$ $(m=1,\ldots,2K_2-N+1)$, $V_{2jm}(z)$ $(m=N-K_2+1,\ldots,N+1)$ are linearly independent solutions of the corresponding homogeneous problem, where $c_{1m}(m=1,\ldots,2K_1-N+1)$, $c_{2m}(m=N-K_2+1,\ldots,N+1)$ are arbitrary real constants. Moreover, we seek a solution $[\tilde{V}_1(z), \tilde{V}_2(z)]$ of the complex equation

$$LV = G(z,w,w_z,\bar{w}_z) \tag{3.60}$$

with the boundary condition

$$\ell_j V = -Re[\beta_j(t)w(t)] + h_j(t)\ ,\ t \in \Gamma\ ,\ j=1,2\ , \tag{3.61}$$

$$Im[\overline{\lambda_j(a_k)}V_j(a_k) + \beta_j(a_k)w(a_k)] = 0\ ,\ k \in \{k\}\ . \tag{3.62}$$

Denote by $\tilde{V} = R_2 w$ a mapping from $w(z)$ onto $\tilde{V} = [\tilde{V}_1(z), V_2(z)]$. It can be verified that R_2 is a linear and bounded operator from

$C^1(\overline{D})$ into $C_\beta(\overline{D}) \times C_\beta(\overline{D})$ $(\beta=1-2/p_0)$. Next, we consider the function

$$R_1\widetilde{V} = \int_0^z \{[V_1(z) + \sum_{m=1}^{N} \frac{d_m}{\zeta-z_m}]dz + \overline{\widetilde{V}_2(\zeta)d\overline{\zeta}}\} . \tag{3.63}$$

It is easy to see that $R_1\widetilde{V}\big|_{z=0} = 0$. Moreover it can be proved that R_1 is a linear, bounded and completely continuous operator from $C_\beta(\overline{D}) \times C_\beta(\overline{D})$ into $C^1(\overline{D})$. Thus we obtain the following linear operator equation

$$w - \varepsilon R_1R_2w = R_1V_0 + R_1V_* + w_0 , \tag{3.64}$$

where w_0 is an arbitrary complex constant, $V_0 = [V_{10}(z), V_{20}(z)]$, $V_* = [V_{1*}(z), V_{2*}(z)]$. It is known that R_1R_2 is a completely continuous operator which maps $C^1(\overline{D})$ into itself, so we can apply the Fredholm theorem to the operator equation. Let

$$\varepsilon_j , 0 < |\varepsilon_j| \le |\varepsilon_{j+1}| \ (j=1,2,\ldots) \tag{3.65}$$

be eigenvalues of the homogeneous operator equation

$$w - \varepsilon R_1R_2w = 0 . \tag{3.66}$$

If $\varepsilon \neq \varepsilon_j$ then the boundary value problem for (3.57) with the boundary condition (3.58) is solvable. Its solution $w(z)$ can be expressed as

$$w(z) = \varepsilon R_1R_2w + R_1V_0 + R_1V_* + w_0 \tag{3.67}$$

which includes $2K_1 + K_2 - N+4$ arbitrary real constants.

Let $w(z)$ of (3.67) and $V_j(z) = \varepsilon\widetilde{V}_j(z) + V_{j0}(z) + V_{j*}(z)$ $(j=1,2)$ be inserted into the boundary condition (3.58) and (3.27). From $h_j(t) = 0$ $(j=2)$ and $d_m = 0$ $(m=1,\ldots,N)$, we obtain $2N-K_2$ algebraic equations

$$\left.\begin{array}{l} Re[\overline{\lambda_j(t)V_j(t)+\varepsilon\beta_j(t)w(t)}] = \tau_j(t)\ ,\ t\in\Gamma\ ,\ j=1,2\ , \\ \int_{\Gamma_m} \{V_1(z)dz+\overline{V_2(z)d\bar{z}}\} = 0\ ,\ m=1,\ldots,N\ . \end{array}\right\} \qquad (3.68)$$

Denoting the rank of the coefficients matrix of the $2K_1 + K_2 - N+4$ arbitrary real constants in (3.68) by S we can determine S constants among the above arbitrary constants and S equalities within the $3N-K_2$ algebraic equations. If the other $3N-K_2-S$ equalities hold, then the function $w(z)$ is a solution for Problem P of (1.22). Hence, Problem P of (1.22) has $3N-K_2-S$ solvability conditions, where $S \le min(3N-K_2, 2K_1+K_2-N+4)$.

Using the same method as above, one can discuss other cases. Thus, we have

Theorem 3.6 Suppose that the constant q_0' in (3.55) is small enough and ε is not an eigenvalue of the corresponding homogeneous operator equation (3.66). Then Problem Q for (3.57) is solvable and the result on the solvability for Problem P for (3.54) is as follows:

(1) If $K_1 \ge N$, $K_2 \ge N$, Problem P of (3.54) has $2N-S$ solvability conditions, $S \le min[2N, 2(K_1+K_2-N+2)]$. Moreover, if these conditions are satisfied, then the solution of the problem includes $2[K_1+K_2-N+2]-S$ arbitrary real constants.

(2) If $0 \le K_1 < N$, $K_2 \ge N$ (or $K_1 \ge N$, $0 \le K_2 < N$) , the total number of solvability conditions for Problem P for (3.54) is not greater than $3N-K_1-S$ (or $3N-K_2-S$) , $S \le min[3N-K_1, 2K_2+K_1-N+4]$, (or $S \le min[3N-K_2, 2K_1+K_2-N+4]$) . Moreover, if the conditions are satisfied, then the solution includes $K_1 + 2K_2 - N+4 - S$ (or $2K_1+K_2-N+4-S$) arbitrary real constants.

(3) If $K_1 < 0$, $K_2 < 0$, under $4N - 2K_1 - 2K_2 - 2 - S$ solvability conditions, Problem P of (3.54) is solvable, $S = 0$, 1 or 2 . Moreover, if the conditions are satisfied, then the general solution includes $2-S$ arbitrary real constants.

In addition, for other cases, we can find the total number of solvability conditions for Problem P for (3.54).

Next, let q be the rank of the eigenvalue ε_j for the operator equation (3.66). We discuss only the case $K_1 < 0$, $0 \le K_2 < N$. According to the Fredholm theorem of linear operator equations, we can write the solvability conditions of the nonhomogeneous operator equation (3.64). Denote by S_1 the rank of the coefficients matrix of $Re\, w_0$, $Im\, w_0$, $c_{2,N-K_2+1},\ldots,c_{2,N+1}$. Obviously, $S_1 \le min[q,K_2+3]$. Thus, we may determine S_1 constants among the above arbitrary constants and simultaneously determine S_1 equalities within the q algebraic equations. This shows that Problem Q for (3.57) has $q - S_1$ solvability conditions and its general solution includes $K_2 + 3 + q - S_1$ arbitrary real constants. Let the solution $[w(z),V_1(z),V_2(z)]$ be substituted into (3.58), (3.27) and take $h_j(t) \equiv 0$ $(j=1,2)$, $d_m = 0$ $(m=1,\ldots,N)$. We obtain $2N + (N-2K_1-1) + (N-K_2) = 4N-2K_1-K_2-1$ algebraic equations. Denoting the rank of the corresponding coefficients matrix by S_2, it is clear that $S_2 \le min\,(4N-2K_1-K_2-1, K_2+3+q-S_1)$. After that, similarly to the proof of Theorem 3.6, we can determine S_2 constants among the $K_2 + 3 + q - S_1$ arbitrary constants. Hence, Problem P has $4N - 2K_1 - K_2 - 1 + q - S$ solvability conditions and its general solution includes $K_2 + 3 + q - S$ arbitrary real constants $S = S_1 + S_2 \le min(4N-2K_1-K_2-1+q, K_2+3+q)$. Thus, we have

<u>Theorem 3.7</u> Suppose that q_0' in (3.55) is small enough and ε_j is an eigenvalue of (3.66) with the rank q, then

(1) if $0 \le K_1 < N$, $0 \le K_2 < N$, then Problem Q for (3.57) has $q - S$ solvability conditions, $S \le min[q, 2(K_1+K_2+2)]$, and Problem P for (3.54) has $4N - K_1 - K_2 + q - S$ solvability conditions, $S \le min[4N-K_1-K_2+q, 2(K_1+K_2+2)+q]$.

(2) If $K_1 < 0$, $K_2 \ge N$, Problem Q for (3.57) has $q - S$ solvability conditions, $S \le min(q, 2K_2-N+3)$, and Problem P for (3.54) has $3N - 2K_1 - 1 + q - S$ solvability conditions, $S \le min[3N-2K_1-1+q, 2K_2-N+3+q]$.

(3) If $K_1 < 0$, $0 \le K_2 < N$, Problem Q has $q - S$ solvability conditions, $S \le min(q, K_2+3)$, and Problem P for (3.54) has $4N - 2K_1 - K_2 - 1 + q - S$ solvability conditions, $S \le min[4N-2K_1-K_2-1+q, K_2+3+q]$.

Furthermore, we can find the total number of solvability conditions for Problem Q and Problem P in other cases.

For the corresponding oblique derivative problem of the nonlinear system (1.24), we can discuss its solvability, too by using a similar method.

4. The oblique derivative problem for nonlinear complex equations of first order in a multiply connected domain

As an application of the above results on the oblique derivative problem of second order complex equations, we shall discuss the oblique derivative problem for the first order complex equation

$$\left.\begin{aligned} &w_{\bar z} = F(z,w,w_z) \ , \ F = Q_1 w_z + Q_2 \bar w_z + A_1 w + A_2 \bar w + A_3 \ , \\ &Q_j = Q_j(z,w,w_z) \ , \ j=1,2 \ , \ A_j = A_j(z,w) \ , \ j=1,2,3 \end{aligned}\right\} \tag{3.69}$$

with the boundary condition

$$Re[\overline{\lambda(t)} w_t + \varepsilon\beta(t)w(t)] = \tau(t) \ , \ t \in \Gamma \ , \ w(1) = 0 \ , \tag{3.70}$$

where $|\lambda(t)| = 1$, $C_\alpha[\lambda,\Gamma] \le \ell$, $C_\alpha[\beta,\Gamma] \le \ell$, $C_\alpha[\tau,\Gamma] \le \ell$ and $\alpha(1/2 < \alpha < 1)$, $\ell(0 \le \ell < \infty)$, $\varepsilon(0 < \varepsilon \le 1)$ are all constants. The above boundary value problem is called Problem P. We assume that for (3.69), together with Condition C as stated in § 1 of Chapter 2, the following conditions are satisfied.

(1) The function $F(z,w,V)$ is continuous for $z \in \bar D$, $w \in \mathbb{C}$, $V \in \mathbb{C}$, and possesses generalized derivatives with respect to $z \in D$, $\bar z$, $w \in \mathbb{C}$, $\bar w$, $V \in \mathbb{C}$, $\bar V$.

(2) The above derivatives are measurable for $z \in D$ and any continuous functions $w(z)$, $V(z)$ in $\bar D$, and are continuous in

$w \in \mathbb{C}$, $V \in \mathbb{C}$ for almost every point $z \in D$, and satisfy the conditions

$$L_p[F_z,\overline{D}] \le k_0 \ , \ L_p[F_{\overline{w}},\overline{D}] \le \varepsilon\, k_0 \ , \ L_p[F_w,\overline{D}] \le k_0 \ . \tag{3.71}$$

(3) F , $F_{\overline{V}}$, and F_V satisfy

$$\left.\begin{aligned} &|F(z,w_1,V_1)-F(z,w_2,V_2)| < \\ &\quad < k_0|z_1-z_2|^{\alpha}+\varepsilon k_0\,|w_1-w_2|+q_0|V_1-V_2| \ , \ z_1,z_2\in D \ , \ w_1,\ldots,V_2\in\mathbb{C} \ , \\ &|F(1,w(1),V(1)| \le k_0<\infty \ , \text{ for any continuous functions } w,V \ , \\ &|F_{\overline{V}}|+|F_V| \le q_0 < 1 \ , \ z\in D \ , \ w,V\in\mathbb{C} \ , \end{aligned}\right\} \tag{3.72}$$

where $\varepsilon(0<\varepsilon\le 1)$, $p(2<p<\infty)$, $k_0(0\le k_0<\infty)$, q_0 are real constants. The above condition for (3.69) is called Condition C_*.

In order to obtain a result on solvability for Problem P for (3.69), we first give an equivalence theorem.

Theorem 3.8 Let $w(z)$ be a solution of Problem P for the nonlinear first order complex equation (3.69) satisfying Condition C_*, and $w(z) \in W^2_{p_0}(D)$, $2 < p_0 < min(p,1/(1-\alpha))$. Then $w(z)$ is a solution of Problem P_* for the second order complex equation

$$\begin{aligned} &w_{z\overline{z}} = f(z,w,w_z,\overline{w}_z,w_{zz},\overline{w}_{\overline{z}z}) \ , \ f = q_1 w_{zz} + q_2\overline{w}_{\overline{z}z} + \\ &+ B_1 w + B_2\overline{w}_z + B_3 \ , \ q_1 = F_{w_z} \ , \ q_2 = F_{\overline{w}_{\overline{z}}} \ , \ B_1 = F_w \ , \ B_2 = F_{\overline{w}} \ , \ B_3 = F_z \end{aligned} \tag{3.73}$$

satisfying the boundary conditions

$$Re[\overline{\lambda(t)}w_t+\varepsilon\beta(t)w(t)] = \tau(t) \ , \ t\in\Gamma \ , \ w(1) = 0 \ , \tag{3.74}$$

$$Re[w_{\overline{t}}] = Re\ F(t,w,w_t)+H(t) \ , \ t\in\Gamma \ , \ w_{\overline{z}}(1) = F(1,w(1)\ ,\ w_z(1)] \ , \tag{3.75}$$

where

$$H(t) = \begin{cases} 0 \ , & t \in \Gamma_0 \ , \\ H_j \ , & t \in \Gamma_j (j=1,\ldots,N) \ , \end{cases}$$

$H_j (j=1,\ldots,N)$ are unknown constants to be determined appropriately. Conversely, if $w(z)$ is a solution to Problem P_* for (3.73), then $w(z)$ is a solution to Problem P for (3.69).

Proof. The first statement in this theorem is obvious and we may take $H(t) = 0$, $t \in \Gamma$. Now we prove the second statement. From (3.73), we can obtain

$$[\bar{w}_z - \overline{F(z,w,w_z)}]_{\bar{z}} = 0 \ , \tag{3.76}$$

which shows that $\bar{w}_z - \overline{F(z,w,w_z)}$ is analytic in D and satisfies the modified Dirichlet boundary condition

$$Re[w_{\bar{t}} - F(t,w,w_t)] = H(t) \ , \ t \in \Gamma \ , \ [\bar{w}_t - F(t,w(t),w_t)]_{t=1} = 0 \ . \tag{3.77}$$

Hence, $\bar{w}_z - \overline{F(z,w,w_z)} = 0$, i.e. $w_{\bar{z}} = F(z,w,w_z)$, $z \in D$. So $w(z)$ is a solution to Problem P for (3.69). □

Theorem 3.9 Let the first order ocmplex equation (3.69) satisfy Condition C_* and the constant ε in (3.70) - (3.72) be sufficiently small. Then the following statements hold.

(1) If $K = 1/2\pi \, \Delta_\Gamma \, arg \, \lambda(t) \geq N$, Problem P for (3.69) has $2N$ solvability conditions.

(2) If $0 \leq K < N$, the total number of solvability conditions for Problem P is not greater than $3N-K$.

(3) If $K < 0$, Problem P has $3N-2K-1$ solvability conditions.

Proof. According to Theorem 3.8, we may only discuss Problem P_* for the second order complex equation (3.73), and consider in detail the case $K < 0$. To apply the Schauder fixed-point theorem, we introduce a bounded, closed and convex set B_M in the Banach space $C^1(\overline{D})$, the elements of which are the functions $W(z)$ satisfying

$$C^1[W,\overline{D}] \le M_{10} \ , \ W(1) = 0 \ , \tag{3.78}$$

where M_{10} is an appropriate constant. Choosing any function $W(z) \in C^1(\overline{D})$ and substituting it into (3.73), we consider the boundary value problem Q for the first order complex equation

$$V_{\bar{z}} = q_1 V_z + q_2 \overline{V}_z + B_1 V + B \ , \ B = B_2 \overline{W}_z + B_3 \tag{3.79}$$

and the boundary conditions

$$Re[\overline{\lambda(t)} V(t)] = \tau(t) + h(t) - Re[\varepsilon\beta(t)W(t)] \ , \ t \in \Gamma \ , \tag{3.80}$$

$$Re[\Phi(t)+T\overline{V}_z] = Re\ F(t,W,V) + H(t) \ , \ t \in \Gamma \ , \tag{3.81}$$

$$[\Phi(t)+T\overline{V}_t]_{t=1} = \overline{F[1,W(t),V(1)]} \ ,$$

where $h(t)$ is as stated in (4.4) of Chapter 2, $\Phi(z)$ is an analytic function in D , continuous on $\overline{D}$. Because (3.79) is equivalent to

$$V_{\bar{z}} = q_1^* V_z + q_2^* \overline{V}_{\bar{z}} + B_1^* V + B_2^* \overline{V} + B_3^* \tag{3.82}$$

where $q_1^* = q_1/q_3$, $q_2^* = \overline{q_1} q_2/q_3$, $B_1^* = B_1/q_3$, $B_2^* = \overline{B_1} q_2/q_3$, $B_3^* = (B+\overline{B}q_2)/q_3$, $q_3 = 1 - |q_2|^2$, and $|q_2^*| + |q_2^*| \le q_0^* < 1$, by Theorem 4.1 and Theorem 4.7 in Chapter 2, it is clear that the above Problem Q has a unique solution $V(z)$. Provided the constant ε in (3.71) is small enough, we can conclude $L_{p_0}[B,\overline{D}] \le 2k_0$, $L_{p_0}[B_3^*,\overline{D}] \le 2k_0/(1-q_0)$, and $C_\alpha\{\tau - Re[\varepsilon\beta W],\Gamma\} \le 2\ \ell$. Consequently, by Theorem 4.3 in Chapter 2, $V(z)$ satisfies the estimates

$$C_\beta[V,\overline{D}] \le M_{11} \ , \ L_{p_0}[|V_{\bar{z}}|+|V_z|,\overline{D}] \le M_{12} \ , \tag{3.83}$$

where $\beta = 1-2/p_0$, $2 < p_0 < min(p,1/(1-\alpha)), M_j = M_j(q_0,p_0,k_0,\alpha,\ell,D)$, $j = 11,12$. Setting $P(z) = V_{\bar{z}}$ and substituting $W(z)$, $V(z) = \Phi_1(z) + T\rho$ into the right-hand side of (3.81), we can find an analytic function $\Phi_2(z)$ in D satisfying the boundary condition

$$\begin{aligned} &Re[\Phi_2(t)] = Re\, F[t,W(t),V(t)] - Re\, T\bar{\rho} + H(t), \\ &\Phi_2(1) = \overline{F[1,W(1),V(1)]} - Re\, T\bar{\rho}\big|_{z=1} . \end{aligned} \tag{3.84}$$

If ε is sufficiently small, from (3.72), (3.78) an (3.83), we have

$$|F[t_1,W(t_1),V(t_1)]-F[t_2,W(t_2),V(t_2)]| \le (2k_0+M_{11})|t_1-t_2|^{\beta} \tag{3.85}$$

and then

$$C_\beta[\Phi_2,\bar{D}] \le M_{13} = M_{13}(q_0,p_0,k_0,\alpha,\ell,D) . \tag{3.86}$$

We construct the single-valued function

$$\left.\begin{aligned} &w(z) = X(z) + Y(z) + H\rho \text{ , } X(z) = \int_1^z \Phi_1(\zeta)d\zeta + \overline{\int_1^z \Phi_2(\zeta)d\zeta} \text{ ,} \\ &Y(z) = \sum_{m=1}^{N} \int_1^z \frac{d_m}{\zeta-z_m}\, d\zeta \text{ , } H\rho = \frac{2}{\pi}\iint_D \ell n\left|\frac{\zeta-z}{\zeta-1}\right| \rho(\zeta)d\sigma_\zeta \end{aligned}\right\} \tag{3.87}$$

where $d_m (m=1,\ldots,N)$ are suitable complex constants. It is not difficult to see that $w(1) = 0$ and $w(z)$ satisfies the estimate

$$C^1_\beta[w,\bar{D}] \le M_{14} = M_{14}(q_0,p_0,k_0,\alpha,\ell,D) . \tag{3.88}$$

We take $M_{10} = M_{14}$ in (3.78) and denote by $w = S(W)$ a mapping from $W(z) \in B_M$ onto $w(z)$. Obviously, $w = S(W)$ maps B_M onto a compact subset of itself. We can prove that $w = S(W)$ is a continuous mapping in B_M . By the Schauder fixed-point theorem, there exists a function $w \in B_M$, so that $w = S(w)$. If $d_m = 0$ $(m=1,\ldots,N)$, then

$w_z = \Phi_1(z) + T\rho = V(z)$, $\bar{w}_z = \Phi_2(z) + T\bar{\rho}$. Let $w(z) = W(z)$, $w_z = V(z)$ be inserted into (3.80), so that $h(t) = 0$, i.e.

$$h_j = 0 \ , \ j=0,1,\ldots,N \ , \ H_{\pm m} = 0 \ , \ m=1,\ldots,|K|-1 \ . \tag{3.89}$$

In this case, the above function $w(z)$ is a solution to Problem P_* for (3.73). By Theorem 3.8, we see that if $K < 0$, Problem P has $3N-2K-1$ solvability conditions. Similarly, we can discuss other cases. This completes the proof (cf. [109]2)). □

Finally, we mention that I.I. Daniljuk [25]1) and Fang Ai-nong [32]2) discussed the oblique derivative problem of the first order complex equation (3.69) in the unit disc.

§ 4. The Riemann-Hilbert problem for complex equations of second order

In this section, we mainly discuss the Riemann-Hilbert problem for the complex equation (1.22) in the unit disc, and simply state the results for the corresponding boundary value problem in a multiply connected domain.

1. Representation and estimation of solutions to the Riemann-Hilbert problem for second order complex equations

Problem A_1. The Riemann-Hilbert problem of the second order complex equation (1.22) in $D = \{|z|<1\}$ is to find a continuously differentiable solution $w(z)$ of (1.22) on $\bar{D}$ satisfying the boundary conditions

$$Re[\bar{t}^{K_1} w_t] = r_1(t) \ , \ Re[\bar{t}^{K_2}\overline{w(t)}] = r_2(t) \ , \ t \in \Gamma = \{|t|=1\} \ , \tag{4.1}$$

where K_1, K_2 are integers and $C_\alpha^{j-1}[r_j,\Gamma] \le \ell < \infty$, $j=1,2$, $1/2 < \alpha < 1$.

If $K_1 < 0$ or $K_2 < 0$, Problem A_1 for (1.22) may not be solvable. Hence we consider the modified boundary value problem B_1 with boundary conditions:

$$
\left.\begin{aligned}
&Re[\bar{t}^{K_1} w_t] = r_1(t) + h_1(t) \;,\; Re[\bar{t}^{K_2}\overline{w(t)}] = r_2(t) + h_2(t) \;,\; t \in \Gamma \;, \\
&h_j(t) = \begin{cases} 0 \;, \text{ for } K_j \geq 0 \;,\; j=1 \text{ or } 2 \;, \\ h_0^{(j)} + Re \displaystyle\sum_{m=1}^{|K_j|-1} (h_m^{(j)} + ih_{-m}^{(j)}) t^m \;, \text{ for } K_j < 0 \;, \end{cases}
\end{aligned}\right\} \tag{4.2}
$$

where $h_{\pm m}^{(j)}$ $(m=0,1,\ldots,|K_j|-1, j=1,2)$ are unknown real constants to be determined appropriately. If $K_1 \geq 0$ or $K_2 \geq 0$, we assume that w_z, $w(z)$ satisfy

$$
\left.\begin{aligned}
&\int_\Gamma [w_t - \Phi_1(t)] e^{-im\theta} d\theta = 0 \; (m=1,\ldots,2K_1) \;, \\
\text{or } &\int_\Gamma [w(t) - \Phi_2(t)] e^{-im\theta} d\theta = 0 \; (m=1,\ldots,2K_2) \;,
\end{aligned}\right\} \tag{4.3}
$$

where

$$
\left.\begin{aligned}
&\frac{z^{K_j}}{2\pi i} \int_\Gamma r_j(t) \frac{t+z}{t-z} \frac{dt}{t} + \sum_{m=0}^{2K_j} c_m^{(j)} z^m \;, \\
&c_{2K_j-m}^{(j)} = -\overline{c_m^{(j)}} \;,\; m=0,1,\ldots,K_j \;, \text{ for } K_j \geq 0 \;, \\
&\frac{1}{\pi i} \int_\Gamma \frac{r_j(t) dt}{t^{-K_j}(t-z)} \;, \text{ for } K_j < 0 \;,\; j=1,2 \;,
\end{aligned}\right\} \tag{4.4}
$$

in which $c_m^{(j)}$ $(m=0,\ldots,K_j-1, j=1,2)$ are arbitrary complex constants, and $ic_{K_j}^{(j)}$ $(j=1,2)$ are arbitrary real constants.

Theorem 4.1 Let the second order complex equation (1.22) satisfy Condition C. Then the solution $w(z) \in W_{p_0}^2(D)$ $(2 < p_0 < p)$ to Problem B_1 for (1.22) can be expressed as

$$
w(z) = U(z) + X(z) \;,\; U(z) = \overline{\Phi_2(z)} + \overline{P_2}\Phi_1 \;,\; X(z) = H_1\rho + \overline{R_2}H_1\rho \;, \tag{4.5}
$$

where for $j=1,2$

$$\left.\begin{aligned}
&H_j = \frac{2}{\pi}\iint_D [\ell n|1-\frac{z}{\zeta}|\,\rho(\zeta)+g_j(z,\zeta)\overline{\rho(\zeta)}d\sigma_\zeta\,,\; P_j\rho = (H_j\rho)_z\,,\\
&g_j(z,\zeta) = \begin{cases} Re\bar{\zeta}^{2-K_j-2}\left[\ell n(1-\bar{\zeta}z) + \sum_{m=1}^{2K_j+1}\frac{1}{m}(\bar{\zeta}z)^m/m\right], & \text{for } K_j \ge 0\,,\\ Re\bar{\zeta}^{-2K_j-2}\,\ell n(1-\bar{\zeta}z)\,, & \text{for } K_j < 0\,,\end{cases}\\
&R_j\bar{\omega} = \begin{cases} \frac{-1}{2\pi i}\int_\Gamma \frac{\overline{\omega(t)}+\bar{t}z^{2K_j+1}\omega(t)}{t-z}dt\,, & \text{for } K_j \ge 0\,,\\ \frac{-1}{2\pi i}\int_\Gamma \frac{\overline{\omega(t)}+\bar{t}^{2|K_j|}\omega(t)}{t-z}dt\,, & \text{for } K_j < 0\,.\end{cases}
\end{aligned}\right\} \quad (4.6)$$

Proof. We substitute the solution $w(z)$ into the right-hand side of (1.22) and denote $\rho(z) = F(z,w,w_z,\bar{w}_z,w_{zz},\bar{w}_{zz})$. Because w_z is a solution of the first order complex equation $(w_z)_{\bar{z}} = \rho(z)$ satisfying the first formula of (4.2), the function w_z can be expressed as

$$w_z = \Phi_1(z) + P_1\rho\,. \quad (4.7)$$

Moreover, $\overline{w(z)}$ is a solution of the first order complex equation $\bar{w}_{\bar{z}} = \overline{\Phi_1(z)} + \overline{P_1\rho}$ satisfying the second formula of (4.2), so that $w(z)$ can be expressed as

$$\overline{w(z)} = \Phi_2(z) + P_2\overline{\Phi_1} + P_2\overline{P_1\rho}\,. \quad (4.8)$$

By the Green and the Pompeiu formulae, we have, if $K_2 \ge 0$,

$$P_2\overline{P_1\rho(z)} = -\frac{1}{\pi}\iint_D \frac{\overline{P_1\rho(t)}}{t-z}\,d\sigma_t - \frac{1}{\pi}\iint_D \frac{z^{2K_2+1}P_1\rho(t)}{1-\bar{t}z}\,d\sigma_t$$

$$= \overline{H_1\rho(z)} - \frac{1}{2\pi i}\int_\Gamma \frac{\overline{H_1\rho(t)}}{t-z}\,dt + \frac{1}{2\pi i}\int_\Gamma \frac{z^{2K_2+1}H_1\rho(t)}{1-\bar{t}z}\,d\bar{t}$$

$$= \overline{H_1\rho(z)} - \frac{1}{2\pi i}\int_\Gamma \frac{\overline{H_1\rho(t)}+\bar{t}z^{2K_2+1}H_1\rho(t)}{t-z}\,dt = \overline{H_1\rho(z)}$$

$$+ R_2\overline{H_1\rho(z)}\ , \text{ for } |z| < 1\ ,$$

and if $K_2 < 0$,

$$P_2\overline{P_1\rho(z)} = \overline{H_1\rho(z)} - \frac{1}{2\pi i}\int_\Gamma \frac{\overline{H_1\rho(t)}}{t-z}\,dt + \frac{1}{2\pi i}\int_\Gamma \frac{\bar{t}^{2|K_j|-1}H_1\rho(t)}{t-z}\,d\bar{t}$$

$$= \overline{H_1\rho(z)} + R_2\overline{H_1\rho(z)}\ , \text{ for } |z| < 1\ .$$

Thus (4.5) is derived.

It is clear that if $w_{z\bar{z}} \in L_{p_0}(\bar{D})$ and $w(z)$ satisfies the boundary condition (4.2), then $w(z)$ possesses the representation (4.5). □

In order to give estimations of solutions to Problem B_1 for (1.22), we need the following lemma.

<u>Lemma 4.2</u> Suppose that the function $\rho(z)$ satisfies $L_{p_0}[\rho,\bar{D}] \le M_1 < \infty$, $2 < p_0 < p$. Then $U(z)$, $X(z)$ in (4.5) satisfy the estimates

$$\left.\begin{aligned} &C^1_\beta[U,\bar{D}] \le M_2\ ,\ L_{p_0}[|U_{zz}|+|\bar{U}_{zz}|,\bar{D}] \le M_3\ ,\\ &C^1_\beta[X,\bar{D}] \le M_4\ .\ L_{p_0}[|X_{zz}|+|\bar{X}_{zz}|,\bar{D}] \le M_5\ ,\end{aligned}\right\} \tag{4.9}$$

where $\beta = 1 - 2/p_0$, $2 < p_0 < min(p, 1/(1-\alpha))$, $M_k = M_k(M_1, \alpha, \ell, D, c_m^{(j)})$ $(k=2,\dots,5)$, $c_m^{(j)}$ are arbitrary constants as stated in (4.4).

Proof. By Theorem 1.10, Theorem 1.11, and Theorem 1.32 in [97]2) we see that $U(z) = \overline{\Phi_2(z)} + \overline{P_2}\Phi_1$ satisfies the first two estimates in (4.9). For an arbitrary function $\rho(z)$ satisfying the condition $L_p[\rho, \overline{D}] \le M_1$, there is no harm in assuming $L_p[\rho, \overline{D}] \neq 0$. We denote $\rho^*(z) = \rho(z)/L_{p_0}[\rho, \overline{D}]$. If $K_2 \ge 0$, by Theorem 4.5 in Chapter 1, we have

$$C_\beta^1[H_1\rho^*, \overline{D}] \le M_6 \ , \ L_{p_0}[|H_1\rho^*)_{zz}| + |\overline{(H_1\rho^*)_{zz}}|, \overline{D}] \le M_7 \tag{4.10}$$

where $M_k = M_k(\alpha, \ell, D)$, $k = 6,7$. Putting

$$T\rho = -\frac{1}{\pi}\iint_D \frac{\rho(\zeta)}{\zeta - z} d\sigma_\zeta \ , \ T_1\rho = -\frac{z^{2K_2+1}}{\pi}\iint_D \frac{\rho(\zeta)}{1-\overline{\zeta}z} d\sigma_\zeta \ ,$$

$$r(t) = \overline{H_1\rho} + \overline{t}\, z^{2K_2+1} H_1\rho \ , \ \frac{\partial r}{\partial \theta} = r_\theta' = it\, r_t - i\overline{t}\, r_{\overline{t}} \ ,$$

we conclude

$$2\pi i(R_2\overline{H_1\rho})_z = \int_\Gamma \frac{i\overline{t}\, r_\theta'}{t-z} dt \ .$$

From

$$(H_1\rho)_t = P_1\rho = \overline{(\overline{H_1\rho})_{\overline{t}}} \ , \ (H_1\rho)_{\overline{t}} = \overline{P_1}\rho = \overline{(\overline{H_1\rho})_t} \ ,$$

we may only compute

$$I_1 = \int_\Gamma \frac{t^m T\rho(t)}{t-z} dt, I_2 = \int_\Gamma \frac{t^m T_1\rho(t)}{t-z} dt, I_3 = \int_\Gamma \frac{\overline{t^m T\rho(t)}}{t-z} dt, I_4 = \int_\Gamma \frac{\overline{t^m T_1\rho(t)}}{t-z} dt \ ,$$

where m is an integer. If $m \ge 0$ and $|z| < 1$, $|\zeta| < 1$, we have

$$I_1 = \int_\Gamma \frac{t^m}{t-z}\left[-\frac{1}{\pi}\iint_D \frac{\rho(\zeta)}{\zeta-t}\,d\sigma_\zeta\right]dt = -\frac{1}{\pi}\iint_D \rho(\zeta)\int_\Gamma \frac{t^m dt}{(t-z)(\zeta-t)}\,d\sigma_\zeta$$

$$= -\frac{1}{\pi}\iint_D \rho(\zeta)\int_\Gamma \frac{t^m}{\zeta-z}\left[\frac{1}{t-z}+\frac{1}{\zeta-t}\right]dt\,d\sigma_\zeta = \int_\Gamma \frac{t^m}{t-z}\,dt\,T\rho$$

$$-\frac{1}{\pi}\iint_D \frac{\rho(\zeta)}{\zeta-z}\left[\int_\Gamma \frac{-t^m}{t-\zeta}\,dt\right]d\sigma_\zeta = 2\pi i[z^m T\rho - T[\zeta^m\rho(\zeta)]] ,$$

and

$$I_2 = \int_\Gamma \frac{t^m}{t-z}\left(-\frac{t^{2K_2+1}}{\pi}\iint_D \frac{\overline{\rho(\zeta)}}{1-\bar{\zeta}t}\,d\sigma_\zeta\right)dt = -\frac{1}{\pi}\int_\Gamma \frac{t^{m+2K_2+1}}{t-z}\left[\iint_G \frac{\rho_1(\zeta_1)}{\zeta_1-t}\,d\sigma_\zeta\right]dt$$

$$= -\frac{1}{\pi}\iint_G \rho_1(\zeta_1)\left[\int_\Gamma \frac{t^{m+2K_2+1}}{(t-z)(\zeta_1-t)}\,dt\right]d\sigma_{\zeta_1}$$

$$= -\frac{1}{\pi}\iint_G \rho_1(\zeta_1)\int_\Gamma \frac{t^{m+2K_2+1}}{\zeta_1-z}\left(\frac{1}{t-z}+\frac{1}{\zeta_1-t}\right)dt\,d\sigma_{\zeta_1}$$

$$= 2\pi i\left[z^{m+2K_2+1}\left(-\frac{1}{\pi}\iint_G \frac{\rho_1(\zeta_1)}{\zeta_1-z}\,d\sigma_{\zeta_1}\right)\right] = 2\pi i\, z^{m+2K_2+1}\, T_1\rho ,$$

where $G = \{|\zeta_1| > 1\}$ and $\rho_1(\zeta_1) = \zeta_1|\zeta_1|^{-4}\,\overline{\rho(1/\overline{\zeta_1})}$. Using the same method we can prove

$$I_3 = \int_\Gamma \frac{\bar{t}^{m-1}}{t-z}\left[-\frac{1}{\pi}\iint_G \frac{\rho_1(\zeta_1)}{t-\zeta_1}\,d\sigma_{\zeta_1}\right]dt = -\frac{1}{\pi}\iint_G \frac{\rho_1(\zeta_1)}{\zeta_1-z}\int_\Gamma\left(\frac{-\bar{t}^{m-1}}{t-z}+\frac{\bar{t}^{m-1}}{t-\zeta_1}\right)dt\,d\sigma_{\zeta_1}$$

$$= -\int_\Gamma \frac{t^{1-m}}{t-z}\,dt\left(-\frac{1}{\pi}\iint_D \frac{\overline{\rho(\zeta)}}{1-\bar{\zeta}z}\,d\sigma_\zeta\right) - \frac{1}{\pi}\iint_D \frac{\bar{\zeta}\,\overline{\rho(\zeta)}}{1-\bar{\zeta}z}\int_\Gamma \frac{t^{1-m}}{\bar{\zeta}t-1}\,dt\,d\sigma_\zeta ,$$

and

$$I_4 = \int_\Gamma \frac{\bar{t}^{m+2K_2}}{t-z}\left(-\frac{1}{\pi}\iint_D \frac{\rho(\zeta)}{t-\zeta}\,d\sigma_\zeta\right)dt = -\frac{1}{\pi}\iint_D \rho(\zeta)\int_\Gamma \frac{\bar{t}^{m+2K_2}}{\zeta-z}\left(\frac{-1}{t-z}+\frac{1}{t-\zeta}\right)dt\,d\sigma_\zeta$$

$$= -\int_\Gamma \frac{\bar{t}^{m+2K_2}}{t-z}\,dt\,T\rho - \frac{1}{\pi}\iint_D \frac{\rho(\zeta)}{\zeta-z}\int_\Gamma \frac{\bar{t}^{m+2K_2}}{t-\zeta}\,dt\,d\sigma_\zeta .$$

By similar computations for $m < 0$ we can obtain the last two estimates, too. In addition we may also deal with the case $K_2 < 0$. □

Theorem 4.3 Let $w(z)$ be a solution for Problem B_1 of (1.22) as stated in (4.5) satisfying

$$L_{p_0}[\rho,\bar{D}] = L_{p_0}[w_{z\bar{z}},\bar{D}] < M_8 , \tag{4.11}$$

where M_8 is an appropriate constant. If the constant ε in (1.20), (1.23) is sufficiently small, then $w(z)$ satisfies the estimates

$$C^1_\beta[w,\bar{D}] \le M_9 , \; L_{p_0}[\rho,\bar{D}] < M_{10} , \; L_{p_0}[w_{zz},\bar{D}] \le M_{11} , \tag{4.12}$$

where β , p_0 are as stated in (4.9), $M_k = M_k(q_0,p_0,k_0,\alpha,\ell,D,c_m^{(j)})$, $k = 9,10,11$.

Proof. Substituting the solution $w(z)$ into the complex equation (1.22), we obtain

$$w_{z\bar{z}} = q_1 w_{zz} + A_1 w_z + A_2 \bar{w}_{\bar{z}} + A , A = q_2 \bar{w}_{zz} + A_3 \bar{w}_z + A_4 w_{\bar{z}} + A_5 w + A_6 \bar{w} + A_7 , \tag{4.13}$$

where

$$q_1 = \begin{cases} [F(z,w,w_z,\bar{w}_z,w_{zz},\bar{w}_{zz}) - F(z,w,w_z,\bar{w}_z,0,\bar{w}_{zz})]/w_{zz} , & \text{for } w_{zz} \neq 0 , \\ 0 , \text{ for } w_{zz} = 0 , \end{cases}$$

$$q_2 = \begin{cases} [F(z,w,w_z\bar{w}_z,0,\bar{w}_{zz}) - F(z,w,w_z,\bar{w}_z,0,0)]/\bar{w}_{zz} , & \text{for } w_{zz} \neq 0 , \\ 0 , \text{ for } \bar{w}_{zz} = 0 . \end{cases}$$

By Lemma 4.2 and (4.11), the solution $w(z)$ satisfies the estimates

$$C^1_\beta[w,\bar{D}] \le M_{12} , \; L_{p_0}[\bar{w}_{zz},\bar{D}] \le M_{13} , \tag{4.14}$$

in which $M_k = M_k(\alpha,\ell,D,c_m^{(j)},M_8)$, $k = 12,13$. It is not difficult to see that no matter how large the constant M_8 in (4.11) is, provided the constant ε in (1.20), (1.23) is small enough, we have

$$L_{p_0}[A,\bar{D}] \le M_{14} = M_{14}(q_0,p_0,k_0,\alpha,\ell,D,c_m^{(j)}) . \tag{4.15}$$

Set $V(z) = w_z$, the second order complex equation (1.22) can be reduced to the first order complex equation

$$V_{\bar{z}} = q_1V_z + A_1V + A_2\bar{V} + A . \tag{4.16}$$

Noting that $q_1(z)$, $A_1(z)$, $A_2(z)$ satisfy the conditions $|q_1(z)| \le q_0 < 1$, $L_{p_0}[A_j,\bar{D}] \le k_0$, $j=1,2$, and according to Theorem 4.3 in Chapter 2, $V(z) = w_z$ satisfies the estimates

$$C_\beta[V,\bar{D}] \le M_{15} , \; L_{p_0}[|V_{z\bar{z}}|+|V_{zz}|,\bar{D}] < M_{16} , \tag{4.17}$$

where $M_k = M_k(q_0,p_0,k_0,\alpha,\ell,D,c_m^{(j)})$, $k=15,16$. Because $\overline{w(z)}$ is a solution of $\bar{w}_{\bar{z}} = \overline{V(z)}$ and satisfies the second formula in (4.2) and (4.3), we also have

$$C_\beta[w,\bar{D}] \le M_{17} , \; L_{p_0}[|w_{\bar{z}}|+|w_z|,\bar{D}] \le M_{18} , \tag{4.18}$$

where $M_k = M_k(q_0,p_0,k_0,\alpha,\ell,D,c_m^{(j)})$, $k=17,18$. From (4.17), (4.18), the last two estimates in (4.12) follow. By the representation (4.5), we conclude the first estimate in (4.12). Moreover, we may choose $M_8 = M_{10}$ in (4.11). □

2. Solvability of the Riemann-Hilbert problem for second order complex equations

Theorem 4.4 Let the second order complex equation (1.22) satisfy Condition C and the constant ε in (1.20), (1.23) be sufficiently small. Then Problem B_1 of (1.22) possesses a solution of the type (4.5).

Proof. We first assume that the coefficients $Q_j (j=1,\ldots,4)$, $A_j (j=1,\ldots,7)$ of (1.22) are equal to zero in the neighbourhood

$\overline{D}\setminus D_n$ of Γ, where $D_n = \{|z| \le 1 - 1/n\}$, n is a positive integer. In this case, the complex equation (1.22) can be written in the form (3.22). As in (3.21), the solution $w(z)$ of Problem B_1 for (3.22) may be expressed as $w(z) = V(z) + H_0\rho$. We consider the integral equation with the parameter $k (0 \le k \le 1)$

$$\rho^*(z) = k\, F_n(z, w, w_z, \bar{w}_z, V_{zz} + \Pi\rho^*, \overline{V}_{zz} + \Pi\bar{\rho}^*), \tag{4.19}$$

where

$$w(z) = U(z) + X(z) = V(z) + H_0\rho,$$

$$w_z = \Phi_1(z) + P_1\rho = V_z + T\rho, \quad T\rho = -\frac{1}{\pi}\iint_{D_n} \frac{\rho(\zeta)}{\zeta - z}\, d\sigma_\zeta,$$

$$w_{zz} = V_{zz} + \Pi\rho, \quad \Pi\rho = -\frac{1}{\pi}\iint_{D_n} \frac{\rho(\zeta)}{(\zeta - z)^2}\, d\sigma_\zeta,$$

and $\rho(z)$ is an arbitrary function in the bounded and open set B_M satisfying the condition

$$L_{p_0}[\rho, D_n] < M_8, \tag{4.20}$$

in which M_8 is the constant in(4.11). By Lemma 4.2, we see that $V_{zz}, \overline{V}_{zz} \in L_{p_0}(\overline{D})$. Applying the principle of contraction the integral equation (4.19) has a unique solution $\rho^*(z) \in L_{p_0}(\overline{D})$. Denote by $\rho^* = S(\rho, k)$ the mapping from $\rho(z) \in B_M$ onto $\rho^*(z)$. Similarly to the proof of Theorem 3.2, we can verify that $\rho^* = S(\rho, k)$ satisfies the conditions of the Leray-Schauder theorem. Hence the integral equation

$$\rho(z) = k\, F_n[z, w, w_z, \bar{w}_z, V_{zz} + \Pi\rho, \overline{V}_{zz} + \Pi\bar{\rho}] \tag{4.21}$$

with $k = 1$ has a solution $\rho(z) \in B_M$, and $w(z) = U(z) + X(z)$ is a solution for Problem B_1 of (3.22). □

Moreover, similarly to the proof of Theorem 3.2, we can derive that Problem B_1 for (1.22) has a solution of the type (4.5), and obtain the following result.

Theorem 4.5 Under the hypothesis in Theorem 4.4, the solvability of Problem A_1 for (1.22) is as follows.

(1) If $K_1 \geq 0$, $K_2 \geq 0$, Problem A_1 has a solution of the type (4.5).

(2) If $K_1 < 0$, $K_2 \geq 0$ (or $K_1 \geq 0, K_2 < 0$), Problem A_1 has $-2K_1-1$ (or $-2K_2-1$) solvability conditions as stated in (3.17), where

$$\tau_j(t) = \begin{cases} r_1(t), & j=1, \\ r_2(t), & j=2, \end{cases} \quad \omega_j(z) = \begin{cases} w_{z\bar{z}}, & j=1, \\ \bar{w}_{\bar{z}}, & j=2. \end{cases}$$

(3) If $K_1 < 0$, $K_2 < 0$, Problem A_1 has $-2(K_1+K_2+1)$ solvability conditions as stated in (2), $j=1,2$.

Finally, we discuss the Riemann-Hilbert Problem A for (1.22) in an (N+1)-connected circular domain D with the boundary conditions

$$\left.\begin{aligned} &Re[\overline{\lambda_1(t)w_t + \varepsilon\beta_1(t)w(t)}] = r_1(t), \quad t \in \Gamma, \\ &Re[\overline{\lambda_2(t)}\ \overline{w(t)}] = r_2(t), \quad t \in \Gamma, \end{aligned}\right\} \tag{4.22}$$

where $|\lambda_j(t)| = 1$, $C_\alpha^{j-1}[\lambda_j,\Gamma] \leq \ell$, $C_\alpha[\beta_1,\Gamma] \leq \ell$, $C_\alpha^{j-1}[r_j,\Gamma] \leq \ell$, $j=1,2;\alpha,\ell$ are as stated in (3.2). We need to propose the corresponding modified boundary value Problem B for (1.22) with the boundary conditions

$$\left.\begin{aligned} &Re[\overline{\lambda_1(t)w_t}] = \tau_1(t) + h_1(t), \quad t \in \Gamma, \\ &Re[\overline{\lambda_2(t)}\ \overline{w(t)}] = \tau_2(t) + h_2(t), \quad t \in \Gamma, \end{aligned}\right\} \tag{4.23}$$

in which $\tau_1(t) = -\varepsilon Re[\beta_1(t)w(t)] + r_1(t)$, $\tau_2(t) = r_2(t)$. A solution $w(z)$ to Problem A for (1.22) can be expressed as

$$w(z) = \overline{\Phi_2(z)} + \tilde{T}_2\overline{\Phi_1} + \tilde{T}_2\overline{\tilde{T}_1\rho}\ ,\ \rho(z) = w_{z\bar{z}}\ , \qquad (4.24)$$

where $\Phi_1(z) + \tilde{T}_1\rho$, $\Phi_2(z) + \tilde{T}_2\bar{w}_{\bar{z}}$ satisfy the first and second conditions in (4.22), respectively, $\Phi_j(z)$, $\tilde{T}_j\omega_j$ $(j=1,2)$ are as stated in (3.29), where $\omega_2(z) = \bar{w}_{\bar{z}}$, $\omega_1(z) = w_{z\bar{z}} = \rho(z)$.

Using the method of a priori estimates for the solutions and the Leray-Schauder theorem, we can obtain the following theorem.

Theorem 4.6 Let the second order complex equation (1.22) satisfy Condition C in the (N+1)-connected circular domain D , and the constant ε in (1.20), (1.23), (4.22) be small enough. Then the following statements hold.

(1) If the indices $K_1 \geq N$, $K_2 \geq N$, then Problem A for (1.22) is solvable.

(2) If $0 \leq K_j < N$ $(j=1,2)$, the total number of solvability conditions for Problem A is not greater than $2N-K_1-K_2$.

(3) If $K_1 < 0$, $K_2 < 0$, Problem A has $2(N-K_1-K_2-1)$ solvability conditions.

For other cases, we can also give corresponding results.

Besides, using the Fredholm theorem for linear operator equations, we can discuss the solvability of Problem A for the linear second order complex equation (1.6).

§ 5. Boundary value problems for another class of second order complex equations

In this section, we discuss the second order complex equation

$$\left.\begin{aligned} & w_{\bar{z}\bar{z}} = F(z,w,w_{\bar{z}},w_z,w_{z\bar{z}},w_{zz})\ , \\ & F = Q_1w_{z\bar{z}} + Q_2\bar{w}_{z\bar{z}} + Q_3w_{zz} + Q_4\bar{w}_{\bar{z}\bar{z}} + A_1w_{\bar{z}} + A_2\bar{w}_z + A_3w_z \\ & \quad + A_4\bar{w}_{\bar{z}} + A_5w + A_6\bar{w} + A_7\ , \end{aligned}\right\} \qquad (5.1)$$

where $Q_j = Q_j(z,w,w_{\bar{z}},w_z,w_{z\bar{z}},w_{zz})$, $j=1,\ldots,4$, $A_j = A_j(z,w,w_{\bar{z}},w_z)$, $j=1,\ldots,7$. For the sake of convenience, we assume that D is an $(N+1)$-connected circular domain and $F(z,w,w_{\bar{z}},w_z,U,V)$ satisfies a corresponding Condition C as stated in § 1. If $Q_j = Q_j(z)$, $j=1,\ldots,4$, $A_j = A_j(z)$, $j=1,\ldots,7$ in (5.1), then the complex equation becomes the linear equation

$$\left.\begin{aligned} w_{\bar{z}\bar{z}} &= Q_1(z)w_{z\bar{z}} + Q_2(z)\bar{w}_{z\bar{z}} + Q_3(z)w_{zz} + Q_4(z)\bar{w}_{\bar{z}\bar{z}} + F_0 , \\ F_0 &= A_1(z)w_{\bar{z}} + A_2(z)\bar{w}_z + A_3(z)w_z + A_4(z)\bar{w}_{\bar{z}} + A_5(z)w + A_6(z)\bar{w} + A_7(z) . \end{aligned}\right\} \tag{5.2}$$

For this case, the condition (1.20) may be replaced by the following condition

$$|Q_1(z)| + |Q_2(z)| \le q_0 , \quad |Q_3(z)| + |Q_4(z)| \le q_0' , \tag{5.3}$$

where $q_0 + q_0' < 1$, $q_0' < \varepsilon$. It is easy to see that condition (1.20) can be derived from the condition (5.3). If $Q_1(z) \ne Q_2(z)$ in (5.2), then we can solve (5.2) for $w_{z\bar{z}}$, and get

$$\begin{aligned} w_{z\bar{z}} &= \frac{1}{|Q_1(z)|^2 - |Q_2(z)|^2} \begin{vmatrix} Q_1(z) & w_{\bar{z}\bar{z}} - Q_3(z)w_{zz} - Q_4(z)\bar{w}_{\bar{z}\bar{z}} - F_0 \\ \overline{Q_2(z)} & \bar{w}_{zz} - \overline{Q_4(z)}w_{zz} - \overline{Q_3(z)}\bar{w}_{\bar{z}\bar{z}} - \overline{F}_0 \end{vmatrix} \\ &= Q_1^*(z)w_{zz} + Q_2^*(z)\bar{w}_{\bar{z}\bar{z}} + Q_3^*(z)\bar{w}_{zz} + Q_4^*(z)w_{\bar{z}\bar{z}} + F_0^*(z,w,w_z,\bar{w}_z) \end{aligned} \tag{5.4}$$

where $Q_1^*(z) = [\overline{Q_2}Q_3 - Q_1\overline{Q_4}]/[|Q_1|^2 - |Q_2|^2]$, $Q_2^*(z) = [\overline{Q_2}Q_4 - Q_1\overline{Q_3}]/[|Q_1|^2 - |Q_2|^2]$, $Q_3^*(z) = Q_1/[|Q_1|^2 - |Q_2|^2]$, $Q_4^* = -\overline{Q_2}/[|Q_1|^2 - |Q_2|^2]$. It is clear that if (5.2) satisfies the condition (5.3), then the coefficients $Q_j^*(z)$ $(j=1,\ldots,4)$ may not satisfy

$$|Q_1^*(z)| + |Q_2^*(z)| \le q_0 , \quad |Q_3^*(z)| + |Q_4^*(z)| \le q_0' , \quad q_0 + q_0' < 1 . \tag{5.5}$$

This shows that the complex equation (5.2) satisfying (5.3) is different from the complex equation (5.4) satisfying the condition (5.5).

B. Bojarski [15]4) divided the elliptic systems of second order equattions (1.1) into three classes E_1 , E_2 , and E_3 . A complex equation of class E_2 can be written in the form (5.4) and essentially the classes E_1 and E_3 correspond to complex equations of the type (5.2).

Problem P. The oblique derivative problem P for the complex equation (5.1) is to find a continuously differentiable solution $w(z)$ on $\overline{D}$ satisfying the boundary condition

$$Re[\overline{\lambda_1(t)}V_j(t) + \varepsilon\beta_j(t)w(t)] = \tau_j(t), V_1(t) = w_{\bar{t}}\,, V_2(t) = w_t\,, t \in \Gamma\,, \tag{5.6}$$

where $\lambda_j(t)$, $\beta_j(t)$, $\tau_j(t)$ $(j=1,2)$ satisfy the condition (3.2). We need to introduce the corresponding modified problem Q for the first order systems of complex equations

$$V_{1\bar{z}} = F(z,w,V_1,V_2,V_{1z},V_{2z})\ , \ V_{2\bar{z}} = V_{1z} \tag{5.7}$$

with the boundary conditions

$$Re[\overline{\lambda_j(t)}V_j(t)] = \tilde{\tau}_j(t) + h_j(t)\ , \tag{5.8}$$

$$\tilde{\tau}_j(t) = -\varepsilon\, Re[\beta_j(t)w_j(t)] + \tau_j(t)\ , \ t \in \Gamma\ , \ j=1,2\ ,$$

$$Im[\overline{\lambda_j(a_k)}V_j(a_k) + \varepsilon\beta_j(a_k)w(a_k)] = b_{jk}\ ,$$

$$k \in \{k\} = \begin{cases} 1,\ldots,2K_j-N+1\ , K_j \geq N\ , \\ N-K_j+1,\ldots,N+1\ , 0 \leq K_j < N\ , \end{cases} \quad j=1,2\ , \tag{5.9}$$

and the relation

$$w(z) = w_0 + \int_0^z \{V_1(\zeta)d\bar{\zeta} + [V_2(\zeta) + \sum_{m=1}^{N} \frac{d_m}{\zeta - z_m}]d\zeta\} \tag{5.10}$$

where $h_j(t)$, a_k , b_{jk} , w_0 , d_m are the same as in (3.25) - (3.28) .

We first give the representation and estimates of solutions for Problem Q of (5.7). Afterwards, we prove the solvability of Problem Q for (5.7) satisfying some conditions.

Theorem 5.1 Let $[w(z), V_1(z), V_2(z)]$ be a solution to Problem Q for the first order system (5.7), and $V_{1\bar{z}} = \rho \in L_{p_0}(\bar{D})$, $2 < p_0 < min(p, 1/(1-\alpha))$. Then $w(z)$ can be expressed as in (5.10), where

$$V_1(z) = \Phi_1(z) + \tilde{T}_1\rho\ ,\ V_2(z) = \Phi_2(z) + \tilde{T}_2[\Phi_1' + \tilde{S}_1\rho]\ , \tag{5.11}$$

where $\tilde{S}_1\rho = (\tilde{T}_1\rho)_z$, $\Phi_1'(z) \in L_{p_0}(\bar{D})$, $\Phi_j(z)$, $\tilde{T}_j\omega_j (j=1,2)$ are as stated in (3.29), $w_1(z) = \rho(z)$, $w_2(z) = \Phi'(z) + \tilde{S}_1\rho$, and $\tilde{T}_j\omega_j$ satisfies the homogeneous boundary condition

$$Re[\overline{\lambda_j(t)}\tilde{T}_j\omega_j] = h_j(t)\ ,\ t \in \Gamma\ ,\ j=1,2\ . \tag{5.12}$$

Besides, $V_1(z)$, $V_2(z)$ can be written in the forms

$$V_1(z) = \Psi_1(z) + T\rho\ ,\ V_2(z) = \Psi_2(z) + T(\Psi_1' + \Pi\rho)\ , \tag{5.13}$$

in which $\Psi_j(z)$ $(j=1,2)$ are analytic in D, $\Psi_j'(z) \in L_{p_0}(\bar{D})$, $j=1,2$, $T\rho = -\frac{1}{\pi}\iint_D \frac{\rho(\zeta)}{\zeta - z} d\sigma_\zeta$, $\Pi\rho = -\frac{1}{\pi}\iint_D \frac{\rho(\zeta)}{(\zeta - z)^2} d\sigma_\zeta$.

Proof. Noting that $V_{1\bar{z}} = \rho(z) \in L_{p_0}(\bar{D})$ and that $V_1(z)$ satisfies the boundary conditions (5.8), (5.9) for $j=1$ and using Theorem 3.6 in Chapter 1, $V_1(z)$ may be expressed as in the first formula in (5.11). We can prove $\Phi_1'(z)$, $\tilde{S}_1\rho = (\tilde{T}\rho)_z \in L_{p_0}(\bar{D})$. Moreover, $V_{2\bar{z}} = V_{1z} = \Phi_1'(z) + \tilde{S}_1\rho$ satisfies the boundary condition (5.8), (5.9) for $j=2$, so that $V_2(z)$ may be expressed as in the second formula in (5.11). Similarly, $V_1(z)$, $V_2(z)$ can be written as in (5.13). □

Theorem 5.2 Suppose that the complex equation (5.1) satisfies Condition C in D, and the constant ε in (5.8), (5.9) and in

Condition C is sufficiently small. Then the solution $[w(z), V_1(z), V_2(z)]$ of Problem Q satisfies the estimates

$$L_j = C_\beta[V_j, \bar{D}] + L_{p_0}[|V_{j\bar{z}}| + |V_{jz}|, \bar{D}] \le M_1 \ , \ j=1,2 \ , \tag{5.14}$$

$$S = C^1_\beta[w, \bar{D}] \le M_2 \ , \tag{5.15}$$

where $\beta = 1-2/p_0$, $2 < p_0 < min(p, 1/(1-\alpha))$, $M_k = M_k(q_0, p_0, k_0, \alpha, \ell, D)$, $k=1,2$.

<u>Proof.</u> From (5.10), it is not difficult to derive

$$S \le M_3(L_1+L_2) \ , \tag{5.16}$$

where $M_3 = M_3(p_0, D)$. Hence, we prove only (5.14). Let the solution $[w(z), V_1(z), V_2(z)]$ be substituted into (5.7) - (5.9). Rewrite (5.7) in the form

$$\left.\begin{aligned} V_{1\bar{z}} &= Q_1 V_{1z} + Q_2 \overline{V}_{1\bar{z}} + A_1 V_1 + A_2 \overline{V_1} + A \ , \\ A &= Q_3 V_{2z} + Q_4 \overline{V}_{2\bar{z}} + A_3 V_2 + A_4 \overline{V_2} + A_5 w + A_6 \bar{w} + A_7 \ . \end{aligned}\right\} \tag{5.17}$$

Similarly to (3.34) and (3.37) we have

$$\begin{aligned} L_{p_0}[A, \bar{D}] &\le q_0' \ L_{p_0}[V_{2z}, \bar{D}] + 2\varepsilon k_0 \{C[V_2, \bar{D}] + C[w, \bar{D}]\} + k_0 \\ &\le \varepsilon(1+2k_0) L_2 + 2\varepsilon \ k_0 M_3 (L_1+L_2) + k_0 = k_1 \ . \end{aligned}$$

and

$$C_\alpha[\varepsilon\beta_j w - \tau, \Gamma] \le \varepsilon\ell C_\alpha[w, \Gamma] + \ell \le \varepsilon\ell M_3(L_1+L_2) + \ell = \ell_1 \ , \ j=1,2 \ .$$

It is easy to see that $V_1^*(z) = V(z)/(k_1+\ell_1)$ is a solution of the complex equation

$$V^*_{1\bar{z}} = Q_1 V^*_{1z} + Q_2 \overline{V^*_{1\bar{z}}} + A_1 V^*_1 + A_2 \overline{V^*_1} + A/(k_1+\ell_1) \tag{5.18}$$

and satisfies the boundary conditions

$$Re[\overline{\lambda_1(t)}V_1^*(t)] = \tilde{\tau}(t)/(k_1+\ell_1) + h_1(t) \ , \ t \in \Gamma \ , \tag{5.19}$$

$$Im[\overline{\lambda_1(a_k)}V_1^*(a_k)] = \{b_{1k} - Im[\varepsilon\beta_1(a_k)w(a_k)]\}/(k_1+\ell_1) \ , k \in \{k\} \ . \tag{5.20}$$

By Theorem 4.3 in Chapter 2, $V_1(z)$ satisfies the estimate

$$L_1 \le M_4[k_1+\ell_1] \le \varepsilon M_4[(1+2k_0)L_2 + M_3(2k_0+\ell)(L_1+L_2)] + M_4(k_0+\ell) \ , \tag{5.21}$$

where $M_4 = M_4(q_0,p_0,k_0,\alpha,\ell,D)$. Provided the constant ε is so small that $1 - \varepsilon M_3M_4(2k_0+\ell) > 0$ holds, we obtain

$$\begin{aligned} L_1 &\le \{\varepsilon M_4[(1+2k_0)+M_3(2k_0+\ell)]L_2+M_4(k_0+\ell)\}/[1-\varepsilon M_3M_4(2k_0+\ell)] \\ &= \varepsilon M_5L_2 + M_6 \ . \end{aligned} \tag{5.22}$$

Noting that $V_{2\bar{z}} = V_{1z}$ $(L_{p_0}[V_{1z},\bar{D}] \le L_1)$ and $V_2(z)$ satisfies the corresponding boundary conditions, similarly we conclude

$$L_2 \le M_7(L_1+\ell_1) = M_7L_1 + \varepsilon\ell M_3M_7(L_1+L_2) + \ell M_7 \ , \tag{5.23}$$

where $M_7 = M_7(p_0,\alpha,\ell,D)$. Choosing the constant ε so small, that $1 - \varepsilon\ell M_3M_7 > 0$ holds,

$$L_2 \le [M_7(1+\varepsilon\ell M_3)L_1+\ell M_7]/[1-\varepsilon\ell M_3M_7] = M_8L_1 + M_9 \ , \tag{5.24}$$

can be derived. Combining (5.22) and (5.24), we have

$$L_1 \le \varepsilon M_5M_8 \ L_1 + \varepsilon M_5M_9 + M_6 \ , \text{ i.e. } L_1 \le [\varepsilon M_5M_9+M_6]/[1-\varepsilon M_5M_8] \ . \tag{5.25}$$

From (5.24), (5.25) and if $1 - \varepsilon M_5M_8 > 0$ it follows that (5.14) holds. □

By Theorem 5.1, Theorem 5.2, and using the Leray-Schauder theorem we can prove the following theorem.

Theorem 5.3 Under the same conditions as in Theorem 5.2, Problem Q for the first order system (5.7) is solvable. Moreover, concerning the solvability of Problem P for the second order complex equation (5.1) there is a similar result as stated in Theorem 3.5.

The linear complex equation (5.2) is rewritten in the form

$$\left.\begin{aligned} w_{\bar z\bar z} &= Q_1(z)w_{z\bar z} + Q_2(z)\bar w_{z\bar z} + Q_3(z)w_{zz} + Q_4(z)\bar w_{\bar z\bar z} + \\ &\quad + A_1(z)w_{\bar z} + A_2(z)\bar w_z + \varepsilon g(z,w,w_z) + A_7(z) , \\ g(z,w,w_z) &= A_3(z)w_z + A_4(z)\bar w_{\bar z} + A_5(z)w + A_6(z)\bar w , \end{aligned}\right\} \tag{5.26}$$

where $Q_j(z)$ $(j=1,\ldots,4)$, $A_j(z)$ $(j=1,\ldots,7)$ satisfy (5.3) and $L_p[A_j,\bar D] \le \ell$, $2 < p < \infty$, and $\varepsilon(-\infty < \varepsilon < \infty)$ is a parameter. The boundary value problem (5.6) for (5.26) is called Problem P.

According to the method of proof for Theorem 3.6 and Theorem 3.7, we can obtain the following theorem.

Theorem 5.4 If the constant q_0' in (5.3) is small enough, then the solvability of Problem P for the second order complex equation (5.26) is the same as in Theorems 3.6 and 3.7 (cf. [103]25)).

VI Boundary value problems for elliptic systems of several equations

In the previous chapters, we considered the boundary value problems for elliptic equations and elliptic systems of two equations. Now, we first transform elliptic systems of several equations of first order and second order into complex forms, and then discuss some boundary value problems for systems of several complex equations of first and second order.

§ 1. Reductions of elliptic systems of several equations of first and second order

1. Reduction of elliptic systems of several first order equations

First of all, we discuss the linear elliptic system of $2n$ first order equations

$$\sum_{k=1}^{2n} (a_{jk}u_{kx}+b_{jk}u_{ky}) = \sum_{k=1}^{2n} c_{jk}u_k+d_j \ , \ j=1,\dots,2n \ , \tag{1.1}$$

where $a_{jk}=a_{jk}(x,y)$, $b_{jk}=b_{jk}(x,y)$, $c_{jk}=c_{jk}(x,y)$, $d_j=d_j(x,y)$ $(j,k=1,\dots,2n)$ are continuous functions on a bounded and closed domain $\overline{D}$. The system (1.1) is called elliptic, if the determinant

$$K(\lambda) = |A+B\lambda| > 0 \quad \text{in} \quad D \tag{1.2}$$

for any real number λ , in which

$$A = (a_{jk}) = \begin{pmatrix} a_{11} & \cdots & a_{12n} \\ \vdots & \ddots & \vdots \\ a_{2n1} & \cdots & a_{2n2n} \end{pmatrix}, \ B = (b_{jk}) = \begin{pmatrix} b_{11} & \cdots & b_{12n} \\ \vdots & \ddots & \vdots \\ b_{2n1} & \cdots & b_{2n2n} \end{pmatrix}.$$

It is evident that there exists a positive number δ , such that

$$|A| \geq \delta \ , \quad \max_{j,k=1,\dots,2n} (|a_{jk}|, |b_{jk}|) \leq \frac{1}{\delta} \quad \text{on } \overline{D} \ . \tag{1.3}$$

We construct a linear transformation

$$\xi = x-ty \ , \ \eta = x+ty \ , \tag{1.4}$$

where t is an appropriate positive constant. One can easily obtain

$$u_{kx} = u_{k\xi} + u_{k\eta} \ , \ u_{ky} = (-u_{k\xi}+u_{k\eta})t \ , \ k=1,\dots,2n \ .$$

Substituting u_{kx} , $u_{ky}(k=1,\dots,2n)$ into (1.1), the system (1.1) reduces to

$$\sum_{k=1}^{2n} [(a_{jk}-b_{jk}t)u_{k\xi} + (a_{jk}+b_{jk}t)u_{k\eta}] = \sum_{k=1}^{2n} c_{jk}u_k + d_j \ , \ j=1,\dots,2n \ . \tag{1.5}$$

Setting $w_k = u_k + iu_{n+k}$, $k=1,\dots,n$, $\zeta = \xi+i\eta$, according to (1.8) in Chapter 2 , we have

$$\left.\begin{aligned} u_{k\xi} &= \frac{1}{2}(w_{k\zeta}+\overline{w_{k\bar\zeta}}+w_{k\bar\zeta}+\overline{w_{k\zeta}}) \ , \ u_{k\eta} = \frac{i}{2}(w_{k\zeta}-\overline{w_{k\bar\zeta}}-w_{k\bar\zeta}+\overline{w_{k\zeta}}) \ , \\ u_{n+k\xi} &= \frac{i}{2}(-w_{k\zeta}+\overline{w_{k\bar\zeta}}-w_{k\bar\zeta}+\overline{w_{k\zeta}}) \ , \ u_{n+k\eta} = \frac{1}{2}(w_{k\zeta}+\overline{w_{k\bar\zeta}}-w_{k\bar\zeta}-\overline{w_{k\zeta}}) \ . \end{aligned}\right\} \tag{1.6}$$

Hence, the system (1.5) can be written in the complex form

$$\left.\begin{aligned} &\sum_{k=1}^{n} [(a_{jk}-ia_{jk}-b_{jk}t-ib_{jk}t-ia_{jn+k}-a_{jn+k}+ib_{jn+k}t-b_{jn+k}t)w_{k\bar\zeta} \\ &\quad + (a_{jk}+ia_{jk}-b_{jk}t+ib_{jk}t+ia_{jn+k}-a_{jn+k}-ib_{jn+k}t-b_{jn+k}t)\overline{w_{k\zeta}}] \\ &= g_j(t,\zeta,w_1,\dots,w_n,w_{1\zeta},\dots,w_{n\zeta}) \ , \ j=1,\dots,2n \ . \end{aligned}\right\} \tag{1.7}$$

Notice that the determinant of coefficients of the above algebraic system is

$$\begin{vmatrix} \frac{a_{11}-b_{1n+1}t}{(1-i)^{-1}} & -\frac{b_{11}t+a_{1n+1}}{(1+i)^{-1}} & \cdots & \frac{a_{1n}-b_{12n}t}{(1+i)^{-1}} & -\frac{b_{1n}t+a_{12n}}{(1-i)^{-1}} \\ \vdots & \vdots & & \vdots & \vdots \\ \frac{a_{2n1}-b_{2nn+1}t}{(1-i)^{-1}} & -\frac{b_{2n1}t+a_{2nn+1}}{(1+i)^{-1}} & \cdots & \frac{a_{2nn}-b_{2n2n}t}{(1+i)^{-1}} & -\frac{b_{2nn}t+a_{2n2n}}{(1-i)^{-1}} \end{vmatrix} \tag{1.8}$$

$$= (2i)^n \begin{vmatrix} a_{11}-b_{11}t-a_{1n+1}-b_{1n+1}t & \cdots & a_{1n}+b_{1n}t+a_{12n}-b_{12n}t \\ \vdots & \ddots & \vdots \\ a_{2n1}-b_{2n1}t-a_{2nn+1}-b_{2nn+1}t & \cdots & a_{2nn}+b_{2nn}t+a_{2n2n}-b_{2n2n}t \end{vmatrix}$$

$$= (4i)^n \begin{vmatrix} a_{11}-b_{1n+1}t & \cdots & a_{12n}+b_{1n}t \\ \vdots & \ddots & \vdots \\ a_{2n1}-b_{2nn+1}t & \cdots & a_{2n2n}+b_{2nn}t \end{vmatrix} = (4i)^n\ |A+Ct|$$

$$= (4i)^n\ [\,|A|+|E|t\,]\ ,$$

where

$$C = \begin{pmatrix} -b_{1n+1} & \cdots & -b_{12n} & b_{11} & \cdots & b_{1n} \\ \vdots & \ddots & \vdots & \vdots & \ddots & \vdots \\ -b_{2nn+1} & \cdots & -b_{2n2n} & b_{2n1} & \cdots & b_{2nn} \end{pmatrix} .$$

By (1.3), we may choose a sufficiently small constant t, so that $|A| + |E|\ t > 0$. Hence, we can solve (1.7) for $w_{\bar\zeta}$ and obtain the system

$$w_{k\bar\zeta} = F_k(\zeta,w_1,\ldots,w_k,w_{1\zeta},\ldots,w_{k\zeta})\ ,\ k=1,\ldots,n \tag{1.9}$$

of complex equations. Moreover, we may write (1.9) in matirx form,

$$w_{\bar\zeta} = F(\zeta,w,w_\zeta)\ ,\ F = Q^{(1)}w_\zeta + Q^{(2)}\bar w_{\bar\zeta} + A^{(1)}w + A^{(2)}\bar w + A^{(3)}\ , \tag{1.10}$$

where

$$w=\begin{pmatrix} w_1 \\ \vdots \\ w_n \end{pmatrix},\ F=\begin{pmatrix} F_1 \\ \vdots \\ F_n \end{pmatrix},\ Q^{(j)}=(Q^{(j)}_{km})=\begin{pmatrix} Q^{(j)}_{11} & \cdots & Q^{(j)}_{1n} \\ \vdots & \ddots & \vdots \\ Q^{(j)}_{n1} & \cdots & Q^{(j)}_{nn} \end{pmatrix},\ A^{(j)}=(A^{(j)}_{km})=\begin{pmatrix} A^{(j)}_{11} & \cdots & A^{(j)}_{1n} \\ \vdots & \ddots & \vdots \\ A^{(j)}_{n1} & \cdots & A^{(j)}_{nn} \end{pmatrix},$$

$$j = 1,2 \ , \ A^{(3)} = \begin{pmatrix} A_1^{(3)} \\ \vdots \\ A_n^{(3)} \end{pmatrix} , \ Q_{km} = Q_{km}(\zeta) \ , \ A_{km}^{(j)} = A_{km}^{(j)}(\zeta) \ ,$$

$$A_k^{(3)} = A_k^{(3)}(\zeta) \ , \ 1 \le k \ , \ m \le n \ , \ 1 \le j \le 2 \ .$$

Secondly, we consider the nonlinear elliptic system of 2n first order equations

$$\Phi_j(x,y,u_1,\ldots,u_{2n},u_{1x},\ldots,u_{2nx},u_{1y},\ldots,u_{2ny}) = 0 \ , j=1,\ldots,2n \ , \qquad (1.11)$$

where $\Phi_j(x,y,\zeta_1,\ldots,\zeta_{6n})(j=1,\ldots,2n)$ are continuous real functions in $(x,y) \in D$ and the real variables $\zeta_1,\ldots,\zeta_{6n}$, and possess continuous partial derivatives with respect to $\zeta_{2n+1},\ldots,\zeta_{6n}$. The nonlinear system (1.11) is called elliptic in D , if the determinant

$$K(\zeta) = |A+B\lambda| > 0 \quad \text{in} \quad D \qquad (1.12)$$

for any real number λ , where

$$A = (\Phi_{ju_{kx}}) \ , \ B = (\Phi_{ju_{ky}}) \ .$$

Putting $w_k = u_k + iu_{n+k}$, $k=1,\ldots,n$, $z = x + iy$ and noting (1.33) in Chapter 2, we have

$$\left.\begin{aligned} \Phi_{ju_{kx}} &= \frac{1}{2}(\Phi_{jw_{kz}} + \Phi_{j\overline{w}_{k\overline{z}}} + \Phi_{jw_{k\overline{z}}} + \Phi_{j\overline{w}_{kz}}) \ , \\ \Phi_{ju_{n+kx}} &= \frac{i}{2}(\Phi_{jw_{kz}} - \Phi_{j\overline{w}_{k\overline{z}}} + \Phi_{jw_{k\overline{z}}} - \Phi_{j\overline{w}_{kz}}) \ , \\ \Phi_{ju_{ky}} &= \frac{i}{2}(-\Phi_{jw_{kz}} + \Phi_{j\overline{w}_{k\overline{z}}} + \Phi_{jw_{k\overline{z}}} - \Phi_{j\overline{w}_{kz}}) \ , \\ \Phi_{ju_{n+ky}} &= \frac{1}{2}(\Phi_{jw_{kz}} + \Phi_{j\overline{w}_{k\overline{z}}} - \Phi_{jw_{k\overline{z}}} - \Phi_{j\overline{w}_{kz}}) \ . \end{aligned}\right\} \qquad (1.13)$$

Hence

$$J = \frac{D(\Phi_1,\dots,\Phi_{2n})}{D(w_{1\bar{z}},\dots,w_{n\bar{z}},\overline{w_{1z}},\dots,\overline{w_{nz}})} =$$

$$= \frac{1}{2^{2n}}$$

$$\times \begin{vmatrix} \Phi_{1u_{1x}} & -i\Phi_{1u_{n+1x}} & -\Phi_{1u_{n+1y}} & -i\Phi_{1u_{1y}} & \cdots & \Phi_{1u_{nx}} & +i\Phi_{1u_{2nx}} & -\Phi_{1u_{2ny}} & +i\Phi_{1u_{ny}} \\ & \vdots & & & \ddots & & \vdots & & \\ \Phi_{2nu_{1x}} & -i\Phi_{2nu_{n+1x}} & -\Phi_{2nu_{n+1y}} & -i\Phi_{2nu_{1y}} & \cdots & \Phi_{2nu_{nx}} & +i\Phi_{2nu_{2nx}} & -\Phi_{2nu_{2ny}} & +i\Phi_{2nu_{ny}} \end{vmatrix}$$

$$= \left(\frac{i}{2}\right)^n \begin{vmatrix} \Phi_{1u_{1x}} & -\Phi_{1u_{n+1y}} & \cdots & \Phi_{1u_{2nx}} & +\Phi_{1u_{ny}} \\ \vdots & & \ddots & & \vdots \\ \Phi_{2nu_{1x}} & -\Phi_{2nu_{n+1y}} & \cdots & \Phi_{2nu_{2nx}} & +\Phi_{2nu_{ny}} \end{vmatrix}$$

$$= \left(\frac{i}{2}\right)^n |A+C| \ , \tag{1.14}$$

in which

$$C = \begin{pmatrix} -\Phi_{1u_{n+1y}} & \cdots & \Phi_{1u_{ny}} \\ \vdots & \ddots & \vdots \\ -\Phi_{2nu_{n+1y}} & \cdots & \Phi_{2nu_{ny}} \end{pmatrix} .$$

If $|A+C| \neq 0$ in D, then (1.11) can be solved for $w_{k\bar{z}}(k=1,\dots,n)$. Thus

$$w_{k\bar{z}} = F_k(z,w_1,\dots,w_n,w_{1z},\dots,w_{nz}) \ , \ k=1,\dots,n \ . \tag{1.15}$$

As in (1.9), we may write (1.15) in matrix form

$$w_{\bar{z}} = F(z,w,w_z) \ , \tag{1.16}$$

where

$$w = (w_1,\dots,w_n)^T \ , \ F = (F_1,\dots,F_n)^T \ ,$$

and T is denoting the transposition of matrices.

Using the linear transformation (1.4) and choosing a sufficiently small posititve constant t, we can also transform (1.11) with conditions similar to (1.3) into a complex system of type (1.9).

2. Reduction of elliptic systems of several second order equations

Now, we discuss the nonlinear system of several second order equations

$$\begin{aligned} &\Phi_j(x,y,u_1,\dots,u_n,u_{1x},\dots,u_{nx},u_{1y},\dots,u_{ny},u_{1xx},\dots,u_{nxx}, \\ &u_{1xy},\dots,u_{nxy},u_{1yy},\dots,u_{nyy}) = 0 \ , \ j=1,\dots,n \ , \end{aligned} \tag{1.17}$$

where $\Phi_j(x,y,\zeta_1,\dots,\zeta_{6n})(j=1,\dots,n)$ are continuous real functions in $(x,y) \in D$ and the real variables $\zeta_1,\dots,\zeta_{6n}$ and possess continuous partial derivatives with respect to $\zeta_{3n+1},\dots,\zeta_{6n}$. If for any real number λ the determinant

$$K(\lambda) = |A+2B\lambda+C\lambda^2| > 0 \quad \text{in} \quad D \ , \tag{1.18}$$

where

$$A = (\Phi_{ju_{kxx}}) \ , \ 2B = (\Phi_{ju_{kxy}}) \ , \ C = (\Phi_{ju_{kyy}}) \ ,$$

then the system of n second order equations is called elliptic in D.

Suppose

$$\frac{D(\Phi_1,\dots,\Phi_n)}{D(u_{1z\bar{z}},\dots,u_{nz\bar{z}})} = \begin{vmatrix} 2(\Phi_{1u_{1xx}}+\Phi_{1u_{1yy}}) & \dots & 2(\Phi_{1u_{nxx}}+\Phi_{1u_{nyy}}) \\ \vdots & \ddots & \vdots \\ 2(\Phi_{nu_{1xx}}+\Phi_{nu_{1yy}}) & \dots & 2(\Phi_{nu_{nxx}}+\Phi_{nu_{nyy}}) \end{vmatrix} \tag{1.19}$$

$$= 2^n |A+C| \neq 0 \ ,$$

then (1.17) can be solved for $u_{kz\bar{z}}$ $(z=x+iy, k=1,\ldots,n)$

$$u_{kz\bar{z}} = F_k(z,u_1,\ldots,u_n,u_{1z},\ldots,u_{nz},u_{1zz},\ldots,u_{nzz}), \quad k=1,\ldots,n . \tag{1.20}$$

Denoting $u = (u_1,\ldots,u_n)^T$, $F = (F_1,\ldots,F_n)^T$, the system (1.20) may be written in vector form

$$u_{z\bar{z}} = F(z,u,u_z,u_{zz}) . \tag{1.21}$$

If (1.17) satisfies the condition

$$\sup_{1\le j,k\le n} \left(\left|\Phi_{ju_{kxx}}\right|, \left|\Phi_{ju_{kxy}}\right|, \left|\Phi_{ju_{kyy}}\right|\right) \le \delta^{-1}, \quad |A| \ge \delta > 0 , \tag{1.22}$$

where δ is a positive constant, then by the linear transformation (1.4), the system (1.17) can be reduced to a complex system written in vector form as

$$u_{\zeta\bar{\zeta}} = F(\zeta,u,u_\zeta,u_{\zeta\zeta}) . \tag{1.23}$$

Here we shall mention that the integer n may not be even. If $n = 2m$ is an even number, and

$$\frac{D(\Phi_1,\ldots,\Phi_n)}{D(w_{1z\bar{z}},\ldots,w_{mz\bar{z}},\overline{w}_{1z\bar{z}},\ldots,\overline{w}_{mz\bar{z}})} = \tag{1.24}$$

$$= \begin{vmatrix} \Phi_{1u_{1xx}}-i\Phi_{1u_{m+1xx}}+\Phi_{1u_{1yy}}-i\Phi_{1u_{m+1yy}} & \cdots & \Phi_{1u_{1xx}}+i\Phi_{1u_{m+1xx}}+\Phi_{1u_{1yy}}+i\Phi_{1u_{m+1yy}} \\ \vdots & \ddots & \vdots \\ \Phi_{nu_{1xx}}-i\Phi_{nu_{m+1xx}}+\Phi_{nu_{1yy}}-i\Phi_{nu_{m+1yy}} & \cdots & \Phi_{nu_{1xx}}+i\Phi_{nu_{m+1xx}}+\Phi_{nu_{1yy}}+i\Phi_{nu_{m+1yy}} \end{vmatrix}$$

$$= (2i)^n |A+C| \neq 0 ,$$

where $w_k = u_k + iu_{m+k}$, $k=1,\ldots,m$, then from (1.17), we can find a system of second order complex equations

$$w_{kz\bar{z}} = F_k(z,w_1,\ldots,w_m,w_{1z},\ldots,w_{mz},\overline{w_{1z}},\ldots,\overline{w_{mz}},w_{1zz},\ldots,w_{mzz}) , \tag{1.25}$$
$$k=1,\ldots,m .$$

Its vector form is as follows:

$$w_{z\bar{z}} = F(z,w,w_z,\bar{w}_z,w_{zz}) \ , \tag{1.26}$$

in which $w = (w_1,\ldots,w_m)^T$, $F = (F_1,\ldots,F_m)^T$.

Again using the transformation (1.4), we may reduce (1.17) with the condition (1.22) to the complex vector form

$$w_{\zeta\bar{\zeta}} = F(\zeta,w,w_\zeta,\bar{w}_\zeta,w_{\zeta\zeta}) \ . \tag{1.27}$$

If the elliptic system (1.11) is linear, i.e.

$$\sum_{k=1}^{n} [a_{jk}(x,y)u_{kxx}+2b_{jk}(x,y)u_{kxy}+c_{jk}(x,y)u_{kyy}] \tag{1.28}$$

$$= \sum_{k=1}^{n} [d_{jk}(x,y)u_{kx}+e_{jk}(x,y)u_{ky}+f_{jk}(x,y)u_k]+g_j(x,y) \ ,j=1,\ldots,n$$

and the coefficients $a_{jk}(x,y)$, $b_{jk}(x,y)$, $c_{jk}(x,y)$, $d_{jk}(x,y)$, $e_{jk}(x,y)$, $f_{jk}(x,y)$, $g_j(x,y)$ $(j,k=1,\ldots,n)$ are continuous on $\bar{D}$, and the condition (1.22) is satisfied, then its complex form (1.23) and (1.27), respectively is also linear.

3. Conditions for elliptic systems of complex equations of first and second order

For the sake of convenience, we discuss only the case of an (N+1)-connected circular domain D as stated in § 1 of Chapter 2.

We rewrite the first order complex system (1.16) in the form

$$w_{\bar{z}} = F(z,w,w_z) \ , \ F = Q^{(1)}w_z + Q^{(2)}\bar{w}_{\bar{z}} + A^{(1)}w + A^{(2)}\bar{w} + A^{(3)} \ , \tag{1.29}$$

where $w = (w_1,\ldots,w_n)^T$, $F = (F_1,\ldots,F_n)^T$, $Q^{(j)} = (Q^{(j)}_{km})$,

$A^{(j)} = (A_{km}^{(j)})$, $j=1,2$, $A^{(3)} = (A_1^{(3)},\ldots,A_n^{(3)})^T$, $Q^{(j)} = Q^{(j)}(z,w,w_z)$, $A^{(j)} = A^{(j)}(z,w)$, $1 \le j \le 3$, and suppose that (1.29) satisfies

<u>Condition C:</u> (1) $Q^{(j)}(z,w,V)$, $A^{(j)}(z,w)$ are continuous in $w \in \mathbb{C}^n$ for almost every point $z \in D$, $V \in \mathbb{C}^n$ and $Q^{(j)} = 0$, $A^{(j)} = 0$, for $z \notin D$.

(2) The above functions are measurable in D for any vectors of functions $w(z) \in C_\beta(\overline{D})$, $V(z) \in L_{p_0}(\overline{D})$ satisfying

$$\left.\begin{aligned} &L_p[A_{km}^{(j)}(z,w(z)),\overline{D}] \le k_0 , \quad 1 \le k , m \le n , \quad 1 \le j \le 2 , \\ &L_p[A_k^{(3)}(z,w(z)),\overline{D}] \le k_0 , \quad 1 \le k \le n , \end{aligned}\right\} \tag{1.30}$$

$$L_p[A_{km}^{(j)}(z,w(z)),\overline{D}] \le \varepsilon k_0 , \quad 1 \le k < m \le n , \quad 1 \le j \le 2 , \tag{1.31}$$

where $p_0, p\,(2<p_0<p<\infty)$, $\beta = 1-2/p_0$, $\ell(0\le \ell < \infty)$, $\varepsilon(0 < \varepsilon \le 1)$ are real constants.

(3) The complex system (1.29) satisfies the following uniform ellipticity condition:

$$\left.\begin{aligned} &|F_k(z,w,V^{(1)})-F_k(z,w,V^{(2)})| \le \sum_{m=1}^{n} q_{km} |V_m^{(1)}-V_m^{(2)}| , \\ &\sum_{m=1}^{n} q_{km} \le q_0 , \quad k=1,\ldots,n , \end{aligned}\right\} \tag{1.32}$$

$$q_{km} \le \varepsilon , \quad 1 \le k < m \le n , \quad q_0 < \frac{1}{n} , \tag{1.33}$$

in which $q_{km}(1\le k,m\le n)$ are nonnegative constants.

If the conditions (1.31), (1.33) are replaced by

$$L_p[A_{km}^{(j)}(z,w(z)),\overline{D}] \le \varepsilon k_0 , \quad 1 \le k , m \le n , \quad 1 \le j \le 2 , \tag{1.34}$$

$$q_0 < \frac{1}{n} , \tag{1.35}$$

respectively, then these modified conditions are called <u>Condition C^*.</u>

If for the linear complex system (1.10), i.e. for

$$w_{\bar{z}} = F(z,w,w_z), F = Q^{(1)} w_z + Q^{(2)} \overline{w}_{\bar{z}} + \varepsilon f(z,w) + A^{(3)}, \quad f = A^{(1)} w + A^{(2)} \overline{w} \qquad (1.36)$$

where $\varepsilon (-\infty < \varepsilon < \infty)$ is a parameter, and

$$Q^{(j)} = (Q_{km}^{(j)}), \; A^{(j)} = (A_{km}^{(j)}), \; j=1,2, \; A^{(3)} = (A_1^{(3)}, \ldots, A_n^{(3)})^T,$$

$$Q_{km}^{(j)} = Q_{km}^{(j)}(z), \; A_{km}^{(j)} = A_{km}^{(j)}(z), \; 1 \le k,m \le n, \; 1 \le j \le 2,$$

$$A_k^{(3)} = A_k^{(3)}(z), \; 1 \le k \le n,$$

the $Q_{km}^{(j)}(z)$ satisfy

$$\sum_{m=1}^{n} [|Q_{km}^{(1)}(z)| + |Q_{km}^{(2)}(z)|] \le q_0 < \frac{1}{n}, \; k=1,\ldots,n, \qquad (1.37)$$

then (1.36) satisfies (1.32). Besides, we assume that $A_{km}^{(j)}(z)$ satisfies

$$L_p[A_{km}^{(j)}, \overline{D}] \le k_0, \; 1 \le k,m \le n, \; 1 \le j \le 2, \; L_p[A_k^{(3)}, \overline{D}] \le k_0, \; 1 \le k \le n, \qquad (1.38)$$

in which p, k_0 are the same as in (1.30), (1.31). The conditions (1.37) and (1.38) are called Condition C_*.

Next, we rewrite the second order complex system (1.21) in the form

$$u_{z\bar{z}} = F(z,u,u_z,u_{zz}), \; F = Re[Q u_{zz} + A^{(1)} u_z + A^{(2)} u + A^{(3)}], \qquad (1.39)$$

where

$$u = (u_1, \ldots, u_n)^T, \; F = (F_1, \ldots, F_n)^T, \; Q = (Q_{km}), \; A^{(j)} = (A_{km}^{(j)}), \; j=1,2,$$

$$A^{(3)} = (A_1^{(3)}, \ldots, A_n^{(3)})^T, \; Q = Q(z,u,u_z,u_{zz}), \; A^{(j)} = A^{(j)}(z,u,u_z), \; 1 \le j \le 3.$$

We suppose that the system (1.39) satisfies

Condition C

(1) $Q(z,u,u_z,V)$, $A^{(j)}(z,u,u_z)$ are continuous in $u \in R^n$, $u_z \in C^n$ for almost every point $z \in D$ and $V \in C^n$.

(2) The above functions are measurable in D for all vectors of Hölder continuously differentiable functions $u(z) \in C_\beta(\overline{D})$ and all vectors of measurable functions $V(z) \in L_{p_0}(\overline{D})$, and satisfy

$$\left.\begin{aligned} &L_p[A_{km}^{(j)}(z,u(z),u_z),\overline{D}] \le k_0 \,,\ 1 \le k,m \le n \,,\ 1 \le j \le 2 \,, \\ &L_p[A_k^{(3)}(z,u(z),u_z),\overline{D}] \le k_0 \,,\ 1 \le k \le n \,, \end{aligned}\right\} \tag{1.40}$$

$$\left.\begin{aligned} &L_p[A_{km}^{(1)}(z,u(z),u_z),\overline{D}] \le \varepsilon k_0 \,,\ 1 \le k < m \le n \,, \\ &L_p[A_{km}^{(2)}(z,u(z),u_z),\overline{D}] \le \varepsilon k_0 \,,\ 1 \le k,m \le n \,, \end{aligned}\right\} \tag{1.41}$$

where p_0 , $p(2<p_0<p<\infty)$, $\beta = 1-2/p_0$, $k_0(0\le k_0<\infty)$, $\varepsilon(0<\varepsilon\le 1)$ are real constants. Here $u(z) \in C_\beta(\overline{D})$, $V(z) \in L_{p_0}(\overline{D})$ mean $u_k(z) \in C_\beta(\overline{D})$, $V_k(z) \in L_{p_0}(\overline{D})(k=1,\dots,n)$, respectively, where $u = (u_1,\dots,u_n)^T$, etc. Besides, we define $C_\beta[u,\overline{D}] = \sum_{k=1}^{n} C_\beta[u_k,\overline{D}]$, $L_{p_0}[V,\overline{D}] = \sum_{k=1}^{n} L_{p_0}[V_k,\overline{D}]$.

(3) The complex system (1.39) satisfies the following uniform ellipticity condition

$$\left.\begin{aligned} &|F_k(z,u,u_z,V^{(1)}) - F_k(z,u,u_z,V^{(2)})| \\ &\le \sum_{m=1}^{n} q_{km}|V_m^{(1)}-V_m^{(2)}| \,,\ \sum_{m=1}^{n} q_{km} \le q_0 \,, \end{aligned}\right\} \tag{1.42}$$

$$q_{km} \le \varepsilon \,,\ 1 \le k < m \le n \,,\ q_0 < \frac{1}{n} \,. \tag{1.43}$$

If the conditions (1.41), (1.43) are replaced by

$$L_p[A_{km}^{(j)}(z,u,u_z),\overline{D}] \le \varepsilon k_0 \,,\ 1 \le k,m \le n \,,\ 1 \le j \le 2 \,, \tag{1.44}$$

$$q_0 < \frac{1}{n} \,, \tag{1.45}$$

respectively, then we say that the complex system (1.39) satisfies

Condition C^*.

In addition, we shall discuss the linear case of (1.26) or (1.27), i.e.

$$\left.\begin{aligned} &u_{z\bar{z}} - Re[Qu_{zz}] = \varepsilon f(z,u,u_z) + A^{(3)}, \ -\infty < \varepsilon < \infty, \\ &f = Re[A^{(1)}u_z + A^{(2)}u], \ Q = (Q_{km}), \ A^{(j)} = (A^{(j)}_{km}), \ j=1,2, \\ &A^{(3)} = (A^{(3)}_1, \ldots, A^{(3)}_n)^T, \ Q = Q(z), \ A^{(j)} = A^{(j)}(z), \ 1 \le j \le 3, \end{aligned}\right\} \tag{1.46}$$

and suppose that (1.46) satisfies Condition C_*, the main conditions of which are

$$\sum_{m=1}^{n} |Q_{km}(z)| \le q_0 < \frac{1}{n}, \ k=1,\ldots,n, \tag{1.47}$$

and

$$L_p[A^{(j)}_{km}, \bar{D}] \le k_0 < \infty \tag{1.48}$$

where $q_0(0 \le q_0 < \frac{1}{n})$, $p(2<p<\infty)$, $k_0(0 \le k_0 < \infty)$ are real constants, $\varepsilon(-\infty<\varepsilon<\infty)$ is a parameter.

For the second order complex system (1.26) or (1.27), we may define Condition C and Condition C_*. We rewrite (1.26) in the form

$$\left.\begin{aligned} &w_{z\bar{z}} = F(z,w,w_z,\bar{w}_z,w_{zz},\bar{w}_{zz}), \ F = Q^{(1)}w_{zz} + Q^{(2)}\bar{w}_{zz} \\ &+ A^{(1)}w_z + A^{(2)}\bar{w}_{\bar{z}} + A^{(3)}\bar{w}_z + A^{(4)}w_{\bar{z}} + A^{(5)}w + A^{(6)}\bar{w} + A^{(7)}, \end{aligned}\right\} \tag{1.49}$$

where $Q^{(j)} = Q^{(j)}(z,w,w_z,\bar{w}_z,w_{zz},\bar{w}_{zz})$, $j=1,2$, $A^{(j)} = A^{(j)}(z,w,w_z,\bar{w}_z)$, $j=1,\ldots,7$. The main conditions in Condition C are as follows:

$$\left.\begin{aligned} &L_p[A^{(j)}_{kn}(z,w(z),w_z,\bar{w}_z),\bar{D}] \le k_0, \ 1 \le k, n \le m, \ 1 \le j \le 7, \\ &L_p[A^{(j)}_{kn},\bar{D}] \le \varepsilon k_0, \ 1 \le k < n \le m, \ j=1,2, \\ &L_p[A^{(j)}_{kn},\bar{D}] \le \varepsilon k_0, \ 1 \le k,n \le m, \ j=3,\ldots,6, \end{aligned}\right\} \tag{1.50}$$

for any vector functions $w(z) \in W_p^{(1)}(D)$, and

$$\left.\begin{array}{l} |F_k(z,w,w_z,\bar{w}_z,U^{(1)},V^{(1)})-F_k(z,w,w_z,\bar{w}_z,U^{(2)},V^{(2)})| \\ \leq \sum_{n=1}^{m} q_{kn} |U_n^{(1)}-U_n^{(2)}| + \sum_{n=1}^{n} q'_{kn} |V_n^{(1)}-V_n^{(2)}| , \\ \sum_{n=1}^{m} (q_{kn}+q'_{kn}) \leq q_0 < \frac{1}{m}, q_{kn} + q'_{kn} < \varepsilon , 1 \leq k < n \leq m . \end{array}\right\} \quad (1.51)$$

§ 2. Boundary value problems for nonlinear complex systems of first order

1. Formulation of the Riemann-Hilbert problem for nonlinear complex systems of first order

Problem A. The Riemann-Hilbert problem for the nonlinear complex system (1.29) may be formulated as follows: Find a vector of continuous functions $w = w(z) = (w_1(z),\ldots,w_n(z))^T$, which is a solution of (1.29) and satisfies the boundary condition

$$Re[\overline{\Lambda(t)}w(t)] = B(t) , t \in \Gamma , \quad (2.1)$$

where $\Lambda(t) = (\Lambda_{km}(t))$, $w(t) = (w_1(t),\ldots,w_n(t))^T$, $B(t) = (B_1(t),\ldots,B_n(t))^T$. The boundary condition (2.1) is the complex form of the real boundary condition

$$\sum_{m=1}^{2n} \sigma_{km}(t)u_m(t) = \psi_k(t) , t \in \Gamma , 1 \leq k \leq n \quad (2.2)$$

(cf. [112]5)). We suppose that $\Lambda(t)$ satisfies the Lopatinski condition

$$Det\ \Lambda(t) = |(\Lambda_{km}(t))| \neq 0 , t \in \Gamma . \quad (2.3)$$

Under this condition, the boundary condition (2.1) can be reduced to the following form

$$Re[\overline{\lambda(t)}w(t)] = r(t,w) , t \in \Gamma , \quad (2.4)$$

where

$$\lambda(t) = \begin{pmatrix} \lambda_1(t) & & 0 \\ & \ddots & \\ 0 & & \lambda_n(t) \end{pmatrix}, \quad r(t,w) = (r_1(t,w),\ldots,r_n(t,w))^T$$

and $Det\ \lambda(t) \neq 0$. This is a nonlinear boundary value problem, which will be considered in Chapter 7. Here we will only discuss the linear boundary condition, i.e.

$$Re[\overline{\lambda(t)}w(t)] = r(t) = (r_1(t),\ldots,r_n(t))^T \ , \ t \in \Gamma \ . \tag{2.5}$$

We assume that $\lambda(t)$, $r(t)$ satisfy the conditions

$$|\lambda_k(t)| = 1 \ , \ k=1,\ldots,n \ , \ C_\alpha[\lambda,\Gamma] \le \ell \ , \ C_\alpha[r,\Gamma] \le \ell \ , \tag{2.6}$$

in which $\alpha(1/2<\alpha<1)$, $\ell(0\le\ell<\infty)$ are real constants. The constant vector $K = (K_1,\ldots,K_n)^T$, $K_k = (1/2\pi)\Delta_\Gamma arg\lambda_k(t)$, $1 \le k \le n$ is called the index of $\lambda(t)$.

Problem B. The boundary condition (2.5) is replaced by

$$Re[\overline{\lambda(t)}w(t)] = r(t) + h(t) \ , \ t \in \Gamma \ , \tag{2.7}$$

in which

$$h(t) = \begin{pmatrix} h_1(t) \\ \vdots \\ h_n(t) \end{pmatrix}, \ h_k(t) = \begin{cases} 0 \ , \ t \in \Gamma \ , \ \text{for } K_k \ge N \ , \\ \left.\begin{array}{l} h_{kj} \ , \ t \in \Gamma_j \ , \ j=1,\ldots,N-K_k \\ 0 \ , \ t \in \Gamma_j \ , \ j=N-K_k+1,\ldots,N+1 \end{array}\right\} \text{for } 0\le K_k \le N \ , \\ h_{kj} \ , \ t \in \Gamma_j \ , \ j=1,\ldots,N \ , \\ h_{k0} + Re \displaystyle\sum_{m=1}^{|K_k|-1} (H_m^{(k)} + iH_{-m}^{(k)})t^m \ , \ \text{for } K_k<0, 1\le k\le n \ , \end{cases} \tag{2.8}$$

and $h_{kj}(j=0,1,\ldots,N)$, $H_{\pm m}^{(k)}(m=1,\ldots,|K_k|-1 \ , \ k=1,\ldots,n)$ are unknown real constants to be determined appropriately. If $K_k \ge 0$, we require the solution $w(z)$ to satisfy the point conditions

$$Im[\overline{\lambda(a_j)}w(a_j)] = b_j = (b_{j1},\ldots,b_{jn})^T , \tag{2.9}$$

$$j \in \{j\} = \begin{cases} 1,\ldots,2K_k-N+1 \text{ , for } K_k \geq N \text{ ,} \\ N-K_k+1,\ldots,N+1 \text{ , for } 0 \leq K_k < N \text{ ,} \end{cases} \quad k=1,\ldots,n ,$$

where $a_j(j=1,\ldots,2K_k-N+1)$ are as stated in (1.3) of Chapter 1, $b_{jk}(j \in \{j\},k=1,\ldots,n)$ are real constants satisfying the condition $|b_{jk}| \leq \ell(j \in \{j\},k=1,\ldots,n)$.

If $\lambda_k(t) = 1$, $k=1,\ldots,n$, then Problem B is the modified Dirichlet problem for the complex system (1.29). This problem under some conditions was studied by W. Tutschke, [96]9).

2. A priori estimates of solutions for the Riemann-Hilbert problem

Let $w(z) \in W^1_{p_0}(D)$, $2 < p_0 < min[p,1/(1-\alpha)]$ be a solution to Problem B represented in the form

$$w(z) = \Phi(z) + T\omega \text{ , i.e. } w_k(z) = \Phi_k(z) + T_k\omega_k \text{ , } k=1,\ldots,n , \tag{2.10}$$

where $\Phi(z)$ is an analytic vector, $\omega(z)$ a vector of functions from $L_{p_0}(\overline{D})$ and T a vector integral operator of the Vekua type so that $T\omega$ and $\phi(z)$ satisfy boundary conditions of type (2.7), (2,9).

Theorem 2.1 Let $w(z)$ be a solution to (1.29), (2.7) and (2.9) of the form (2.10), where (2.6) and Condition C are satisfied, but $C_\alpha[r,\Gamma] \leq \ell$ in (2.6), $|b_{jk}| \leq \ell$ in (2.9) and $L_p[A_k^{(3)},\overline{D}] \leq k_0$ in (1.30) are replaced by $C_\alpha[r,\Gamma] \leq \ell_1$, $|b_{jk}| \leq \ell_1$ and $L_p[A_k^{(3)},\overline{D}] \leq k_1$, respectively, in which ℓ_1 and k_1 are nonnegative constants. If $\varepsilon > 0$ is sufficiently small, then there exists a constant $M_1 = M_1(q_0,p_0,k_0,\alpha,\ell,D,\varepsilon)$ such that

$$S = C_\beta[w,\overline{D}] + L_{p_0}[|w_{\bar z}|+|w_z|,\overline{D}] = \tag{2.11}$$
$$= \sum_{k=1}^{n} \{C_\beta[w_k,\overline{D}] + L_{p_0}[|w_{k\bar z}|+|w_{kz}|,\overline{D}]\} \leq (k_1+\ell_1)M_1 .$$

In particular, if $k_1 = k_0$, $\ell_1 = \ell$, then this estimate is

$$S \le (k_0+\ell_0)\, M_1 = M_2 = M_2(q_0,p_0,k_0,\alpha,\ell,D,\varepsilon) \,. \qquad (2.12)$$

Proof. The components of solution $w(z) = \Phi(z) + T\omega$ of Problem B satisfy the complex equations of first order

$$w_{k\bar{z}} = \sum_{m=1}^{n} [Q_{km}w_{mz}+A_{km}^{(1)}w_m+A_{km}^{(2)}\overline{w_m}] + A_k^{(3)} \,, \qquad (2.13)$$

i.e.

$$w_{k\bar{z}} - Q_{kk}w_{kz} = A_{kk}^{(1)}w_k + A_{kk}^{(2)}\overline{w_k} + A_k \,,$$

$$A_k = A_k^{(3)} + \sum_{m\neq k} [Q_{km}w_{mz}+A_{km}^{(1)}w_m+A_{km}^{(2)}\overline{w_m}] \,,\ k=1,\dots,n \,,$$

and the boundary and point conditions

$$Re[\overline{\lambda_k(t)}w_k(t)] = r_k(t) + h_k(t) \,,\ t \in \Gamma \,, \qquad (2.14)$$

$$Im[\overline{\lambda_k(a_j)}w_k(a_j)] = b_{kj} \,,\ j \in \{j\} \,,\ k=1,\dots,n \,. \qquad (2.15)$$

Let us consider the first component $w_1(z)$ of $w(z)$. Because for $2 \le m \le n$

$$|Q_{1m}| \le q_{1m} \le \varepsilon \,,\ L_{p_0}[A_{1m}^{(1)},\overline{D}] + L_{p_0}[A_{1m}^{(2)},\overline{D}] \le d_{1m} \le 2\varepsilon k_0 \,,$$

we have

$$L_{p_0}[A_1,\overline{D}] \le L_{p_0}[A_1^{(3)},\overline{D}] +$$

$$+ \sum_{m=2}^{n} \{q_{1m}L_{p_0}[w_{mz},\overline{D}]+[L_{p_0}(A_{1m}^{(1)},\overline{D})+L_{p_0}[A_{1m}^{(2)},\overline{D}]]C[w_m,\overline{D}]\}$$

$$\le k_1+\varepsilon(1+2k_0)[L_{p_0}[w_z,D]+C_\beta[w,\overline{D}]] \le k_1+\varepsilon(1+2k_0)S = k_2 \,.$$

Putting $W_1 = w_1/(\ell_1+k_2)$, which satisfies

$$W_{1\bar{z}} - Q_{11}W_{1z} = A_{11}^{(1)}W_1 + A_{11}^{(2)}\overline{W_1} + A_1/(\ell_1+k_2) \ , \tag{2.16}$$

$$Re[\overline{\lambda_1(t)}W_1(t)] = r_1(t)/(\ell_1+k_2)+h_1(t) \ , \ t \in \Gamma \ , \tag{2.17}$$

$$Im[\overline{\lambda_1(a_j)}W_1(a_j)] = b_{1j}/(\ell_1+k_2) \ , \tag{2.18}$$

$$1 \le j \le 2K_1-N+1 \ , \text{ if } N \le K_1 \ , \ N-K_1+1 \le j \le N+1 \ , \text{ if } 0 \le K_1 \le N \ ,$$

where

$$L_{p_0}[A_1/(\ell_1+k_2),\overline{D}] \le 1 \ , \ C_\beta[r_1/(\ell_1+k_2),\Gamma] \le 1 \ , \ |b_{1j}/(\ell_1+k_2)| \le 1 \ ,$$

so that

$$C_\beta[W_1,\overline{D}] + L_{p_0}[|W_{1z}|+|W_{1\bar{z}}|,\overline{D}] \le M_3 = M_3(q_0,p_0,k_0,\alpha,\ell,D) \ ,$$

i.e.

$$S_1=C_\beta[w_1,\overline{D}]+L_{p_0}[|w_{1\bar{z}}|+|w_{1\bar{z}}|,\overline{D}] \le (\ell_1+k_2)M_3=[\ell_1+k_1+\varepsilon(\ell+2k_0)S]M_3 \ . \tag{2.19}$$

Let us now consider (2.13) - (2.15) for $k=2$. Because

$$\begin{aligned} L_{p_0}[A_2,\overline{D}] &\le L_{p_0}[A_2^{(3)},\overline{D}] + \sum_{m\neq 2}\{q_{2m}L_{p_0}[w_{mz},\overline{D}] + [L_{p_0}[A_{2m}^{(1)},\overline{D}] \\ &+ L_{p_0}[A_{2m}^{(2)},\overline{D}]]C[w_m,\overline{D}]\} \le k_1 + L_{p_0}[w_{1z},\overline{D}] \\ &+ 2\varepsilon k_0 \sum_{m=3}^{n} L_{p_0}[w_{mz},\overline{D}] + 2k_0C[w_1,\overline{D}] + 2\varepsilon k_0 \sum_{m=3}^{n} C[w_m,\overline{D}] \\ &= k_1 + (1+2k_0)[L_{p_0}[w_{1z},\overline{D}]+C[w_1,\overline{D}]] + 2\varepsilon k_0[L_{p_0}[w_z,\overline{D}] \\ &+ C[w,\overline{D}]] \le k_1 + (1+2k_0)S_1 + 2\varepsilon k_0 S \ . \end{aligned}$$

From here similarly to (2.19) one has

$$\begin{aligned} S_2 &= C_\beta[w_2,\overline{D}] + L_{p_0}[|w_{2\bar{z}}|+|w_{2z}|,\overline{D}] \\ &\le (\ell_1+k_1+(1+2k_0)S_1 + 2\varepsilon k_0 S)M_3 \end{aligned} \tag{2.20}$$

$$\le \{\ell_1+k_1+(1+2k_0)[\ell_1+k_1+\varepsilon(1+2k_0)S]M_3 + 2\varepsilon k_0 S\}M_3$$
$$\le [\ell_1+k_1+\varepsilon(1+2k_0)S][1+(1+2k_0)M_3]M_3 \ .$$

Similarly, for $1 \le k \le n$ one has

$$S_k = C_\beta[w_k,\bar{D}] + L_{p_0}[|w_{k\bar{z}}|+|w_{kz}|,\bar{D}] \tag{2.21}$$
$$\le [\ell_1+k_1+(1+2k_0)\sum_{m=1}^{k-1} S_m+2\varepsilon k_0 S]M_3$$
$$\le [\ell_1+k_1+\varepsilon(1+2k_0)S][1+(1+2k_0)M_3]^{k-1}M_3 \ .$$

Thus

$$S = \sum_{k=1}^{n} S_k \le \sum_{k=1}^{n} [\ell_1+k_1+\varepsilon(1+2k_0)S]M_3[1+(1+2k_0)M_3]^{k-1} =$$
$$= [\ell_1+k_1+\varepsilon(1+2k_0)S]M_3\{[1+(1+2k_0)M_3]^n-1\}/(1+2k_0)M_3 \ .$$

If now

$$\varepsilon\{[1+(1+2k_0)M_3]^n-1\} < 1 \ ,$$

then

$$S \le \frac{\ell_1+k_1}{(1+2k_0)} \ \frac{[1+(1+2k_0)M_3]^n-1}{1-\varepsilon\{[1+(1+2k_0)M_3]^n-1\}} = (\ell_1+k_1)M_1 \ .$$

This proves the estimate (2.11). □

Secondly, we discuss Problem B with the condition (3.1) from § 3.

Theorem 2.2 Suppose that the complex system (1.29) satisfies Condition C^* and the constant ε in (1.34) is sufficiently small. Then the solution $w(z)$ to Problem B for (1.29) with the condition (3.1) and $K_j \le 0 (j=1,\dots,n)$ satisfies the estimate (2.12).

Proof. The solution $w(z)$ can be expressed as in (3.2) of the next section. We subsitute the solution $w(z) = \Phi(z) + \tilde{T}\omega$ into (1.29), (2.7) and (2.9), and rewrite the complex system (1.29) in the following form

$$\left.\begin{aligned} &w_{\bar{z}} - Q^{(1)} w_z - Q^{(2)} \overline{w_z} = A \,,\ A^{(1)} w + A^{(2)} \bar{w} + A^{(3)} \quad \text{or} \\ &\omega(z) - Q^{(1)} \tilde{S}\omega - Q^{(2)} \overline{\tilde{S}\omega} = B \,,\ B = A^{(1)} \tilde{T}\omega + A^{(2)} \overline{\tilde{T}\omega} + B_0 \\ &B_0 = Q^{(1)} \Phi' + Q^{(2)} \overline{\Phi'} + A^{(1)} \Phi + A^{(2)} \bar{\Phi} + A^{(3)} . \end{aligned}\right\} \tag{2.22}$$

Taking the properties of $\Phi(z)$, $\tilde{T}\omega$ in Theorem 3.1 of the next section into the account, we have

$$L_{p_0}[B_0,\bar{D}] \le M_4 = M_4(q_0,p_0,k_0,\alpha,\ell,D) \,, \tag{2.23}$$

$$L_{p_0}[A^{(1)}\tilde{T}\omega + A^{(2)}\tilde{T}\omega,\bar{D}] \le 2\varepsilon n^2 k_0 M_5 L_{p_0}[\omega,\bar{D}], \tag{2.24}$$

where $M_5 = M_5(p_0,D)$. Choosing the constant ε so small, that $1-q_0\Lambda_{p_0} - 2\varepsilon n^2 M_5 k_0 < 1$, from (2.22), we obtain

$$L_{p_0}[\omega,\bar{D}] \le M_4/(1-q_0\Lambda_{p_0} - 2\varepsilon n^2 M_5 k_0) \,. \tag{2.25}$$

By (3.5), (3.6), (3.8) and the above formula follows (2.12). □

3. Solvability of the Riemann-Hilbert problem for nonlinear complex systems of first order

Theorem 2.3 Let the complex system (1.29) satisfy Condition C and ε be sufficiently small. Then Problem B is solvable.

Proof. Let

$$D_m = \{z\colon z \in D \,,\ d(z,\Gamma) > 1/m\} \,, \tag{2.26}$$

where m is a positive integer and

$$F^{(m)}(z,w,w_z) = \begin{cases} F(z,w,w_z) , & z \in D_m , \\ 0 , & z \in D \setminus D_m . \end{cases} \tag{2.27}$$

Instead for (1.29), Problem B now will be considered for

$$w_{\bar{z}} = F^{(m)}(z,w,w_z) . \tag{2.28}$$

For any element $\omega(z)$ of the Banach space $[L_{p_0}(\overline{D_m})]^n$ with norm

$$L_{p_0}[\omega,D_m] = \sum_{k=1}^{n} L_{p_0}[\omega_k,D_m] < M = 1+M_2 ,$$

where M_2 is the constant in (2.12), the vector function

$$\Psi(z) = T\omega = -\frac{1}{\pi} \iint_{D_m} \frac{\omega(\zeta)}{\zeta - z} d\sigma_\zeta \tag{2.29}$$

is an analytic vector in $\bar{D} \setminus D_m$. For a given $\omega(z)$, let $\Phi(z)$ be the analytic vector uniquely defined by

$$Re\{\overline{\lambda(t)}[\Phi(t)+\Psi(t)]\} = r(t) + h(t) , \quad t \in \Gamma , \tag{2.30}$$

$$Im\{\overline{\lambda(a_j)}[\Phi(a_j)+\Psi(a_j)]\} = b_j , \quad j \in \{j\} . \tag{2.31}$$

To make $w(z) = \Phi(z) + \Psi(z)$ a solution to (2.28), $w(z)$ has to satisfy

$$\omega(z) = F^{(m)}(z,w(z),\Phi'(z)+\Pi\omega) , \quad \Pi\omega = (T\omega)_z . \tag{2.32}$$

To show that this integral equation is solvable, the Leray-Schauder theorem will be applied. Instead of (2.32) the equation

$$\omega(z) = \tau F^{(m)}(z,w(z),\Phi'(z)+\Pi\omega) \tag{2.33}$$

with a real parameter $\tau \in [0,1]$ is considered. The mapping

$$(\omega,\tau) \longrightarrow \Omega , \quad \Omega(z) = \tau F^{(m)}(z,w(z),\Phi'(z)+\Pi\Omega)$$

satisfies the following conditions.

(1) It is completely continuous on $\overline{B_M} \times [0,1]$ and uniformly continuous on $\overline{B_M}$ with respect to $\tau \in [0,1]$, where $B_M = \{\omega(z) \,|\, L_{p_0}[\omega, \overline{D}_m] < M\}$.

(2) For $\tau = 0$, there is only the solution $\omega(z) = 0$ in B_M.

(3) For any number $\tau \in [0,1]$, there is no solution $\omega(z)$ of (2.33) on ∂B_M.

Property (1) follows from the assumptions in Condition C, property (3) from the a priori estimate (2.12), property (2) is obvious. Thus, there exists a solution to (2.33) for any $\tau \in [0,1]$, especially for $\tau = 1$ a solution $\omega(z)$ to (2.32) and therefore

$$w(z) = \Phi(z) + \Psi(z) = \Phi(z) + T\omega$$

is a solution to Problem B for (2.28) satisfying the a priori estimate (2.12).

Denote now by $w_m(z)$ a solution to Problem B for (2.28) and consider the sequence $\{w_m(z)\}$ for $m \geq m_0$. Because (2.12) holds for any $w_m(z)$ the sequence $\{w_m(z)\}$ may be assumed to converge uniformly on $\overline{D}$ by choosing a proper subsequence. Let the limit vector be $w_0(z)$. It satisfies (2.7), (2.9) because the $w_m(z)$ do. From the a priori estimate (2.12) it can be seen that $\omega_m(z) = w_{m\overline{z}}$ as well as w_{mz} may be assumed to converge in the L_p-norm. Let the limit functions be $\omega_0(z)$ and $Y_0(z)$, respectively. From $w_m(z) = \Phi_m(z) + T\omega_m$, we see because $\Phi_m(z)$, $T\omega_m$ and $\Phi_m'(z)$ tend to $\Phi_0(z)$, $T\omega_0$ and $\Phi_0'(z)$, respectively as m tends to infinity

$$w_0(z) = \Phi_0(z) + T\omega_0 \,, \; w_{0z} = Y_0(z) = \Phi_0(z) = \Phi_0(z) + \Pi\omega_0 \,.$$

It remains to show $w_0(z)$ satisfies (1.29). From the continuity of $F^{(m)}(z,w,w_z)$ in the second variable it follows that

$$\lim_{n\to\infty} L_{p_0}[F^{(m)}(z,w_m,w_{0z}) - F(z,w_0,w_{0z}),\overline{D}] = 0 \ ,$$

because

$$F^{(m)}(z,w_m,w_{0z}) - F(z,w_0,w_{0z}) = -F(z,w_0,w_{0z}) \ , \ z \in \overline{D}\setminus D_m \ .$$

Moreover, from Condition C

$$\lim_{m\to\infty}[F^{(m)}(z,w_m,w_{mz}) - F(z,w_m,w_{0z}),\overline{D}] = 0 \ ,$$

for almost every point $z \in D$. By using the method of the proof for Theorem 2.3 of Chapter 5 we have

$$\lim_{m\to\infty} L_{p_0}[F^{(m)}(z,w_m,w_{mz})-F(z,w_0,w_{0z}),\overline{D}] = 0 \ ,$$

i.e. $\lim_{m\to\infty} L_{p_0}[\omega_m-F(z,w_0,w_{0z}),\overline{D}] = 0$, so that in D

$$w_{0\bar{z}} = \omega_0(z) = F(z,w_0,w_{0z}) \ .$$

This proves the theorem. □

A consequence is

Theorem 2.4 Let (1.29) satisfy Condition C. If $K_k < 0$ for $1 \le k \le n_0$, $0 \le K_k < N$ for $n_0 + 1 \le k \le n_1$, and $N \le K_k$ for $n_1+1 \le k \le n$, then Problem A for (1.29) is solvable, if

$$(n_1-n_0)N-2\sum_{k=1}^{n_1} K_k + n_0(N-1) + \sum_{k=n_0+1}^{n_1} K_k$$

solvability conditions are satisfied. For other cases, the total number of solvability conditions for Problem A of (1.29) can be given, too.

Finally, we discuss the solvability of Problem A for (1.29) with condition (3.1) of the next section.

Theorem 2.5 Under the hypothesis in Theorem 2.2, Problem B ($K_j \le 0$, $j=1,\dots,n$) for (1.29) with the condition (3.1) is solvable, and Problem A has J solvability conditions, where

$$J = \begin{cases} -2\sum_{j=1}^{n} K_j + nN - n, & \text{for } K_j < 0, \ j=1,\dots,n, \\ nN, & \text{for } K_j = 0, \ j=1,\dots,n. \end{cases}$$

Proof. We denote by B_M a bounded and open set in the Banach space $L_{p_0}(\overline{D})^n$, the elements of which are the vectors $\omega(z)$ satisfying the inequality

$$L_{p_0}[\omega,D] < M = 1+M_2, \tag{2.34}$$

where M_2 is the constant in (2.12). Choosing any vector $\omega(z) \in \overline{B_M}$ and using the double integral vector $\tilde{T}\omega$ as stated in (3.2), (3.3) of the next section, one can find an analytic vector satisfying the boundary condition

$$Re\{\overline{\lambda(t)}[\Phi(t)+\tilde{T}\omega]\} = r(t) + h(t), \quad t \ \Gamma, \tag{2.35}$$

$$Im\{\overline{\lambda(t)}[\Phi(t)+\tilde{T}\omega]\}|_{t=a_j} = b_j, \quad j \in \{j\}. \tag{2.36}$$

Next, using Condition C^* and the principle of cantraction, we can find a solution $\omega^*(z) \in L_{p_0}(\overline{D})$ of the integral system with the parameter $\tau(0 \le \tau \le 1)$:

$$\omega^*(z) = \tau F(z,\Phi+T\omega,\Phi'+\tilde{S}\omega^*), \quad \tilde{S}\omega^* = (\tilde{T}\omega^*)_z. \tag{2.37}$$

Denote by $\omega^* = S(\omega,\tau)(0 \le \tau \le 1)$ a mapping from $\omega(z)$ onto $\omega^*(z)$, we can verify that $\omega^* = S(\omega,\tau)$ satisfies the conditions of the Leray-Schauder theorem. Hence, the system of integral equations

$$\omega(z) = \tau F(z,\Phi+T\omega,\Phi'+\tilde{S}\omega) \tag{2.38}$$

has a solution $\omega(z) \in B_M$. Let $\omega(z)$ be a solution of $\omega = S(\omega,1)$.

Then $w(z) = \Phi(z) + \tilde{T}\omega$ is just a solution for Problem B of (1.29).

Substituting the above solution $w(z)$ into the boundary condition (2.7) with $h(t) = 0$, the total number of solvability conditions for Problem A for (1.29) is derived (cf. [103]14)). □

§ 3. Boundary value problems for complex systems of several linear equations of first and second order

We first discuss the Riemann Hilbert boundary value problem for complex systems fo linear first order equations and then consider the oblique derivative boundary value problem for complex systems of linear second order equations.

1. The Riemann-Hilbert problem for complex systems of linear first order equations

By using the method in (3.4) - (3.6) of Chapter 1, we may transform the boundary condition (2.5) into the canonical form, i.e.

$$\lambda(t) = \begin{pmatrix} \lambda_1(t) & & 0 \\ & \ddots & \\ 0 & & \lambda_n(t) \end{pmatrix}, \quad \overline{\lambda_k(t)} = \begin{cases} t^{-K_k}, & t \in \Gamma_0, \ k=1,\ldots,n. \\ e^{-i\theta_{kj}}, & t \in \Gamma_j, \ j=1,\ldots,N. \end{cases} \tag{3.1}$$

Here, we discuss the canonical boundary condition (2.5), (3.1) for the linear complex system (1.36) still denoted by Problem A. The modified boundary value problem for (1.36) with the boundary condition (2.7), (3.1) is called Problem B.

We now give an integral representation for solutions to Problem B for (1.36) and its properties.

Theorem 3.1 Let $w(z)$ be a solution of Problem B for the complex system (1.36) and $w(z) \in W^1_{p_0}(D)$, $2 < p_0 < min(p,1/(1-\alpha))$. Then

(1) $w(z)$ can be expressed as

$$w(z) = \Phi(z) + \tilde{T}\omega, \tag{3.2}$$

where $\omega(z) = w_{\bar{z}} = (\omega_1(z),\ldots,\omega_n(z))^T$, $\Phi(z) = (\Phi_1(z),\ldots,\Phi_n(z))^T$ is a vector of analytic functions, $\tilde{T}\omega = (\tilde{T}_1\omega_1,\ldots,\tilde{T}_n\omega_n)^T$, $\Phi_k(z)$ and $\tilde{T}_k\omega_k$ possess the integral representations

$$\left.\begin{aligned}\Phi_k(z) &= \frac{1}{2\pi}\int_\Gamma P_k(z,t)r(t)d\theta + \Phi_{k0}(z) \ , \\ \tilde{T}_k\omega_k &= -\frac{1}{\pi}\iint_D [G_{k1}(z,\zeta)Re\,\omega_k(\zeta)+G_{k2}(z,\zeta)\,i\,Im\,\omega_k(\zeta)]d\sigma_\zeta \ ,\end{aligned}\right\} \tag{3.3}$$

in which $\Phi_{k_0}(z)$ is an analytic function in D , $P_k(z,t)$, $G_{k1}(z,\zeta)$, $G_{k2}(z,\zeta)$ are the Schwarz kernel and the Green functions as stated in (3.26), (3.35) of Chapter 1.

(2) The double integral vector $\tilde{T}$ possesses the properties:

$$(\tilde{T}\omega)_{\bar{z}} = \omega(z) \ , \ \tilde{S}\omega = (\tilde{T}\omega)_z \ , \tag{3.4}$$

$$L_{p_0}[\tilde{S}\omega,\bar{D}] = \sum_{k=1}^{n} L_{p_0}[\tilde{S}_k\omega,\bar{D}] \le \lambda_{p_0} L_{p_0}[\omega,\bar{D}] \ , \ \lambda_{p_0} < \infty \ , \ 2 < p_0 < p \ , \tag{3.5}$$

$$\lambda_2 \le 1 \ , \text{ for } K_k \le 0 \ , \ k=1,\ldots,n \ , \tag{3.6}$$

$$C_\beta[\tilde{T}\omega,\bar{D}] = \sum_{k=1}^{n} C_\beta[\tilde{T}_k\omega_k,\bar{D}] \le M_1 L_{p_0}[\omega,\bar{D}] \ , \tag{3.7}$$

where $\beta = 1-2/p_0$, $M_1 = M_1(p_0,D)$; $\Phi(z)$ satisfies the condition (2.7) and the estimates

$$C_\beta[\Phi,\bar{D}] \le M_2 \ , \ L_{p_0}[\Phi',\bar{D}] \le M_3 \ , \tag{3.8}$$

where $2 < p_0 < min(p,1/(1-\alpha))$, $M_j = M_j(p_0,\alpha,\ell,D)$, $j=2,3$. Moreover, if the constant $nq_0 < 1$, then there exists a positive constant $p_0 (2<p_0<p)$ so that

$$nq_0\lambda_{p_0} < 1 \ , \text{ for } K_k \le 0 \ , \ k=1,\ldots,n \ . \tag{3.9}$$

On the basis of the results in Theorem 3.3 and Theorem 3.5 of Chapter 1, it is not difficult to prove this theorem.

Next, using the Fredholm theorem for linear operator equations, we can show the solvability of Problem B and Problem A of (1.36).

If $K_k = (1/2\pi)\Delta_\Gamma arg\lambda_k(t) \le 0$, $k=1,\ldots,n$, and the vector $w(z)$ of functions in (3.2) is substituted into the linear complex system (1.36), we obtain the system of integral equations

$$\omega(z) - Q^{(1)}\tilde{S}\omega - Q^{(2)}\overline{\tilde{S}\omega} = \varepsilon f(z,\tilde{T}\omega) + g(z,\varepsilon) , \qquad (3.10)$$

where

$$g(z,\varepsilon) = Q^{(1)}\Phi' + Q^{(2)}\overline{\Phi'} + \varepsilon f(z,\Phi) + A^{(3)}(z) . \qquad (3.11)$$

From (3.9), we see that $I - Q^{(1)}\tilde{S} - Q^{(2)}\overline{\tilde{S}}$ has an inverse operator R and obtain the system of integral equations

$$\omega(z) = \varepsilon R[f(z,\tilde{T}\omega)] + R[g(z,\omega)] . \qquad (3.12)$$

Because $\tilde{T}$ is a completely continuous operator from $L_{p_0}(\overline{D})$ into $C_\beta(\overline{D})$ $(\beta=1-2/p_0$, $2 < p_0 < min(p,1/(1-\alpha)))$, the inverse operator R is also completely continuous. By the Fredholm theorem for operator equations, the homogeneous system of integral equations

$$\omega(z) = \varepsilon R[f(z,\tilde{T}\omega)] \qquad (3.13)$$

possesses discrete eigenvalues

$$\varepsilon_j(j=1,2,\ldots) , \ 0 < |\varepsilon_j| \le |\varepsilon_{j+1}| , \ j \ge 1 . \qquad (3.14)$$

Hence, if $\varepsilon \neq \varepsilon_j(j=1,2,\ldots)$, the nonhomogeneous system of integral equations (3.12) has a solution $\omega(z) \in L_{p_0}(\overline{D})$, and then Problem B of the complex system (1.36) is solvable. Besides it can be seen that if $K_j = 0(j=1,\ldots,n)$, the general solution of Problem B includes n arbitrary real constants. If $\varepsilon = \varepsilon_j$, where ε_j is an eigenvalue of (3.13) with the rank q , then if $K_j < 0(j=1,\ldots,n)$, Problem B has q solvability conditions, and if $K_j = 0(j=1,\ldots,n)$, using a similar method as in § 3 of Chapter 5, we find that Problem B has q-S solvability conditions and the general solution includes n-S arbitrary real constants, $S \leq min(q,n)$. □

<u>Theorem 3.2</u> Suppose that the linear complex system (1.36) satisfies Condition C_*. If the indices $K_j \leq 0 (j=1,...,n)$ and there are p of them equal to zero, then for $\varepsilon \neq \varepsilon_j (j=1,2,...)$, where $\varepsilon_j (j=1,2...)$ are the eigenvalues of (3.13), Problem B of (1.36) is solvable and its general solution includes p arbitrary real constants. If ε is an eigenvalue of rank q as stated in (3.14), then Problem B has q-S solvability conditions, and its general solution includes p-S arbitrary real constants, $S \leq min(q,p)$.

Besides, the solvability of Problem A for (1.36) is as follows.

If $K_j < 0 (j=1,...,n)$ and $\varepsilon \neq \varepsilon_j (j=1,2,...)$, then Problem A of (1.36) has

$$-2 \sum_{k=1}^{n} K_k + nN-n-S$$

solvability conditions. If $\varepsilon = \varepsilon_j$ is an eigenvalue with rank q, then Problem A has

$$-2 \sum_{k=1}^{n} K_k + nN-n+q-S$$

solvability conditions, and the general solution includes q-S arbitrary real constants,

$$S \leq min(-2 \sum_{k=1}^{n} K_k + nN-n+q, q) .$$

If $K_j = 0 (j=1,...,n)$ and $\varepsilon \neq \varepsilon_j (j=1,2,...)$, then Problem A has nN-S solvability conditions, and the general solution includes n-S arbitrary real constants, $S \leq min(nN,n)$. If $\varepsilon = \varepsilon_j$ is an eigenvalue with the rank q, then Problem A has nN+q-S solvability conditions, and its general solution includes n+q-S arbitrary real constants, $S \leq min(nN+q,n+q)$.

<u>Proof.</u> We only discuss Problem A for (1.36) and substitute the general solution $w(z)$ of Problem B into the boundary condition (2.7). Let

$h(t) = 0$ in (2.7), and if $K_j = 0 (j=1,\ldots,n)$, $\varepsilon \neq \varepsilon_j$, we obtain nN algebraic equations. Denoting the rank of the coefficients matrix of the n arbitrary real constants by S, we can determine S constants among the above arbitrary constants and S equations within the nN albegraic equations. If the other nN-S equations hold, then the vector w(z) is a solution for Problem A of (1.36). Hence, Problem A has nN-S solvability conditions. Let $K_j = 0(j=1,\ldots,n)$ and $\varepsilon = \varepsilon_j$ be an eigenvalue of (3.13) with the rank q. From the Fredholm theorem of linear operator equations, we can find the solvability conditions for the nonhomogeneous operator equations (3.12). Denoting by S_1 the rank of the coefficients matrix of the arbitrary constants we have $S_1 \leq \min(nN,q)$. We can determine S_1 constants among the above arbitrary constants and S_1 equations within the q equations. This shows that Problem B of (1.36) has $q-S_1$ solvability conditions and its general solution includes $n+q-S_1$ arbitrary real constants. Let us substitute the solution w(z) into the boundary condition (2.7) and take $h(t) = 0$. We obtain nN algebraic equations. Denoting the rank of the corresponding coefficients matrix by S_2 we have $S_2 \leq \min(nN,n+q-S_1)$. Therefore, we can determine S_2 constants among the $n+q-S_1$ arbitrary constants. Hence, Problem A has nN+q-S solvability conditions, and its general solution includes n+q-S arbitrary real constants, $S = S_1 + S_2 \leq \min(nN+q,n+q)$.

Similarly, we may discuss the solvability of Problem A for (1.36). □

Next, we assume that the linear complex system (1.36) satisfies Condition C. We rewrite the representation formula (3.2) in the form

$$w(z) = \hat{\Phi}(z) + \hat{T}\omega \; , \; \hat{T}\omega = \tilde{T}\omega + \tilde{\Phi}(z) \; , \tag{3.15}$$

where $\tilde{\Phi}(z)$ is an analytic function so that $\hat{T}\omega$ satisfies the homogeneous boundary condition of (2.5) and

$$Im[\overline{\lambda(t)}\hat{T}\omega]\Big|_{t=a_j} = 0, j \in \{j\} = \begin{cases} 1,\ldots,2K_k-N+1, \text{for } K_k \geq N, \\ N-K_k+1,\ldots,N+1, \text{for } 0 \leq K_k < N, k=1,\ldots,n . \end{cases} \tag{3.16}$$

Let $w(z) = \hat{\Phi}(z) + \hat{T}\omega$ be substituted into (1.36). Corresponding to (3.10), we obtain

$$\omega(z) - Q^{(1)}\hat{S}\omega - Q^{(2)}\overline{\hat{S}\omega} = \varepsilon f(z,\hat{T}\omega) + g(z,\varepsilon) , \tag{3.17}$$

where

$$g(z,\varepsilon) = Q^{(1)}\hat{\Phi}' + Q^{(2)}\overline{\hat{\Phi}'} + \varepsilon f(z,\hat{\Phi}) + A^{(3)} .$$

Let us choose any function $\omega^*(z) \in L_{p_0}(\overline{D})$, and replace $\hat{T}\omega$ on the right-hand side of (3.17) by $\hat{T}\omega^*$. From Theorem 2.3, we know that the system of integral equations

$$(\hat{T}\omega)_{\bar{z}} - Q^{(1)}\hat{S}\omega - Q^{(2)}\overline{\hat{S}\omega} = f(z,\hat{T}\omega^*) + g(z,\varepsilon)$$

has a solution $\hat{T}\omega$. This shows that $I - Q^{(1)}\hat{S} - Q^{(2)}\overline{\hat{S}}$ possesses an inverse operator R . Henc, we obtain the system of integral equations

$$\omega(z) = \varepsilon R[f(z,\hat{T}\omega)] + R[g(z,\varepsilon)] \tag{3.18}$$

Denoting the discrete eigenvalues of the homogeneous system of the integral equation

$$\omega(z) = \varepsilon R[f(z,\hat{T}\omega)] , \tag{3.19}$$

as in (3.14), by a similar method as in the proof of Theorem 3.2, we can prove

Theorem 3.3 Let the linear complex system (1.36) satisfy Condition C. If $\varepsilon \neq \varepsilon_j$, $\varepsilon_j (j=1,2,\ldots)$ are eigenvalues of (3.19), then Problem B of (1.36) is solvable. If ε is an eigenvalue with rank q , then Problem B has $q - S(S \leq q)$ solvability conditions. Moreover, the solvability of Problem A for (1.36) is as follows:

(1) If $K_k \geq N(k=1,\ldots,n)$ and $\varepsilon \neq \varepsilon_j$, $\varepsilon_j(j=1,2,\ldots)$ are as stated in (3.14), the Problem A of (1.36) is solvable, and the general solution $w(z)$ includes

$$2\sum_{k=1}^{n} K_k - nN+n$$

arbitrary real constants. If $\varepsilon = \varepsilon_j$, ε_j is an eigenvalue of rank q, then Problem A of (1.36) has q-S solvability conditions and the general solution includes

$$2\sum_{k=1}^{n} K_k - nN+n-S$$

arbitrary real constants,

$$S \leq \mathit{min}(q, 2\sum_{k=1}^{n} K_k - nN+n+q) .$$

(2) If $0 \leq K_k < N(k=1,\ldots,n)$, and $\varepsilon \neq \varepsilon_j(j=1,2,\ldots)$, then Problem A has

$$nN - \sum_{k=1}^{n} K_k - S$$

solvability conditions, the general solution includes

$$\sum_{k=1}^{n} K_k + n-S$$

arbitrary real constants,

$$S \leq \mathit{min}(nN - \sum_{k=1}^{n} K_k , \sum_{k=1}^{n} K_k + n) .$$

If $\varepsilon = \varepsilon_j$ is an eigenvalue of rank q, then Problem B has

$$nN - \sum_{k=1}^{n} K_k + q-S$$

solvability conditions, and the general solution includes

$$\sum_{k=1}^{n} K_k + n+q-S$$

arbitrary real constants,

$$S \le \min(nN - \sum_{k=1}^{n} K_k + q \;, \; \sum_{k=1}^{n} K_k + n+q) \;.$$

For other cases one can give the total number of solvability conditions for Problem A, too.

2. The oblique derivative problem for complex systems of linear second order equations

Here we consider the complex system of linear second order equations (1.46), and denote by Problem P the oblique derivative problem for the system (1.46) with the boundary condition

$$Re[\overline{\lambda(t)}u_t] = -\varepsilon\sigma(t)u(t) + \tau(t) \;, \; t \in \Gamma \;, \tag{3.20}$$

where $\lambda(t)$ is as stated in (3.1), and $\sigma(t)$, $\tau(t)$ satisfy the conditions

$$C_\alpha[\sigma,\Gamma] \le \ell \;, \; C_\alpha[\tau,\Gamma] \le \ell \tag{3.21}$$

in which $\alpha(1/2 < \alpha < 1)$, $\ell(0 \le \ell < \infty)$ are real constants.

Moreover, we denote by Problem Q the following boundary value problem:

$$Lw = w_{\bar z} - Re[Q(z)w_z] = \varepsilon f(z,u,u_z) + A^{(3)}(z) \;, \tag{3.22}$$

$$\ell w = Re[\overline{\lambda(t)}w(t)] = -\varepsilon\sigma(t)u(t) + \tau(t) + h(t) \;, \; t \in \Gamma \;, \tag{3.23}$$

and the relation

$$u(z) = u_0 + Re\int_0^z [w(\zeta) + \sum_{m=1}^{n} \frac{id_m}{\zeta - z_m}]d\zeta \;, \tag{3.24}$$

where $h(t)$ is an unknown vector as stated in (2.8);

$u_0 = (u_{01},\ldots,u_{0n})^T$, is a real constant vector;

$d_m = (d_{m1},\ldots,d_{mn})^T (m=1,\ldots,N)$ are appropriate real constant vectors, such that the function vector determined by the integral in (3.24) is single-valued in D.

Theorem 3.4 Let $[w(z),u(z)]$ be a solution of Problem Q for the complex system (1.46) and $u(z) \in W^2_{p_0}(D)$, $2 < p_0 < min(p,1/(1-\alpha))$.

Then

(1) $[w(z),u(z)]$ possesses the representation (3.24) in which $w(z)$ can be written in the following form

$$\left.\begin{aligned} w(z) &= \Phi(z) + \tilde{T}\rho \ , \ \tilde{T}\rho = -\frac{1}{\pi}\iint_D G(z,\zeta)\rho(\zeta)d\sigma_\zeta \ , \\ \rho(z) &= u_{z\bar{z}} \ , \ \Phi(z) = \frac{1}{2\pi}\int_\Gamma P(z,t)r(t)d\theta + \Phi_0(z) \ , \end{aligned}\right\} \tag{3.25}$$

where $r(t) = -\varepsilon\sigma(t)u(t) + \tau(t)$, $P(z,t)$ and $G(z,\zeta)$ are the Schwarz kernel and the Green function for Problem Q, respectively, $\Phi(z)$, $\Phi_0(z)$ are analytic vectors, and $G(z,\zeta)$, $\Phi_0(z)$ satisfy the homogeneous boundary condition

$$Re[\overline{\lambda(t)}\phi(t)] = h(t) \ , \ t \in \Gamma \ . \tag{3.26}$$

(2) The double integral vector $\tilde{T}\rho$ possesses the following properties

$$(\tilde{T}\rho)_{\bar{z}} = \rho(z) \ , \ \tilde{S}\rho = (\tilde{T}\rho)_z \ , \tag{3.27}$$

$$L_{p_0}(\tilde{S}\rho,\bar{D}) = \sum_{k=1}^n L_{p_0}[\tilde{S}_k\rho_k,\bar{D}] \le \Lambda_{p_0} L_{p_0}(\rho,\bar{D}) \ , \ \Lambda_{p_0} < \infty \ , \ 2 < p_0 < p \ , \tag{3.28}$$

$$\Lambda_2 \le 1 \ , \ \text{for} \ K_k \le 0 \ , \ k=1,\ldots,n \ , \tag{3.29}$$

$$C_\beta[\tilde{T}\rho,\bar{D}] = \sum_{k=1}^n C_\beta[\tilde{T}_k\rho_k,\bar{D}] \le M_4 L_{p_0}(\rho,\bar{D}) \ , \tag{3.30}$$

where $\beta = 1-2/p_0$, $M_4 = M_4(p_0,D)$. The analytic vector $\Phi(z)$ with condition (2.9) satisfies

$$C_\beta[\Phi,\overline{D}] \le M_5 \ , \ L_{p_0}[\Phi',\overline{D}] \le M_6 \ , \tag{3.31}$$

where $M_j = M_j(p_0,\alpha,\ell,D)$, $j=5,6$. In addition, for the constant $nq_0 < 1$ we are able to find a constant $p_0(>2)$ such that

$$nq_0\Lambda_{p_0} < 1 \ , \ \text{for} \ K_k \le 0 \ , \ k=1,\ldots,n \ . \tag{3.32}$$

Proof. Using a method similar to that of § 3 of Chapter 5 and in the proof of Theorem 3.1, we can obtain the integral representation (3.24) for Problem Q of (1.46), and its properties (3.27) - (3.32). □

Finally, we apply the Fredholm theorem for linear operator equations to discuss the solvability for Problem Q and Problem P. In fact, if $K_k \ge N(k=1,\ldots,n)$, by Theorem 2.3 we find a general solution $w(z) = w_0(z) + w_*(z)$ for the boundary value problem Q_1:

$$Lw = A^{(3)}(z) \ , \ z \in D \ , \tag{3.33}$$

$$\ell w = \tau(t) + h(t) \ , \ t \in \Gamma \ , \tag{3.34}$$

where $w_0(z)$ is a special solution of Problem Q_1, and

$$w_*(z) = \sum_{m=1}^{2K-N+1} c_m w_m(z) \ , \ c_m = (c_{1m},\ldots,c_{2K_m-N+1m}) \ , \tag{3.35}$$

$$w_m = (w_{1m},\ldots,w_{2K_m-N+1m})^T$$

is the general solution of the corresponding homogeneous problem. Next, for any vector $u(z) \in C^1(\overline{D})$, we are able to find a solution $w(z) \in C_\beta(\overline{D})$ for

$$Lw = f(z,u,u_z) \ , \ z \in D \ , \tag{3.36}$$

$$\ell w = -\sigma(t)u(t) + h(t) \ , \ t \in \Gamma \ , \tag{3.37}$$

$$Im[\overline{\lambda(a_j)}w(a_j)]=0, j \in \{j\} \begin{cases} 1,\dots,2K_k-N+1 \text{ , for } K_k \ge N \text{ ,} \\ N-K_k+1,\dots,N+1, \text{for } 0 \le K_k < N \text{ , } k=1,\dots,n. \end{cases} \tag{3.38}$$

We denote by $w = R_2 u$ the mapping from $u(z) \in C^1(\overline{D})$ onto $w(z) \in C(\overline{D})$, and by $u = R_1 w$ the mapping (3.24) from $w(z) \in C(\overline{D})$ onto $u(z) \in C^1(\overline{D})$. It is clear that R_1 is a linear and bounded operator, and R_2 is a linear bounded and completely continuous operator. Thus, we can obtain the following linear operator equation

$$u - \varepsilon R_1 R_2 u = R_1 w_0 + R_1 w_* + u_0 \text{ ,} \tag{3.39}$$

where u_0 is an arbitrary constant vector, $w_0(z)$, $w_*(z)$ are as stated before. Because R_1R_2 is a completely continuous operator from $C^1(\overline{D})$ onto itself, we can apply the Fredholm theorem for the linear operator equation (3.39). Let

$$\varepsilon_j (j=1,2,\dots) \text{ , } |\varepsilon_j| \le |\varepsilon_{j+1}| \text{ , } j \ge 1 \tag{3.40}$$

be the eigenvalues of the homogeneous system

$$u - \varepsilon R_1 R_2 u = 0 \text{ .} \tag{3.41}$$

If $\varepsilon \neq \varepsilon_j (j=1,2,\dots)$, Problem Q is solvable, and its solution

$$u(z) = \varepsilon R_1 R_2 u + R_1 w_0 + R_1 w_* + u_0 \tag{3.42}$$

includes

$$2\sum_{k=1}^{n} K_k - nN+2n$$

arbitrary real constants. If $\varepsilon = \varepsilon_j$ is an eigenvalue of (3.41) with the rank q, we denote by S the rank of the coefficients matrix of the arbitrary constants,

$$S \le min(q, 2\sum_{k=1}^{n} K_k - nN+2n) \text{ .}$$

Then we can determine S constants within the above arbitrary constants, and simultaneously determine S equalities within the q algebraic equations. This shows that Problem Q of (3.22) has q-S solvability conditions, and its general solution includes

$$2\sum_{k=1}^{n} K_k - nN+2n-S$$

arbitrary real constants.

As for Problem P of (1.46), we substitute the solution $[w(z),u(z)]$ of Problem Q into the relation (3.24), and set $d_m = 0$, $m=1,\ldots,n$. If ε is not an eigenvalue of (3.41), using the above method and denoting the rank of the coefficients matrix of the arbitrary constants by S, and

$$S \le \min(nN, 2\sum_{k=1}^{n} K_k - nN+2n),$$

then Problem P of (1.46) has nN-S solvability conditions, and its general solution includes

$$2\sum_{k=1}^{n} K_k - nN+2n-S$$

arbitrary real constants. If ε is an eigenvalue of (3.41) with rank q, similarly, we can deal with the solvability of Problem P for (1.46).

Thus, we have the following theorem.

Theorem 3.5 Suppose that the linear complex system (1.46) satisfies Condition C. If $K_k \le 0 (k=1,\ldots,n)$, Condition C may be replaced by Condition C_*. The solvability of Problem Q for (3.22) is as follows:

(1) If ε is not an eigenvalue of (3.41), then Problem Q of (3.22) is solvable.

(2) If ε is an eigenvalue with the rank q, then Problem Q of (3.22) has q-S solvability conditions, $S \le q$.

Moreover, the result of the solvability of Problem P for (1.46) is as follows:

(1) If $K_k \geq N(k=1,\ldots,n)$ and $\varepsilon \neq \varepsilon_j$, $\varepsilon_j(j=1,2,\ldots)$ are as stated in (3.40), Problem P of (1.46) has nN-S solvability conditions,

$$S \leq min(nN, 2\sum_{k=1}^{n} K_k - nN+2n),$$

and its general solution $u(z)$ includes

$$2\sum_{k=1}^{n} K_k - nN+2n-S$$

arbitrary real constants. If $\varepsilon \neq \varepsilon_j$ is an eigenvalue of (3.41) with the rank q , then Problem P of (1.46) has nN+q-S solvability conditions and the general solution includes

$$2\sum_{k=1}^{n} K_k - nN+2n+q-S$$

arbitrary real constants,

$$S \leq min(nN+q\ ,\ 2\sum_{k=1}^{n} K_k - nN+2n+q)\ .$$

(2) If $0 \leq K_k < N(k=1,\ldots,n)$, and $\varepsilon \neq \varepsilon_j$, then Problem P of (1.46) has

$$2nN - \sum_{k=1}^{n} K_k - S$$

solvability conditions, the general solution includes

$$\sum_{k=1}^{n} K_k + 2n-S$$

arbitrary real constants,

$$S \leq min(2nN - \sum_{k=1}^{n} K_k\ ,\ \sum_{k=1}^{n} K_k + 2n)\ .$$

If $\varepsilon = \varepsilon_j$ is an eigenvalue with rank q , then Problem P has

$$2nN - \sum_{k=1}^{n} K_k + q-S$$

solvability conditions, and the general solution includes

$$\sum_{k=1}^{n} K_k + 2n+q-S$$

arbitrary real constants,

$$S \le min(2nN-\sum_{k=1}^{n} K_k + q \ , \ \sum_{k=1}^{n} K_k + 2n+q) \ .$$

(3) If $K_k < 0(k=1,\ldots,n)$, and $\varepsilon \neq \varepsilon_j$, then Problem P of (1.46) has

$$-2\sum_{k=1}^{n} K_k + 2nN-n-S$$

solvability conditions; and the general solution includes n-S arbitrary real constants,

$$S \le min(-2\sum_{k=1}^{n} K_k + 2nN-n,n) \ .$$

If $\varepsilon = \varepsilon_j$ is an eigenvalue with rank q , then Problem P of (1.46) has

$$-2\sum_{k=1}^{n} K_k + 2nN-n+q-S$$

solvability conditions, and the general solution includes n+q-S arbitrary real constants,

$$S \le min(-2\sum_{k=1}^{n} K_k + 2nN-n+q \ , \ n+q) \ .$$

For other cases, we can also give the total number of solvability conditions for Problem P and Problem Q.

§ 4. The oblique derivative problem for complex systems of nonlinear second order equations

1. Formulation of the oblique derivative problem and the corresponding modified problem

Problem P. The oblique derivative boundary value problem of (1.39) is to find a solution $u(z)$ of (1.39) in D, which satisfies the boundary condition

$$Re[\overline{\Lambda(t)}u_t] = -\varepsilon B^{(1)}(t)u(t) + B^{(2)}(t) \ , \ t \in \Gamma \ , \tag{4.1}$$

where $\Lambda(t) = (\Lambda_{km}(t))$, $B^{(j)} = (B_1^{(j)}(t),\ldots,B_n^{(j)}(t))^T$, $j=1,2$, and $|\Lambda(t)| \neq 0$. We can reduce (4.1) to the following form

$$Re[\overline{\lambda(t)}u_t] = -\varepsilon\sigma(t)u(t) + Re\beta(t,u_t) \ , \ t \in \Gamma \ , \tag{4.2}$$

in which

$$\lambda(t) = \begin{pmatrix} \lambda_1(t) & & 0 \\ & \ddots & \\ 0 & & \lambda_n(t) \end{pmatrix}, \ \sigma(t) = (\sigma_{km}(t)) \ , \ \beta(t,u_t) = (\beta_{km}(t,u_t)) \ .$$

This is a nonlinear boundary value problem, which will be discussed in Chapter 7.*) Here, we only consider the linear case, i.e. $Re\beta(t,u_t) = \tau(t)$, and suppose that $\lambda(t)$, $\sigma(t)$, $\tau(t)$ satisfy the following conditions

$$|\lambda(t)| = 1 \ , \ C_\alpha[\lambda,\Gamma] \le \ell \ , \ C_\alpha[\sigma,\Gamma] \le \ell \ , \ C_\alpha[\tau,\Gamma] \le \ell \ , \tag{4.3}$$

where $\alpha (1/2 < \alpha < 1)$, $\ell (0 \le \ell < \infty)$ are real constants.

In order to obtain the result on solvability of Problem P, we need the corresponding boundary value problem (Problem Q) for the nonlinear system of first order complex equations

*) A second volume to this book is in preparation.

$$\left.\begin{aligned}
&w_{\bar z} = f(z,u,u_z,w,w_z)\ ,\ f = Re[Qw_z+\varepsilon(A^{(1)}w+A^{(2)}u)+A^{(3)}]\ ,\\
&Q = (Q_{km})\ ,\ A^{(j)} = (A^{(j)}_{km})\ ,\ A^{(3)} = (A^{(3)}_1,\dots,A^{(3)}_n)^T\ ,\\
&Q_{km}=Q_{km}(z,u,u_z,w_z),A^{(j)}_{km}(z,u,u_z),A^{(3)}_k=A^{(3)}_k(z,u,u_z),\\
&1 \le k\ ,\ m \le n\ ,\ 1 \le j \le 2
\end{aligned}\right\} \tag{4.4}$$

with the boundary conditions (3.23) and

$$Im[\overline{\lambda(a_j)}w(a_j)] = b_j = (b_{j1},\dots,b_{jn})^T\ ,$$

$$j \in \{j\} = \begin{cases} 1,\dots,2K_k-N+1\ , & \text{for } K_k \ge N\ ,\\ N-K_k+1,\dots,N+1\ , & \text{for } 0 \le K_k < N\ ,\end{cases} \tag{4.5}$$

and the relation (3.24), where $a_j(j=1,\dots,2K_k-N+1)$ are as stated in (1.3) of Chapter 1, u_{0k} , $b_{jk}(j\in\{j\},k=1,\dots,n)$ are real constants satisfying the conditions

$$|u_0| \le \ell\ ,\ |b_{jk}| \le \ell(j\in\{j\},k=1,\dots,n)\ ,$$

and $d_m = (d_{m1},\dots,d_{mn})^T$, $d_{mk}(m=1,\dots,N,k=1,\dots,n)$ are appropriately selected constants such that $u_k(z)$ $(k=1,\dots,n)$ in (3.24) are single-valued functions in D .

The oblique derivative boundary value problem (Problem P) of (1.49) may be formulated as follows. Find a solution $w(z)$ of (1.49) satisfying the boundary condition

$$Re[\overline{\lambda^{(j)}(t)}V^{(j)}(t)+\varepsilon\sigma^{(j)}(t)w(t)] = \tau^{(j)}(t),t\in\Gamma\ ,\ j=1,2\ , \tag{4.6}$$

where $V^{(1)} = U = w_z$, $V^{(2)} = V = \bar w_z$, and $\lambda^{(j)}(t)$, $\sigma^{(j)}(t)$, $\tau^{(j)}(t)$ $(j=1,2)$ satisfy the conditions

$$|\lambda^{(j)}(t)| = 1,C_\alpha[\lambda^{(j)},\Gamma]\le\ell,C_\alpha[\sigma^{(j)},\Gamma]\le\ell,C_\alpha[\tau^{(j)},\Gamma]\le\ell\ , \tag{4.7}$$

where $j=1,2$, $\alpha(1/2<\alpha<1)$, $\ell(0<\ell<\infty)$, $(0<\varepsilon\le 1)$ are constants.

The corresponding modified problem (Problem Q) is to find a solution $[w(z),U(z),V(z)]$ of the complex system of first order

$$\left.\begin{aligned}&U_{\bar z}=f(z,w,w_z,U,V,U_z,V_z)\ ,\ f=Q^{(1)}U+Q^{(2)}V+A^{(1)}U+A^{(2)}\overline{U}\\&\quad+A^{(3)}V+A^{(4)}\overline{V}+A^{(5)}w+A^{(6)}\overline{w}+A^{(7)}\ ,\ V_{\bar z}=\overline{U}_z\ ,\\&Q^{(j)}=Q^{(j)}(z,w,w_z,\overline{w}_z,U,V,U_z,V_z)\ ,\ j=1,2\ ,\\&A^{(j)}=A^{(j)}(z,w,w_z,\overline{w}_z)\ ,\ j=1,\ldots,7\ ,\end{aligned}\right\}\qquad(4.8)$$

satisfying the boundary conditions

$$Re[\overline{\lambda^{(j)}(t)V^{(j)}(t)+\varepsilon\sigma^{(j)}(t)w(t)}]=\tau^{(j)}(t)+h^{(j)}(t)\ ,\ t\in\Gamma\ ,\ j=1,2\ ,\qquad(4.9)$$

and point conditions

$$Im[\overline{\lambda^{(j)}(a_k)V^{(j)}(a_k)+\varepsilon\sigma^{(j)}(a_k)w(a_k)}]=b_k^{(j)}\ ,\ k\in\{k\}\ ,\ j=1,2\ ,\qquad(4.10)$$

and the relation

$$w(z)=w_0+\int_0^z\left\{\left[V^{(1)}(\zeta)+\sum_{k=1}^N\frac{d_k}{\zeta-z_k}\right]d\zeta+\overline{V^{(2)}(\zeta)d\bar\zeta}\right\}\ ,\ |w_0|\le\ell,\qquad(4.11)$$

where $U=V^{(1)}$, $V=V^{(2)}$, $h^{(j)}(t)(j=1,2)$ are unknown vectors similar to (3.28) in Chapter 5 and to (2.8); a_k, $b_k^{(j)}$, $\{k\}$ are defined as in (4.5) with conditions $|b_k^{(j)}|\le\ell$, and $d_k(k=1,\ldots,N)$ are appropriate complex constant vectors so that the function vector $w(z)$ is singlevalued in D.

In order to discuss the solvability of Problem P and Problem Q, we need to give a priori estimates of solutions to Problem Q for (4.4) and (4.8).

2. A priori estimates of solutions for the oblique derivative problem

Theorem 4.1 Let (1.39) satisfy Condition C and the constant ε in (1.41), (1.43) and (4.4) be small enough. Then the solution $[w(z),u(z)]$ for Problem Q of (4.4) satisfies the estimates

$$Rw = C_\beta[w,\overline{D}] + L_{p_0}[|w_{\bar{z}}| + |w_z|,\overline{D}] < M_1 , \tag{4.12}$$

$$Su = C_\beta^1[u,\overline{D}] < M_2 , \tag{4.13}$$

where $p_0 (2 < p_0 < p)$, $\beta = 1-2/p_0$, M_1 , M_2 are constants, $M_j = M_j(q_0,p_0,k_0,\alpha,\ell,D)$, $j=1,2$.

Proof. From (4.4) and (3.23), we have

$$L_{p_0}\left[\varepsilon \sum_{m=1}^{n} A_{km}^{(2)} u_m + A_k^{(3)},\overline{D}\right] \le (n\varepsilon Su+1)k_0 , \tag{4.14}$$

$$C_\alpha[-\varepsilon\sigma u+\tau,\Gamma] \le (\varepsilon Su+1)\ell . \tag{4.15}$$

With $H = \varepsilon Su(n+1)+1$ and $w^*(z) = w(z)/H$, $u^*(z) = u(z)/H$, it is obvious that $w^*(z)$ is a solution of Problem B for the complex system of first order

$$w_{\bar{z}}^* = f^*(z,u,u_z,w^*,w_z^*) , \quad f^* = Re[Qw_z^*+\varepsilon A^{(1)}w^*+A^*] ,$$

and the boundary conditions

$$Re[\overline{\lambda(t)}w^*(t)] = r(t) , \quad r(t) = -\varepsilon\sigma(t)u^*(t) + [\tau(t)+h(t)]/H , \quad t \in \Gamma ,$$

$$Im[\overline{\lambda(a_j)}w^*(a_j)] = b_j/H , \quad j \in \{j\} .$$

If the constant ε is sufficiently small, then $A^* = \varepsilon A^{(2)}u^* + A^{(3)}/H$ and $r(t) = -\varepsilon\sigma(t)u^*(t) + \tau(t)/H$ is satisfying the following conditions

$$L_{p_0}[A_k^*,\overline{D}] \le 2k_0 , \quad C_\alpha[r,\Gamma] \le 2\ell , \quad |b_j/H| \le \ell ,$$

in which $A^* = (A_1^*,\dots,A_n^*)^T$.

By Theorem 2.1, we see that the solution $w^*(z)$ satisfies the following estimate

$$Rw^* \le M_3 = M_3(q_0,p_0,k_0,\alpha,\ell,D) , \tag{4.16}$$

and then

$$Rw \le M_3 H = M_3[\varepsilon Su(n+1)+1] \ . \qquad (4.17)$$

From (3.24), it follows that

$$Su \le M_4 C_\beta[w,\overline{D}] + \ell \le M_4 Rw + \ell \ , \qquad (4.18)$$

where $M_4 = M_4(p_0,D)$. Combining (4.17) and (4.18), we obtain

$$Su \le M_4 M_3[\varepsilon Su(n+1)+1] + \ell \ . \qquad (4.19)$$

Choosing the constant ε so small that $\varepsilon(n+1)M_3M_4 < 1$, we conclude

$$Su \le [M_3M_4+\ell]/[1-\varepsilon(n+1)M_3M_4] = M_2 \ . \qquad (4.20)$$

Moreover, we have

$$Rw \le M_3[\varepsilon M_2(n+1)+1] = M_1 \ . \qquad (4.21)$$

This completes the proof. □

If (1.39) satisfies Condition C_*, we can obtain a similar result.

Theorem 4.2 Suppose that (1.39) satisfies Condition C_* and the constant ε in (1.44) and (4.4) is sufficiently small, then the solution $[w(z),u(z)]$ to Problem Q with the condition (3.1) and $K_j \le 0$ $(j=1,\ldots,n)$ satisfies the estimates (4.12) and (4.13).

Proof. Let the solution $[w(z),u(z)]$ of Problem Q be substituted into (4.4), (4.5), (3.23) and (3.24). Noting that the solution $w(z)$ is representable by (3.25), the complex system (4.8) can be written in the form

$$w_{\bar{z}} - \mathit{Re}[Qw_z] = A \ , \ A = \mathit{Re}[\varepsilon(A^{(1)}w+A^{(2)}u)+A^{(3)}] \ , \qquad (4.22)$$

where A satisfies the inequality

$$L_{p_0}[A_j,\overline{D}] \leq \varepsilon n k_0(Rw+Su) + k_0 \ , \ j=1,\dots,n \ . \tag{4.23}$$

With $H = \varepsilon(n+1)(Rw+Su)+1$, and $w^*(z) = w(z)/H$, $u^*(z) = u(z)/H$, the function $w^*(z)$ is a solution of the complex system

$$w^*_{\bar{z}} - \mathit{Re}[Qw^*_z] = A^* \ , \ A^* = \mathit{Re}[\varepsilon(A^{(1)}w^*+A^{(2)}u^*)+A^{(3)}/H] \ .$$

Provided the constant ε is small enough, by Theorem 2.2 and Theorem 3.4, we see that the estimate (4.16) is satisfied. Hence we have

$$Rw \leq M_3 \ H \leq M_3[\varepsilon(n+1)(Rw+Su)+1] \ .$$

The remaining part of the proof is similar to the proof of Theorem 4.1. So we obtain the estimates (4.12) and (4.13). □

Using a similar method to that in the proof of Theorem 4.2, one can show

Theorem 4.3 Let (1.49) satisfy Condition C, and the constant ε in (1.50), (1.51) and (4.9), (4.10) be small enough. Then the solution $[w(z),V^{(1)}(z),V^{(2)}(z)]$ for Problem Q of (4.8) satisfies the estimates

$$RV^{(j)} = C_\beta[V^{(j)},\overline{D}] + L_{p_0}[|V^{(j)}_{\bar{z}}|+|V^{(j)}_z|,\overline{D}] < M_5 \ , \ j=1,2 \ , \tag{4.24}$$

$$Sw = C^1_\beta[w,\overline{D}] < M_6 \ , \tag{4.25}$$

where $2 < p_0 < \mathit{min}(p,1/(1-\alpha)), \beta = 1-2/p_0$, $M_j = M_j(q_0,p_0,k_0,\alpha,\ell,D)$, $j=5,6$.

3. Solvability of the oblique derivative problem for nonlinear complex systems of second order

Theorem 4.4 Under the assumption of Theorem 4.2, Problem Q for (4.4) with the condition (3.1) and $K_j \leq 0(j=1,\dots,n)$ is solvable, and the solvability of Problem P is as follows.

(1) If $K_k < 0$, $k=1,\ldots,n$, Problem P of (1.39) has

$$-2\sum_{k=1}^{n} K_k + n(2N-1)$$

solvability conditions.

(2) If $K_k = 0$, $k=1,\ldots,n$, Problem P of (1.39) has 2nN solvability conditions.

(3) If $K_k < 0$ $(k=1,\ldots,m)$ and $K_k = 0$ $(k=m+1,\ldots,n)$, Problem P of (1.39) has

$$-2\sum_{k=1}^{m} K_k + 2nN-m$$

solvability conditions.

Proof. Let us introduce a bounded and open set B_M in the Banach space $B = C^1(\overline{D}) \times L_{p_0}(\overline{D})$, the elements of which are the vectors of functions $\omega(z) = [u(z),\rho(z)]$ satisfying the condition

$$C^1[u,\overline{D}] < M_2 \ , \ L_{p_0}[\rho,\overline{D}] < M_1 \ , \tag{4.26}$$

where M_1 , M_2 are constants as stated in (4.12), (4.13). We arbitrarily choose $[u(z),\rho(z)] \in \overline{B_M}$ and consider the double integral vector $\widetilde{T}\rho$. Substituting $u(z)$ into proper positions of the boundary conditions (3.23) and (4.5), we can find an analytic vector $\Phi(z)$ in D satisfying the boundary condition

$$Re\{\overline{\lambda(t)}[\Phi(t)+\widetilde{T}\rho]\} = \varepsilon\sigma(t)u(t) + \tau(t)+h(t) \ , \ t \in \Gamma \ , \tag{4.27}$$

$$Im\{\overline{\lambda(z)}[\Phi(z)+\widetilde{T}\rho]\}\Big|_{z=a_j} = b_j \ , \ j \in \{j\} \ , \tag{4.28}$$

one can also prove that $\Phi(z)$, $w(z) = \Phi(z) + \widetilde{T}\rho$ satisfies the esitmates

$$C_\beta[w,\overline{D}] \le M_7 \ , \ L_{p_0}[\Phi',\overline{D}] \le M_8 \ , \tag{4.29}$$

where $M_j = M_j(p_0,\alpha,\ell,D,M_1,M_2)$, $j=7,8$. Now, after substituting $u(z)$, $\Phi'(z)$ into the appropriate positions of the complex system (4.4), we consider the system of integral equations with the parameter $t(0 \le t \le 1)$

$$\rho^*(z) = tf(z,u,u_z,\Phi+\widetilde{T}\rho,\Phi'+\widetilde{S}\rho^*) \ , \tag{4.30}$$

where $\widetilde{S}\rho^* = (\widetilde{T}\rho^*)_z$. By Condition C and the principle of contracting mappings, the system (4.30) has a unique solution $\rho^*(z) \in L_{p_0}(\overline{D})$. Next, we can find a corresponding analytic vector $\Phi^*(z)$ in D similar to the above analytic vector $\Phi(z)$, and determine the corresponding single-valued vector

$$u^*(z) = u_0 + Re\int_0^z [\Phi^*(\zeta)+\widetilde{T}\rho^* + \sum_{m=1}^N \frac{id_m}{\zeta-z_m}]d\zeta \ . \tag{4.31}$$

Let us denote the mapping from ω onto $\omega^* = [u^*(z),\rho^*(z)]$ by $\omega^* = S(\omega,t)$ $(0 \le t \le 1)$. If the constant ε is sufficiently small, then we can show that the solution $\omega = [u(z),\rho(z)]$ of the system

$$\rho(z) = tf(z,u,u_z,\Phi+\widetilde{T}\rho,\Phi'+\widetilde{S}\rho) \ , \ 0 \le t \le 1 \tag{4.32}$$

satisfies (4.26), i.e. $\omega = [u(z),\rho(z)] \in B_M$. Furthermore, we can verify that the mapping $\omega^* = S(\omega,t)$ $(0 \le t \le 1)$ satisfies the conditions of the Leray-Schauder theorem. Therefore, the system (4.32) with $t = 1$ has a solution $\omega = [u(z),\rho(z)] \in B_M$, and then $[w(z),u(z)]$ is a solution to Problem Q for (4.4) with $w(z) = \Phi(z)+\widetilde{T}\rho$.

If we substitute the above solution $[w(z),u(z)]$ into the boundary conditions (3.23), (4.5) and into the relation (3.24) with

$$d_m = 0 \ , \text{ i.e. } Re\int_{\Gamma_m} [\Phi(\zeta)+\widetilde{T}\rho]d\zeta = 0 \ , \ m=1,\dots,N \ , \tag{4.33}$$

$$h(t) = 0 \ , \ t \in \Gamma \ , \tag{4.34}$$

then the solution $u(z)$ for Problem Q is also a solution to Problem P for (1.39), so that the total number of solvability conditions of Problem P for (1.39) are as stated in the theorem. □

Secondly, we prove the general result for Problem Q and Problem P under different conditions.

Theorem 4.5 Under the hypothesis of Theorem 4.1, Problem Q for (4.4) is solvable, and the solvability of Problem P for (1.39) is as follows.

(1) If $K_k \geq N$, $k=1,\ldots,m$, $0 \leq K_k < N$, $k=m+1,\ldots,n$, then Problem P of (1.39) has

$$2nN - \sum_{k=m+1}^{n} K_k - mN$$

solvability conditions.

(2) If $K_k \geq N$, $k=1,\ldots,m$, $K_k < 0$, $k=m+1,\ldots,n$, then Problem P has

$$-2\sum_{k=m+1}^{n} K_k + (n-m)(N-1) + nN$$

solvability conditions.

(3) If $0 \leq K_k < N$, $k=1,\ldots,m$, $K_k < 0$, $k=m+1,\ldots,n$, then Problem P has

$$-\sum_{k=1}^{m} K_k - 2\sum_{k=m+1}^{n} K_k + 2nN - (n-m)$$

solvability conditions.

In addition, we can give the results in the other cases.

Proof. We first consider the complex system

$$w_{\bar{z}} = f^{(m)}(z,u,u_z,w,w_z) \ , \ f^{(m)} = \sigma_m(z)f(z,u,u_z,w,w_z) \ , \tag{4.35}$$

where $\sigma_m(z)$, $f(z,u,u_z,w,w_z)$ are as stated in (2.34) of Chapter 2 and (4.4), respectively. Let B_M be a bounded and open set as stated in (4.26), and we arbitrarily select $[u(z),\rho(z)] \in \overline{B_M}$. According to the proof of Theorem 2.3, we consider the double integral vector

$$T\rho = -\frac{1}{\pi}\iint \frac{\rho(\zeta)}{\zeta - z}\, d\sigma_\zeta .$$

The remaining proof is similar to the proof of Theorem 4.4, but we have to replace $\tilde{T}\rho$, $\tilde{S}\rho$ by $T\rho$, $\Pi\rho$ in (4.27) - (4.33), respectively. Moreover, $\Phi(z)$, $\Phi^*(z)$ are vectors of analytic functions in D such that $\Phi(z) + T\rho$, $\Phi^*(z) + T\rho^*$ satisfy the boundary conditions (4.27), (4.28). Thus, we can find a solution $[w(z),u(z)]$ for Problem Q of (4.35). According to the method in the proof of Theorem 2.5 of Chapter 5, we can eliminate the assumption that the coefficients are equal to zero in the neighbourhood of the boundary Γ .

In addition, we can also derive the result for solvability for Problem P of (1.39). □

<u>Theorem 4.6</u> Under the hypothesis in Theorem 4.3, Problem Q for (4.8) is solvable, and the solvability of Problem P for (1.49) can be obtained. For instance, if

$$K_k^{(j)} = \frac{1}{2\pi}\Delta_\Gamma \arg \lambda_k^{(j)}(t)$$

satisfies the condition $0 \le K_k^{(j)} < N$, $k=1,\ldots,n$, $j=1,2$, then Problem P has

$$4nN - \sum_{j=1}^{2}\sum_{k=1}^{n} K_k^{(j)}$$

solvability conditions.

<u>Proof.</u> Using a similar method to that in the proof of Theorem 4.5, we introduce a bounded and open set B_M , the elements of which are the vectors of functions $\omega = [w(z),\rho^{(1)}(z),\rho^{(2)}(z)]$ satisfying the conditions

$$C^{(1)}[w,\overline{D}] < M_6 \ , \ L_{p_0}[\rho^{(j)},\overline{D}] < M_5 \ , \ j=1,2 \ , \tag{4.36}$$

where M_6 , M_5 are nonnegative constants as stated in (4.24), (4.25). Choosing any vector $[w(z),\rho^{(1)}(z),\rho^{(2)}(z)] \in \overline{B_M}$, and considering the double integral vector

$$T\rho^{(j)} = -\frac{1}{\pi}\iint_D \frac{\rho^{(j)}(\zeta)}{\zeta-z}\, d\sigma_\zeta \ ,$$

we can find two vecotrs of analytic functions $\Phi^{(j)}(z)$ satisfying the boundary conditions

$$Re\{\overline{\lambda^{(j)}(t)}[\Phi^{(j)}(t)+T\rho^{(j)}]+\varepsilon\sigma^{(j)}(t)w(t)\} = \tau^{(j)}(t) + h^{(j)}(t) \ , \tag{4.37}$$

$$Re\{\overline{\lambda^{(j)}(z)}[\Phi^{(j)}(z)+T\rho^{(j)}]+\varepsilon\sigma^{(j)}(z)w(z)\}\Big|_{z=a_k} = b_k^{(j)} , k \in \{k\}, j=1,2. \tag{4.38}$$

Let $V^{(j)}(z) = \Phi^{(j)}(z) + T\rho^{(j)}$, $j=1,2$. Substituting $w(z)$, $V^{(j)}(z)$, $\Phi^{(j)\prime}(z) + \Pi\rho^{*(j)} (\Pi\rho^{*(j)} = (T\rho^{*(j)})_z)$ into proper positions of the complex system

$$\left.\begin{aligned}&\rho^{*(1)}(z) = tf^{(n)}(z,w,w_z,\bar{w}_z,\Phi^{(1)}+T\rho^{(1)},\Phi^{(2)}+T\rho^{(2)} , \Phi^{(1)\prime} + \\ &\Pi\rho^{*(1)},\Phi^{(2)\prime} + \Pi\rho^{*(2)}), 0 \le t \le 1, f^{(n)}=\sigma_n(z)f, \rho^{*(2)}(z)=\overline{\rho^{*(1)}(z)},\end{aligned}\right\} \tag{4.39}$$

by the principle of contraction we find $\rho^{*(j)}(z)(j=1,2)$ and the corresponding vectors of analytic functions $\Phi^{*(j)}(z)(j=1,2)$ in D . Next, we can determine a single-valued function vector

$$w^*(z) = w_0 + \int_0^z \left\{\left[V^{*(1)}(\zeta)+\sum_{k=1}^{N} \frac{d_k}{\zeta-z_k}\right]d\zeta+\overline{V^{*(2)}(\zeta)d\zeta}\right\} , \tag{4.40}$$

and denote by $\omega^* = [w^*(z),\rho^{*(1)}(z),\rho^{*(2)}(z)] = S(\omega,t)$ $(0 \le t \le 1)$ a mapping from ω onto ω^* , which satisfies the conditions of the Leray-Schauder theorem. So we can conclude that Problem Q for (4.8) is solvable. Moreover, we can derive a result for the solvability of Problem P for (1.49). □

References

[1] Agmon, S., Douglis, A., Nirenberg, L.
Estimates near the boundary for solutions of elliptic partial differential equations satisfying general boundary conditions. I, II. Comm. Pure Appl. Math. 12 (1959), 623-727; 17 (1964), 35-92.

[2] Ahlfors, L.
1) Lecture on quasiconformal mappings. Princeton, 1966.
2) Conformality with respect to Riemannian metrics. Ann. Acad. Sci. Fenn. A.I. 206 (1955), 22 pp.

[3] Begehr, H.
1) Randwertaufgaben für elliptische und für zusammengesetzte Systeme partieller fastlinearer Differentialgleichungen erster Ordnung. Komplexe Analysis und ihre Anwendung auf partielle Differentialgleichungen. Martin-Luther-Univ., Halle-Wittenberg, Wiss. Beiträge 27 (1977), 6-10.
2) Boundary value problems for mixed kind systems of first order partial differential equations. 3. Roumanian-Finnish Seminar on Complex Analysis, Bucharest 1976, Lecture Notes in Math. 743, Springer-Verlag, Berlin etc., 1979, 600-614.
3) An approximation method for the Dirichlet problem of nonlinear elliptic systems in $\mathbb{R}^r$. Rev. Roumaine Math. Pure Appl. 27 (1982), 927-934.
4) Boundary value problems for analytic and generalized analytic functions. Complex Analysis - Methods, Trends, and Applications. Akademie-Verlag, Berlin 1983, 150-165.
5) Boundary value problems for systems with Cauchy Riemannian main part. Complex Analysis. Fifth Roumanian-Finnish Seminar, Bucharest 1981. Lecture Notes in Math. 1014, Springer-Verlag, Berlin etc., 1983, 265-279.
6) Remark on Hilbert's boundary value problem for Beltrami systems. Proc. Roy. Soc. Edinburgh 98A (1984), 305-310.

[4] Begehr, H., Gilbert, R.P.
1) Über das Randwert-Normproblem für ein nichtlineares elliptisches System. Function Theoretic Methods Part. Diff. Equ., Darmstadt 1976, Lecture Notes in Math. 561, Springer-Verlag, Berlin etc., 1976, 112-122.
2) Das Randwert-Normproblem für ein fastlineares elliptisches System und eine Anwendung. Ann. Acad. Sci. Fenn. A.I. 3 (1977), 179-184.

3) Randwertaufgaben ganzzahliger Charakteristik für verallgemeinerte hyperanalytische Funktionen. Appl. Anal. 6 (1977), 189-205.
4) On Riemann boundary value problems for certain linear elliptic systems in the plane. J. Diff. Equ. 32 (1979), 1-14.
5) Boundary value problems associated with first order elliptic systems in the plane. Contemporary Math. 11 (1982), 13-48.

[5] Begehr, H., Hile, G.N.
1) Nonlinear Riemann boundary value problems for a semilinear elliptic system in the plane. Math. Z. 179 (1982), 241-261.
2) Riemann boundary value problems for nonlinear elliptic systems. Complex Variables, Theory Appl. 1 (1983), 239-261.

[6] Begehr, H., Hsiao, G.C.
1) Nonlinear boundary value problems for a class of elliptic systems. Komplexe Analysis und ihre Anwendungen auf partielle Differentialgleichungen. Martin-Luther-Univ., Halle-Wittenberg. Wiss. Beiträge 1980, 90-102.
2) On nonlinear boundary value problems of elliptic systems in the plane. Ord. Part. Diff. Equ. Proc. Dundee 1980. Lecture Notes in Math. 846, Springer-Verlag, Berlin etc., 1981, 55-63.
3) Nonlinear boundary value problems of Riemann-Hilbert type. Contemporary Math. 11 (1982), 139-153.
4) The Holbert boundary value problem for nonlinear elliptic systems. Proc. Roy. Soc. Edinburgh 94A (1983), 97-112.
5) A priori estimates for elliptic systems. Z. Anal. Anw. 6 (1987), 1-21.

[7] Begehr, H., Wen Guo-chun
1) The discontinuous oblique derivative problem for nonlinear elliptic systems of first order. Rev. Roumaine Math. Pures Appl. 33 (1988), 7-19.
2) A priori estimates for the discontinuous oblique derivative problem for elliptic systems. Math. Nachr. 142 (1989), 307-336.

[8] Bergman, S.
Integral operators in the theory of linear partial differential equations. Springer-Verlag, Berlin etc., 1961.

[9] Bergman, S., Schiffer, M.
1) Kernel functions and elliptic differential equations in mathematical physics. Academic Press, New York, 1953.
2) Potential-theoretic methods in the theory of functions of two complex variables. Compositio Math. 10 (1952), 213-240.

[10] Bers, L.
1) Theory of pseudoanalytic functions. Courant Institute, New York, 1953.
2) Univalent solutions of linear elliptic systems. Comm. Pure Appl. Math 6 (1953), 513-526.
3) Mathematical aspects of subsonic and transonic gas dynamics. Wiley, New York, 1958.

4) Uniformizations by Beltrami equations. Comm. Pure Appl. Math. 14 (1961), 215-228.
5) Quasiconformal mappings, with applications to differential equations, function theory and topology. Bull. Amer. Math. Soc. 83 (1977), 1083-1100.

[11] Bers, L., John, F., Schechter, M.
Partial differential equations. Interscience, New York etc., 1964.

[12] Bers, L., Nirenberg, L.
1) On a representation theorem for linear elliptic systems with discontinuous coefficients and its application. Conv. Intern. Eq. Lin. Derivate Partiali, Trieste. Cremonense, Roma, 1954, 111-140.
2) On linear and nonlinear elliptic boundary value problems in the plane. Conv. Intern. Eq. Lin. Derivate Partiali, Trieste. Cremonense, Roma, 1954, 141-167.

[13] Beyer, K.
1) Nichtlineare Randwertprobleme für elliptische Systeme 1. Ordnung für zwei Funktionen von zwei Variablen. Beiträge zur Analysis 4 (1972), 31-34.
2) Abschätzungen für elliptische Systeme erster Ordnung für zwei Funktionen in Außengebieten des $\mathbb{R}^2$. Math. Nachr. 57 (1973), 1-13.

[14] Bicadze, A.V.
1) On the problem of mixed type equations. Trudy Mat. Inst. Akad. Nauk SSSR 41 (1953) (Russian).
2) Boundary value problems for elliptic equations of second order. Nauka, Moscow, 1966 (Russian); Engl. transl. North Holland Publ. Co., Amsterdam, 1968.
3) The theory of non-Fredholm elliptic boundary value problems. Am. Math. Soc. Transl. (2) 105 (1976), 95-103.
4) Some classes of partial differential equations. Gordon a. Breach, New York etc., 1988.

[15] Bojarski, B.
1) Generalized solutions of a system of differential equation of first order and elliptic type with discontinuous coefficients. Mat. Sb. N. S. 43 (85) (1957), 451-563 (Russian).
2) The general representation of solutions to elliptic systems of 2m equations in the plane. Dokl. Akad. Nauk SSSR 122 (1958), 543-546 (Russian).
3) Some boundary value problems for systems of 2m equations of elliptic type in the plane. Dokl. Akad. Nauk SSSR 124 (1958), 15-18 (Russian).
4) On the first boundary value problem for elliptic systems of second order in the plane. Bull. Acad. Polon. Sci., Sér. Sci. Math. Astr. Phys. 7 (1959), 565-570.
5) Riemann-Hilbert problem for a holomorphic vector. Dokl. Akad. Nauk SSSR 126 (1959), 695-698 (Russian).

6) On the Dirichlet problem for a system of elliptic equations in space. Bull. Acad. Polon. Sci., Sér. Sci. Math. Astr. Phys. 8 (1960), 19-23.
7) Theory of generalized analytic vectors. Ann. Polon. Math. 17 (1966), 281-320 (Russian).
8) Subsonic flow of compressible fluid. Arch. Mech. Stos. 18 (1966), 497-520; Mathematical Problems in Fluid Mechanics, Polon. Acad. Sci. Warsaw, 1967, 9-32.

[16] Bojarski, B., Iwaniec, T.
Quasiconformal mappings and non-linear elliptic equations in two variable, I, II. Bull. Acad. Polon. Sci., Sér. Sci. Math. Astr. Phys. 22 (1974), 473-478; 479-484.

[17] Brackx F., Delanghe, R., Sommen, F.
Clifford analysis. Pitman,London etc., 1982.

[18] Buchanan, J.
1) The Hilbert and Riemann-Hilbert problems for systems of Pascali type. Doctoral Dissertation, University of Delaware, Newark, 1980.
2) Bers-Vekua equations of two complex variable. Contemporary Math. 11 (1982), 71-88.

[19] Calderon, A.P., Zygmund, A.
1) On existence of certain integrals. Acta Math. 88 (1952), 85-139.
2) On singular integrals. Amer. J. Math. 78 (1956), 289-309.

[20] Chen Jian-gong
The Hölder continuity of the general solutions of the linear elliptic system of partial differential equations. Sci. Sinica 10 (1961), 153-159.

[21] Ciechanowicz-Halka, B.
Ein nichtlineares Randproblem der flachen Elastizitätstheorie. Demonstratio Math. 11 (1978), 583-590.

[22] Colton, D.
1) Partial differential equations in the complex domain. Pitman, London etc., 1976.
2) Solution of boundary value problems by the method of integral operators. Pitman, London etc., 1976.

[23] Colton, D., Gilbert, R.P. (Ed.)
Constructive and computational methods for differential and integral equations. Lecture Notes in Math. 430, Springer-Verlag, Berlin etc., 1974.

[24] Courant, R., Hilbert D.
Methods of mathematical physics II. Interscience, New York, 1962.

[25] Daniljuk, I.I.
1) On the oblique derivative problem for the general quasilinear system of the first order. Dokl. Akad. Nauk SSSR 127 (1959), 953-956 (Russian).
2) Nonregular boundary value problems in the plane. Izdat. Nauka, Moscow, 1975 (Russian).

[26] Delanghe, R.
1) On regular-analytic functions with values in a Clifford algebra. Math. Ann. 185 (1970), 91-111.
2) On regular points and Liouville's theorem for functions with values in a Clifford algebra. Simon Stevin 44 (1970), 55-66.
3) Sur les solutions de l'equation $\Delta u+a_0\partial u/\partial x_0=0$. Bull. Math. Soc. Sci. Math. R.S. Roumainie (N.S.) 14 (62) (1970), 147-151.
4) On the singularities of functions with values in a Clifford algebra. Math. Ann. 196 (1972), 293-319.
5) On regular functions with values in topological modules over a Clifford algebra. Bull. Soc. Math. Belg. 25 (1973), 131-138.

[27] Ding Shio-kuai, Wang Kan-ting, Ma Ju-nien, Shun Chia-lo, Zhang Tong
Definition of ellipticity of a system of second order partial differential equations with constant coefficients. Acta Math. Sinica 10 (1960), 276-287 (Chinese).

[28] Douglis, A.A.
1) A function-theoretic approach to elliptic systems of equations in two variables. Comm. Pure Appl. Math. 6 (1953), 259-289.
2) On uniqueness in Cauchy problems for elliptic systems of equations. Comm. Pure Appl. Math. 13 (1960), 593-607.

[29] Douglis, A., Nirenberg, L.
Interior estimates for elliptic systems of partial differential equations. Comm. Pure Appl. Math. 8 (1953), 503-538.

[30] Dressel, F.G., Gergen, J.J.
Mapping for elliptic equations. Trans. Amer. Math. Soc. 77 (1954), 151-178.

[31] Dzurajev, A.
1) On properties of some degenerate elliptic systems of first order in the plane. Dokl. Akad. Nauk SSSR 223 (1975) (Russian); Sov. Math. Dokl. 16 (1975), 914-918.
2) Study of partial differential equations by the means of generalized analytical functions. Function Theoretic Methods for Partial Differential Equations. Lecture Nores in Math. 561, Springer-Verlag, Berlin etc., 1976, 29-38.
3) Systems of equations of composite type. Moscow, 1972 (Russian); Engl.transl. Longman, Essex, 1989.

[32] Fan Ai-nong
1) On the differentiability and the existence theorem of a solution of the problem of oblique derivative, J. Hunan Univ., 1979, no. 4, 1-13 (Chinese).

2) Quasiconformal mappings and the theory of functions for systems of nonlinear elliptic partial differential equations of first order. Acta. Math. Sinica 23 (1980), 280-292 (Chinese).
3) On integral operators and (nonlinear mixed) boundary value problems. Sci. Sinica Ser. A, 25 (1982), 225-236.

[33] Gakhov, F.D.
Boundary value problems. Pergamon, Oxford, 1962.

[34] Gilbert, R.P.
1) Function theoretic methods in partial differential equations. Academic Press, New York, 1969.
2) Pseudohyperanalytic function theory. Ber. Ges. Math. Datenverarb. Bonn 77 (1973), 53-63.
3) Nonlinear boundary value problems for elliptic systems in the plane. Nonlinear Systems and Applications, ed. V. Lakshmikantham. Academic Press, New York, 1977, 97-124.
4) Verallgemeinerte hyperanalytische Funktionentheorie. Komplexe Analysis und ihre Anwendungen auf partielle Differentialgleichungen. Martin-Luther-Univ. Halle-Wittenberg, Wiss. Beiträge 1980, 124-145.

[35] Gilbert, R.P., Hile, G.N.
1) Generalized hypercomplex function theory. Trans. Amer. Math. Soc. 195 (1974), 1-29.
2) Hypercomplex function theory in the sense of L. Bers. Math. Nachr. 72 (1976), 187-200.
3) Hilbert function modules with reproducing kernels. Nonlinear Anal. 1 (1977), 135-150.
4) Degenerate elliptic systems whose coefficient matrix has a group inverse. Complex Variables, Theory Appl. 1 (1982), 61-88.

[36] Gilbert, R.P., Buchanan, J.L.
First order elliptic systems: A function theoretic approach. Academic Press, New York, 1983.

[37] Gilbert, R.P., Lin Wei
Algorithms for generalized Cauchy kernels. Complex Variables, Theory Appl. 2 (1983), 103-124.

[38] Gilbert, R.P., Hsiao, G.C.
Constructive function theoretic methods for higher order pseudoparabolic equations. Lecture Notes in Math. 561, Springer-Verlag, Berlin etc., 1976, 51-67.

[39] Gilbert, R.P., Kukral, D.K.
A function theoretic method for $\Delta_4^2 u+Q(x)u=0$. Ann. Math. Pure Appl. 104 (1975), 31-42.

[40] Gilbert, R.P., Linz, P.
The numerical solution of some elliptic boundary value problems by integral operator methods. Lecture Notes in Math. 430, Springer-Verlag, Berlin etc., 1974, 237-252.

[41] Gilbert, R.P., Roach, G.F.
Contructive methods for meta- and pseudoparabolic systems.
Bull. Math. Soc. Sci. Math. R.S. Roumaine, N
97-109.

[42] Gilbert, R.P., Schneider, M.
Generalized meta- and pseudoparabolic equations in the plane.
Complex Analysis and its Applications. Akad. Nauk SSSR, Izd.
Nauka, Moscow, 1978, 160-172.

[43] Gilbert, R.P., Wendland, W.
Analytic, generalized hyperanalytic function theory and an
application to elasticity. Proc. Roy. Soc. Edinburgh 73A (1975),
317-331.

[44] Goldschmidt, B.
1) Funktionentheoretische Eigenschaften verallgemeinerter
analytischer Vektoren. Math. Nachr. 90 (1979), 57-90.
2) Verallgemeinerte analytische Vektoren in R^n. Habilitations-
schrift, Martin-Luther-Univ., Halle-Wittenberg, 1980.
3) Eigenschaften der Lösungen elliptischer Differentialgleichungen
erster Ordnung in der Ebene. Math. Nachr. 95 (1980), 215-221.
4) Generalized analytic functions in $\mathbb{R}^n$. Komplexe Analysis und
ihre Anwendungen auf partielle Differentialgleichungen.
Martin-Luther-Univ., Halle-Wittenberg, Wiss. Beiträge 1980,
175-178.
5) Regularity properties of generalized analytic vectors in $\mathbb{R}^n$.
Math. Nachr. 103 (1981), 245-254.
6) Series expansions and maximum principles for generalized
analytic vectors in the space $\mathbb{R}^n$. Math. Nachr. 107 (1982),
241-251.

[45] Goluzin, G.M.
Geometric theory of functions of a complex variable. Amer. Math.
Soc., Providence, 1969.

[46] Haack, W., Wendland, W.
Lectures on Pfaffian and partial differential equations.
Pergamon Press, Oxford, 1972.

[47] Habetha, K.
1) Über lineare elliptische Differentialgleichungssysteme mit
analytischen Koeffizienten. Ber. Ges. Math. Datenverab.
Bonn 77 (1973), 65-89.
2) On zeros of elliptic systems of first order in the plane.
Function Theoretic Methods in Differential Equations. Pitman,
London, 1976, 42-62.

[48] Hellwig, G.
Partial differential equaitons. Ginn (Blaisdell), Boston,
Massachusetts, 1964.

[49] Hile, G.N.
1) Hypercomplex function theory applied to partial differential equations. Ph. D. Dissertation, Indiana University, Bloomington, 1972.
2) Representation of solutions of a special class of first order systems. J. Diff. Eq. 25 (1977), 410-424.
3) Elliptic systems in the plane with first order terms and constant coefficients. Comm. Part. Diff. Eq. 3 (10) (1978), 949-977.
4) Function theory for a class of elliptic systems in the plane. J. Diff. Eq. 32 (3) (1979), 369-387.

[50] Hile, G.N., Protter, M.H.
1) Unique continuation and the Cauchy problem for first order systems of partial differential equations. Comm. Part. Diff. Eq. 1 (1976), 437-465.
2) Maximum principles for a class of first order elliptical systems. J. Diff. Eq. 24 (1) (1977), 136-151.

[51] Hilbert, D.
Grundzüge einer allgemeinen Theorie der linearen Integralgleichungen. Teubner, Leipzig, 1912; 2. Ausg. 1924.

[52] Hörmander, L.
1) Differentiability properties of solutions of systems of differential equations. Ark. Mat. 3 (1958), 527-535.
2) Linear partial differential operators. Springer-Verlag, Berline etc., 1964.

[53] Hsiao, G.C., Wendland, W.
A finite element method for some integral equations of the first kind. J. Math. Anal. Appl. 58 (1977), 449-481.

[54] Hua Loo-keng, Lin Wei, Wu Tzu-chien,
Second order systems of partial differential equations in the plane. Pitman, London, 1985.

[55] Hou Zong-yi
1) Dirichlet problem for a class of linear elliptic second order equations with parabolic degeneracy on the boundary of the domain. Sci. Record (N.S.) 2 (1958), 244-249 (Chinese).
2) A Carleman boundary value problem for elliptic systems of first order equations. Sci. Sinica 12 (1963), 1237 (Russian).

[56] Iwaniec, T.
1) Green's function of multiply connected domain and Dirichlet problem for systems of second order in the plane. Lecture Notes in Math. 561, Springer-Verlag, Berlin etc., 1976, 261-276.
2) Quasiconformal mapping problem for general nonlinear systems of partial differential equations. Symp. Math. 18 (1976), 501-517.

[57] Iwaniec, T., Mamourian, A.
On the first order nonlinear differential systems with degeneration of ellipticity. Proceedings of the Second Finnish-Polish Summer School in Complex Analysis at Jyväskylä 1984. Univ. Jyväskylä, Dep. Math. Report 28 (1984), 41-52.

[58] Krushkal, S.L.
Quasiconformal mappings and Riemann surfaces, Wiley, N.Y., 1979.

[58a] Krushkal, S.L., Kühnau, R.
Quasikonforme Abbildungen. Neue Methoden und Anwendungen. Teubner-Verlag, Leipzig, 1983.

[59] Ladyshenskaja, O.A., Uraltseva, N.N.
Linear und quasilinear elliptic equations. Academic Press, New York, 1968.

[60] Lavrent'ev, M.A.
1) Sur une classe de représentations continués. Mat. Sbornik 42 (1935), 407-424.
2) Quasiconformal mappings and their derived systems. Dokl. Akad. Nauk SSSR 52 (1946), 287-290 (Russian).
3) The general problem of the theory of quasiconformal mappings of plane regions. Mat. Sbornik 21 (63) (1947), 285-320 (Russian).
4) Certain boundary value problems for systems of elliptic type. Sibirsk Mat. Z. 3 (1962), 715-728 (Russian).

[61] Lavrent'ev, M.A., Shabat, B.V.
A geometric property of the solutions of nonlinear systems of equations with partial derivatives. Dokl. Akad. Nauk SSSR 112 (1957), 810-811 (Russian).

[62] Lehto, O.
1) Remarks on generalized Beltrami equations and conformal mappings. Proc. Romanian-Finnish Seminar on Teichmüller Spaces and Quasiconformal Mappings, Brasov, 1969. Sém. Inst. Math. Acad. R. S. Roumanie, Publ. House Acad. S.R. Roumanie, 1971, 203-214.
2) Univalent functions and Teichmüller Spaces. Graduate Texts in Math. 109, Springer-Verlag, Berlin etc., 1987.

[63] Lehto, O., Virtanen, K.I.
Quasikonforme Abbildungen. Springer-Verlag, Berlin etc., 1965.

[64] Leray, J., Schauder, J.
Topologie et équations fonctionelles. Ann. Sci. École Norm. Sup. 51 (1934), 45-78 (YMH 1 (1946), 71-95).

[65] Levinson, N.
Dirichlet problem for $\Delta u = f(x,y,u)$. J. Math. Mech. 12 (1963), 567-575.

[66] Li Zhong
1) On the existence of homeomorphic solutions of a system of quasilinear partial differential equations of elliptic type. Acta Math. Sinica 13 (1963), 454-461 (Chinese Math. 4 (1964), 493-500).
2) The modified Dirichlet problem for a system of quasilinear partial differential equations of elliptic type. Beijingdaxue Xuebao (Acta Sci. Nat. Univ. Peking) 1964, 319-341 (Chinese).

[67] Li Zhong, Wen Guo-chun
1) On Cauchy's formula for elliptic systems of linear partial differential equations of the first order. Acta Math. Sinica 14 (1964), 23-32 (Chinese Math. 5 (1964), 25-35).
2) On Riemann-Hilbert boundary value problems of elliptic systems of linear partial differential equations of the first order. Acta Math. Sinica 15 (1965), 599-613 (Chinese Math. 7 (1965), 323-338).

[68] Li Ming-zhong
1) An existence theorem and a representation formula for generalized solutions of second order elliptic differential equations. Acta Math. Sinica 14 (1964), 7-22 (Chinese).
2) Hausdorff solvability of the Dirichlet problem for a class of second order elliptic systems. Chinese Ann. Math. 3 (1982), 319-328.

[68a] Lions, J.L., Magenes, E.
Non-homogeneous boundary value problems and applictions. Springer-Verlag, Berlin etc., 1972.

[69] Litvinchuk, G.S.
Boundary value problems and singular integral equations with shift. Nauka, Moscow, 1977 (Russian).

[70] Lopatinski, Y.B.
On a method of reducing boundary problems for a system of differential equations of elliptic type to regular equations. Ukrain. Mat. Z. 5 (1953), 123-151.

[71] Lu Chien-ke
1) The periodic Riemann boundary value problem and its applications to the theory of elasticity. Chinese Math. 4 (1964), 372-422.
2) On compound boundary problems. Sci. Sinica 14 (1965), 1545-1555.
3) The approximation of Cauchy-type integrals by some kinds of interpolatory splines. J. Approx. Theory 36 (1982), 197-212.

[72] Mikhailov, L.G.
1) Special cases in the theory of generalized analytic functions. Revue Roumaine Math. Pures Appl. 13 (1968), 1403-1408.
2) A new class of singular integral equations and its applications to differential equations with singular coefficients. Wolters-Noordhoff, Groningen, 1970.

3) On the analytic functions method in the theory of partial differential equations with singular coefficients . Lecture Notes in Math. 561, Springer-Verlag, Berlin etc., 1976, 510-520.

[73] Miranda, C.
Partial differential equations of elliptic type. Springer-Verlag, Berlin etc., 1970.

[74] Morrey, C.B.
On the solution of quasilinear elliptic partial differential equations. Trans. Am. Math. Soc. 43 (1938), 126-166.

[75] Monakhov, V.N.
1) Transformations of multiply connected domains by the solutions of nonlinear L-elliptic systems of equations. Dokl. Akad. Nauk SSSR 220 (1975), 520-523 (Russian).
2) Boundary value problems with free boundaries for elliptic systems. Amer. Math. Soc.,Providence, R.I., 1983.

[76] Muskhelishvili, N.I.
1) Singular integral equations. Noordhoff, Groningen, 1953.
2) Some basic problems of the mathematical theory of elasticity. Nauka, Moscow, 1966 (Russian); Noordhoff, Groningen, 1953.

[77] Naas, J., Tuschke, W.
1) Some probabilistic aspects in partial complex differential equations. Complex Analysis and its Applications. Akad. Nauk SSSR, Izd. Nauka, Moscow, 1978, 409-412.
2) On the error in the approximate solution of boundary value problems of nonlinear first order differential equations in the plane. Appl. Anal. 7 (1978), 239-246.

[78] Nitsche, J., Nitsche, J.J.
1) Das zweite Randwertproblem der Differentialgleichung $\Delta u = e^u$. Arch. Math. 3 (1952), 460-464.
2) Bemerkungen zum zweiten Randwertproblem der Differentialgleichung $\Delta\phi = \phi_x^2 + \phi_y^2$. Math. Ann. 126 (1953), 69-74.

[79] Nirenberg, L.
1) On nonlinear elliptic partial differential equations and Hälder continuity. Comm. Pure Appl. Math. 6 (1953), 103-156.
2) An application of generalized degree to a class of nonlinear problems. Coll. Analyse Fonct. (Liége, 1970), Vander, Louvain, 1971, 57-74.

[80] Orton, M.,
Hilbert boundary value problems - a distributional approach. Proc. Roy. Soc. Edinburgh 77A (1977), 193-208.

[81] Oskolkov, A.P.
Interior estimates for the first order derivatives for a class of quasilinear elliptic systems. Proc. Steklov Inst. Math. 110 (1970), 116-121.

[82] Parter, S.
On mappings of multiply connected domains by solutions of partial differential equations. Comm. Pure Appl. Math. 13 (1960), 167-182.

[83] Polozhy, G.N.
Generalization of the theory of analytic functions of a complex variable. p-analytic and (p,q)-analytic functions and some of their applications. Izdat. Kiev. Univ.,Kiev, 1965 (Russian).

[84] Protter, M.H., Weinberger, H.F.
Maximum principles in differential equations. Prentice-Hall, Englewood Cliffs, M.J., 1967.

[85] Rodin, Yu.L.
Generalized analytic functions on Riemann surfaces. Lecture Notes in Math. 1288, Springer-Verlag, Berlin etc., 1987.

[86] Schauder, L.
Der Fixpunktsatz in Funktionalräumen. Studia Math. 2 (1930), 171-181.

[87] Schapiro, Z.
Sur l'existence des representations quasiconformes. C. R. (Doklady) Acad. Sci. USSR (N.S.), 30 (1941), 690-692.

[88] Schleiff, M.
Über geschlossene Lösungen des Poincaréschen Randwertproblems mit Hilfe einer komplexen Differentialgleichung. Beiträge zur Analysis 6 (1974), 73-85.

[89] Simader, C.
Another approach to the Dirichlet problem for very strongly nonlinear elliptic equations. Ordinary and Partial Differential Equations, Lecture Notes in Math. 564, Springer-Verlag, Berlin etc., 1976, 425-437.

[90] Sobolev, S.L.
Applications of functional analysis in mathematical physics. Amer. Math. Soc., Providence, R.I., 1963.

[91] Stein, E.M.
Singular integrals and differentiability properties of functions. Princeton Univ. Press, Princeton,N. J., 1970.

[92] Tai Chung-wei
The oblique derivative problem for various elliptic complex equations in a multiply connected domain. Appl. Math. Num. Math. 1982, no. 1, 69-73 (Chinese).

[93] Teichert-Gaida, A.
1) On some compound nonlinear boundary value problem. Demonstratio Math. 9 (1976), 409-421.

2) A nonlinear problem of Sjöstrand-v. Wolfersdorf's type. Demonstratio Math. 11 (1978), 395-413.

[94] Tian Mao-ying
The mixed boundary value problem for a class of nonlinear degenerate elliptic equations of second order in the plane. Beijingdaxue Xuebao, 1982, no. 5, 1-13 (Chinese).

[95] Tjurikov, E.V.
The nonlinear Riemann-Hilbert boundary value problem for quasilinear elliptic systems. Soviet Math. Dokl. 20 (1979), 863-866.

[96] Tutschke, W.
1) Die neuen Methoden der komplexen Analysis und ihre Anwendung auf nichtlineare Differentialgleichungssysteme. Sitzungsber. Akad. Wiss. DDR 17N (1976).
2)Lösung nichtlinearer partieller Differentialgleichungssysteme erster Ordnung in der Ebene durch Verwendung einer komplexen Normalform. Math. Nachr. 75 (1976), 283-298.
3) Partielle komplexe Differentialgleichungen in einer und in mehreren komplexen Variablen. VEB Deutscher Verlag der Wissenschaften, Berlin, 1977.
4) The Riemann-Hilbert problem for nonlinear systems of differential equations in the plane. Complex Analysis and its Applications. Akad. Nauk SSSR, Izd. Nauka, Moscow, 1978, 537-542.
5) Vorlesungen über partielle Differentialgleichungen. Teubner, Leipzig, 1978.
6) Solutions with prescribed periods on the boundary components for nonlinear elliptic systems of first order in multiply connected domains in the plane. Martin-Luther-Univ., Halle, Preprint Nr. 27 (1979), 3-9.
7) A general method for approximate solving of mixed problems in the theory of partial differential equations. Complex Variables, Theory Appl. 1 (1982), 89-96.
8) Partielle Differentialgleichungen: Klassische, funktional-analytische und komplexe Methoden. Teubner, Leipzig, 1983.
9) Solutions with prescribed periods on the boundary components of nonlinear elliptic systems of first order in multiply connected domains in the plane. Complex Analysis. Banach Center Publ. 11 (1983), 347-351.

[97] Vekua, I.N.
1) Systems of differential equations of the first order of elliptic type and boundary value problems; application to the theory of shells. Mat. Sbornik 31 (1952), 217-314 (Russian).
2) Generalized analytic functions. Pergamon, Oxford, 1962.
3) Theory of thin shallow shells of variables thickness. Metsniereba, Tbilisi, 1965.
4) New methods for solving elliptic equations. North-Holland Publ., Amsterdam, 1967.
5) Foundations of Tensor Analysis. Izd. Tbil. Gos. Univ., Tbilisi, 1967.

6) On two ways of contructing the consistent theory of elastic shells. Mater. perv. vsesoyuz. shk. po. teor. i chisl. met. rasch. obol. i plast.. Metsniereba, Tbilisi, 1974.
7) Foundations of Tensor analysis and theory of covariants. Nauka, Moscow, 1978.
8) Shell theory: General methods of construction. Pitman, London etc., 1985.

[98] Vinogradov, V.S.
1) On a boundary value problem for linear elliptic systems of differential equations of the first order on the plane. Dokl. Akad. Nauk SSSR 118 (1958), 1059-1062 (Russian).
2) On the boundedness of solutions of boundary value problems for linear elliptic systems of first order in the plane. Dokl. Akad. Nauk SSSR 121 (1958), 399-402 (Russian).
3) On some boundary value problems for quasilinear elliptic systems of first order in the plane. Dokl. Akad. Nauk SSSR 121 (1958), 579-581 (Russian).
4) A certain boundary value problem for an elliptic system of special form. Differential'nye Uravnenija 77 (1971), 1226-1234 (Russian).
5) A boundary value problem for a first order elliptic equation on the plane. Differential'nye Uravnenija 77 (1971), 1440-1448 (Russian).
6) On a method of solution of a boundary value problem for a first order elliptic system on the plane. Dokl. Akad. Nauk SSSR 201 (1971) (Soviet Math. Dokl. 12 (1971), 1699-1703).
7) On the solvability of a singular integral equation. Soviet Math. Dokl. 19 (1978), 827-829.
8) The geometric meaning of the condition of ellipticity of a system of first order differential equations in the plane and its reduction to a complex special form. Soviet Math. Dokl. 24 (1981), 364-367.
9) Elliptic systems in the plane. Complex Analysis-Methods, Trends, and Applications. Akademie-Verlag, Berlin, 1983, 166-173.

[99] Volkovyskii, L.I.
On differentiability of quasi-conformal mappings. L'vov. Gos. Univ. Uc. Zap. 29, Ser. Meh.-Mat. 6 (1954), 50-57 (Russian).

[100] Wagner, E., Wolfersdorf, L.v.
A regularization method in the Cauchy problem for holomorphic functions. Z. Anal. Anw. 1 (6) (1982), 35-44.

[101] Warowna-Dorau, G.
1) Application of the method of successive approximations to a nonlinear Hilbert problem in the class of generalized analytic functions. Demonstratio Math. 2 (1970), 101-116.
2) A compound boundary value problem of Hilbert-Vekua type. Demonstratio Math. 7 (1974), 337-352.

[102] Warschawski, W.
On differentiability at the boundary in conformal mapping. Proc. Amer. Math. Soc. 12 (1961), 614-620.

[103] Wen Guo-chun
1) The existence theorems of continuously differentiable and homeomorphic solutions for nonlinear elliptic systems of first order. Beijingdaxue Xuebao (Acta Sci. Natur. Univ. Peking), 1979, no. 2, 60-72 (Chinese).
2) Modified Dirichlet problem and quasiconformal mappings for nonlinear elliptic systems of first order. Kuxue Tongbao 25 (1980), 449-453.
3) The Riemann-Hilbert boundary problem for nonlinear elliptic systems of first order in the plane. Acta. Math. Sinica 23 (1980), 244-255 (Chinese).
4) Boundary value problems of Riemann type for nonlinear elliptic systems of first order. Beijingdaxue Xuebao, 1980, no. 4, 1-13 (Chinese).
5) On representation theorems and existence theorems of solutions for nonlinear elliptic complex equations of first order. Hebei Huagong Xueyuan Xuebao (Shuxue Zhuanji), 1980, 41-61 (Chinese).
6) The oblique derivative boundary value problem for nonlinear elliptic systems of second order (I). Hebei Huagong Xueyuan Xuebao (Shuxue Zhuanji), 1980, 119-144 (Chinese).
7) The singular case of Riemann-Hilbert boundary value problem. Beijingdaxue Xuebao, 1981, no. 4, 1-14 (Chinese).
8) The Poincaré problem with negative index for linear elliptic systems of second order in a multiply connected domain. Jour. of Math. Res. and Expos., 1981, no. 3, 61-76 (Chinese).
9) Integral representations of solutions for boundary value problems of first order complex equations. Hebeishida Xuebao (Ziran Kuxue), 1981, no. 1,2, 130-145 (Chinese).
10) The mixed boundary value problem for nonlinear elliptic equations of second order in the plane. Proc. of 1980 Beijing Sym. on Diff. Geom. and Diff. Eq., 1982, 1543-1557.
11) On compound boundary value problem with shift for nonlinear elliptic complex equations of first order. Complex Variables, Theory Appl. 1 (1982), 39-59.
12) Green functions in a multiply connected domain and integral representation of solutions for elliptic boundary value problems. Yingyong Shuxue Yu Jisuan Shuxue, 1982, no. 1, 55-60 (Chinese).
13) The third boundary value problem for elliptic systems of second order. Chinese Ann. Math. 4 (1983), 1-12 (Chinese).
14) Nonlinear boundary value problems for elliptic systems of several equations of first order in a multiply connected domain. Beijingdaxue Xuebao, 1983, no. 4, 1-12 (Chinese).
15) The boundary value problems with shift for linear elliptic equations of second order. Chinese Ann. Math. 4 (1983), 465-473 (Chinese).
16) On representation theorem of solutions and mixed boundary value problems for second order nonlinear elliptic equations with unbounded measurable coefficients. Acta. Math. Sinica 26 (1983), 533-537 (Chinese).

17) Oblique derivative boundary value problems for nonlinear elliptic systems of second order. Scientia Sinica, Ser. A 26 (1983), 113-124.
18) Nonlinear quasiconformal glue theorem. Analytic Functions. Lecture Notes in Math. 1039, Springer-Verlag, Berlin etc., 1983, 458-463.
19) Function theoretic methods for nonlinear elliptic complex equations. Bull. Math. Soc. Sci. Math. Roumanie, 28 (76) (1984), 87-90.
20) Boundary value problems with shift for nonlinear elliptic equations of second order. Jour. of Math. Res. Expos., 1984, no. 3, 107-113.
21) Some boundary value problems for a class of nonlinear elliptic systems of several second order equations. Beijingdaxue Xuebao, 1984, no. 2, 1-11 (Chinese).
22) Some nonlinear boundary value problems for nonlinear elliptic equations of second order in the plane. Complex Variables, Theory Appl. 4 (1985), 189-210.
23) Nonlinear discontinuous boundary value problems for nonlinear elliptic systems of first order in a multiply connected domain. Beijingdaxue Xuebao, 1985, no. 3, 1-10 (Chinese).
24) Conformal mappings and boundary value problems. Higher Education Press, 1985 (Chinese).
25) Linear and nonlinear elliptic complex equations. Shanghai Science Techn. Publ. House, 1986 (Chinese).
26) The irregular oblique derivative problem for nonlinear elliptic equations of second order. Beijingdaxue Xuebao, 1986, no. 5, 15-25 (Chinese).
27) Applications of complex analysis to nonlinear elliptic systems of partial differential equations. Contemporary Math. 48 (1985), 217-234.

[104] Wen Guo-chun, Fang Ai-nong
1) Complex function theory for nonlinear elliptic systems of first order. Hanshulun Zhuanji, Chongqing 1978, 98-105 (Chinese).
2) The complex form and some boundary value problems for nonlinear elliptic systems of second order. Chinese Ann. Math. 2 (1981), 201-216 (Chinese).

[105] Wen Guo-chun, Gao Shan-zhi, Huang Sha
1) The Dirichlet problem for nonlinear elliptic systems of second order. Hebeishida Xuebao (Ziran Kuxue), 1979, 1-11 (Chinese).
2) The Neumann boundary value problem for elliptic systems of second order. Heibeishida Xuebao (Ziran Kuxue), 1981, 13-24 (Chinese).

[106] Wen Guo-chun, Li Hong-zhen, Li Zi-shi
1) The Haseman problem for nonlinear elliptic complex equations of first order. Hebeidaxue Xuebao, 1980, 65-85 (Chinese).
2) Properties of solutions for first order elliptic complex equations in the whole plane. Hebeidaxue Xuebao, 1981, no. 2, 28-35 (Chinese).

[107] Wen Guo-chun, Li Sheng-xun, Xu Ke-ming
1) On basic theorems of quasiconformal mappings. Hebei Huagong Xueyuan Xuebao (Shuxue Zhuanji) , 1980, 20-40 (Chinese).
2) Nonlinear quasiconformal mappings for the univalent Riemann surface. Neimenggudaxue Xuebao (Ziran Kuxue), 1984, no. 2, 18-26 (Chinese).

[108] Wen Guo-chun, Li Qi-feng
1) The Riemann-Hilbert problem for elliptic systems of fourth order. Sichuan Schiyuan Xuebao (Shuxue Zhuanji), 1981, 90-103 (Chinese).
2) The oblique derivative boundary value problem for elliptic systems of fourth order. Sichuan Schiyuan Xuebao (Ziran Kuxue), 1983, no. 2, 53-64 (Chinese).
3) Nonlinear Riemann-Hilbert problems for nonlinear elliptic complex equations of first order. Sichuan Schiyuan Xuebao (Ziran Kuxue),1983, no. 4, 36-47 (Chinese).

[109] Wen Guo-chun, Tai Chung-wei
1) The oblique derivative boundary value problem for elliptic complex equations of first order in a multiply connected domain. Beijingdaxue Xuebao, 1981, no. 3, 19-29 (Chinese).
2) The Poincaré boundary value problem for the linear elliptic complex equation of second order in a multiply connected domain. Beijingdaxue Xuebao, 1983, no. 2, 1-10 (Chinese).
3) Nonlinear boundary value problems for elliptic complex equations of second order. Hebeishida Xuebao (Ziran Kuxue), 1983,no. 1, 6-18 (Chinese).

[110] Wen Guo-chun, Tian Mao-ying
Solutions for elliptic equations of second order in the whole plane. J. Math. 2 (1982), no. 1, 23-36 (Chinese).

[111] Wen Guo-chun, Yang Guang-wu
1) The properties of solutions and first boundary problem for nonlinear elliptic equation of second order in the plane. Hebei Huagong Xueyuan Xuebao (Shuxue Zhuanji), 1980, 84-103 (Chinese).
2) The Riemann-Hilbert boundary problem for uniformly elliptic complex equation of second order. Hebei Huagong Xueyuan Xuebao, 1980, no. 2, 49-57 (Chinese).
3) The Poincaré boundary value problem for linear equations of second order. Advances in Math. 10 (1981), 157-160 (Chinese).

[112] Wendland, W.
1) Bemerkungen über die Fredholmschen Sätze. Meth. Verf. Math. Physik, 3, BI Hochschulskripten,722/722a, Mannheim, 1970,141-176.
2) An integral equation method for generalized analytic functions. Lecture Notes in Math. 430, Springer-Verlag, Berlin etc., 1974, 414-452.
3) On a class of semilinear boundary value problems for certain elliptic systems in the plane. Complex Analysis and its Applications. Akad. Nauk SSSR, Izd. Nauka, Moscow, 1978,108-119.

4) On the imbedding method for semilinear first order elliptic systems and related finite element methods. Continuation Methods. Academic Press, New York etc., 1978, 277-336.
5) Elliptic systems in the plane. Pitman, London etc., 1979.
6) Zur Behandlung elliptischer Randwertaufgaben mit Integralgleichungen. Wiss. Schriftenreihe TH Karl-Marx-Stadt, 1979, 143-153.
7) Numerische Methoden bei Randwertproblemen elliptischer Systeme in der Ebene. Kpmplexe Analysis und ihre Anwendungen auf partielle Differentialgleichungen. Martin-Luther-Univ., Halle-Wittenberg, Wiss. Beiträge 1980, 369-373.

[113] Wendland, W., Stephan, E.
Remarks to Galerkin and least squares methods with finite elements for general elliptic problems. Ordinary and Partial Differential Equations. Lecture Notes in Math. 564, Springer-Verlag, Berlin etc., 1976, 461-471.

[114] Wolfersdorf, L. v.
1) Monotonicity methods for a class of first order semilinear elliptic systems. Komplexe Analysis und ihre Anwendungen auf partielle Differentialgleichungen, Martin-Luther-Univ., Halle-Wittenberg, Wiss. Beiträge 1980, 369-373.
2) On a boundary value problem for a special class of first order semilinear elliptic system in the plane. Z. Anal. Anw. 2 (1983), 37-40.
3) A class of nonlinear Riemann-Hilbert problems for holomorphic functions. Math. Nachr. 116 (1984), 89-107.
4) On strongly nonlinear Poincaré boundary value problems for harmonic functions. Z. Anal. Anw. 3 (1984), 385-399.

[115] Wolfersdorf, L. v., Wolska-Bochenek, J.
A compound Riemann-Hilbert problem for holomorphic functions with nonlinear boundary condition. Demonstratio Math. 17 (1984), 545-556.

[116] Wolska-Bochenek, J.
1) Problème nonlinéaire à dérivée oblique. Ann. Polon. Math. 9 (1960/61), 253-264.
2) Sur un problème nonlinéaire d'Hilbert dans le théorie des fonctions pseudo-analytiques. Zeszyty Naukowe Politechniki Warszawskiej 172 Mat. 11 (1968), 145-157.
3) On some generalized nonlinear problem of the Hilbert type. Zeszyty Nauk. Politech. Warszawsk. 183 Mat. 14 (1968), 15-32.
4) A compound nonlinear boundary value problem in the theory of pseudo-analytic functions. Demonstratio Math. 4 (1972), 105-117.

[117] Wu Su-ming
A free boundary problem in filtration theory of an earth dam. Beijingdaxue Xuebao, 1987, no. 3.

[118] Xu Zhen-yuan
1) Two-dimensional singular integral equations with analytical coefficients. Chinese Ann. Math. 5A (1984), 455-460 (Chinese).
2) Nonlinear Poincaré problem for a system of first-order elliptic equations in the plane. Complex Variables, Theory Appl. 7 (1987), 363-381.

[119] Yuan Yi-rang,
The boundary value problem of Riemann type for generalized analytic functions. Kuxue Tongbao 26 (1981), 58-61.

[120] Zhao Zhen
The sufficient and necessary conditions for Noether solvability of singular integral equations with two Carleman shifts. Chinese Ann. Math. 2 (1982), 91-100 (Chinese).

Index